T0235377

Lecture Notes in Computer Science   9106

Commenced Publication in 1973
Founding and Former Series Editors:
Gerhard Goos, Juris Hartmanis, and Jan van Leeuwen

More information about this series at http://www.springer.com/series/7408

Weizhong Qiang · Xianghan Zheng
Ching-Hsien Hsu (Eds.)

Cloud Computing
and Big Data

Second International Conference, CloudCom-Asia 2015
Huangshan, China, June 17–19, 2015
Revised Selected Papers

 Springer

Editors

Weizhong Qiang
School of Computer Science
  and Technology
Huazhong University of Science
  and Technology
Wuhan
China

Ching-Hsien Hsu
Department of Computer Science
  and Information Engineering
Chung Hua University
Hsinchu
Taiwan

Xianghan Zheng
College of Mathematics and Computer
  Science
Fuzhou University
Fuzhou
China

ISSN 0302-9743   ISSN 1611-3349   (electronic)
Lecture Notes in Computer Science
ISBN 978-3-319-28429-3   ISBN 978-3-319-28430-9   (eBook)
DOI 10.1007/978-3-319-28430-9

Library of Congress Control Number: 2015959928

LNCS Sublibrary: SL2 – Programming and Software Engineering

This Springer imprint is published by SpringerNature
The registered company is Springer International Publishing AG Switzerland

Preface

The International Conference on Cloud Computing and Big Data (CloudComAsia) is an international conference aimed at providing an exciting platform and forum for researchers and developers from academia and industry to exchange information regarding advancements in the state of the art and practice of cloud computing and big data, as well as to identify emerging research topics and define the future directions of cloud computing and big data.

This year CloudComAsia 2015 received 106 submissions from authors in 14 countries/regions. The papers were reviewed by a 51-member Program Committee, with 24 members from mainland China, seven members from USA, and the rest from Norway, UK, Australia, Taiwan, Romania, Spain, Hong Kong, India, Singapore, and Ireland. Each paper received three to five reviews. Based on a total of 348 reviews, the program co-chairs accepted 29 papers for the LNCS proceedings. The acceptance rate is 27.4 %.

For the conference, we had invited two distinguished keynote speakers:

- "Business Transformation with the Cloud and Big Data," by Hui Lei (IBM T. J. Watson Research Center, USA)
- "Emerging Memories Advancing Big Data," by Nong Xiao (National University of Defense Technology, China)

We would like to thank all the authors of submitted papers for their work and their interest in the conference. We would like to express our sincere appreciation to all members of the Program Committee. Of the 369 reviews we assigned, 348 (94 %) were completed. From these reviews we identified a set of submissions that were clear, relevant, and described high-quality work in cloud computing and big data. Hundreds of authors received objective and often detailed feedback from a diverse group of experts.

In addition, we would like to thank the general chairs, Drs. Hai Jin and Chunming Rong, and the Steering Committee members, Drs. Martin Gilje Jaatun, Thomas J. Hacker, and Guolong Chen, for their invaluable advice and guidance as well as Alfred Hofmann, executive editor, and Anna Kramer of the LNCS editorial office for their prompt and patient response to our questions and requests. The conference proceedings would not have been possible without the support of these individuals and organizations. In closing, it is our hope that all of these efforts have helped to improve and promote cloud computing and big data in China, other Asian countries, the US, and beyond.

July 2015

Weizhong Qiang
Xianghan Zheng
Ching-Hsien Hsu

Organization

Executive Committee

General Co-chairs

Hai Jin	Huazhong University of Science and Technology, China
Chunming Rong	University of Stavanger, Norway

Program Chairs

Weizhong Qiang	Huazhong University of Science and Technology, China
Xianghan Zheng	Fuzhou University, China
Ching-Hsien Hsu	Chung Hua University, Taiwan, China

Steering Committee Chair

Chunming Rong	University of Stavanger, Norway
Hai Jin	Huazhong University of Science and Technology, China
Martin Gilje Jaatun	SINTEF, Norway
Thomas J. Hacker	Purdue University, USA
Guolong Chen	Fuzhou University, China

Advisory Committee

Rajkumar Buyya	University of Melbourne, Australia
Stephen Diamond	IEEE CCI, USA
Albert Zomaya	University of Sydney, Australia
Kai Hwang	University of Southern California, USA
Hai Jin	Huazhong University of Science and Technology, China
Jiannong Cao	Hong Kong Polytechnic University, Hong Kong, SAR China
Siani Pearson	HP Labs, UK
Jianying Zhou	Institute for Infocomm Research, Singapore
Pavan Balaji	Argonne National Lab, USA
Victor Leung	UBC, Canada
Chung-Ming Hwang	National Cheng-Kun University, Taiwan, China
Sadie Creese	Oxford, UK
Ake Edlund	KTH, Sweden
Huanguo Zhang	Wuhan University, China

Publicity Chair

Wenbing Jiang	Huazhong University of Science and Technology, China

Program Committee

Tomasz Wiktor Wlodarczyk	University of Stavanger, Norway
Geng Yang	Nanjing University of Posts Telecommunications, China
Dingyi Pei	Guangzhou University, China
Vladimir Oleshchuk	University of Agder, Norway
Lei Jiao	University of Agder, Norway
Xuan Zhang	University of Agder, Norway
Yi Ren	National Chiao Tung University, Taiwan, China
Xinyi Huang	Fujian Normal University, China
Guohua Liu	Donghua University, China
Chunming Tang	Guangzhou University, China
Zhou Quan	Guangzhou University, China
Dehua Chen	Donghua University, China
Xin Hong	Hua Qiao University, China
Tor-Morten Gronli	Norwegian School of Information Technology, Norway
Kyrre Begnum	Oslo University College, Norway
Zhen Chen	Tsinghua University, China
Hui Cheng	University of Bedfordshire, UK
Andrzej M. Goscinski	Deakin University, Australia
Bo Hong	Georgia Institute of Technology, USA
Liusheng Huang	University of Science and Technology of China, China
Ching-Hsien Hsu	Chung Hua University, Taiwan
Lingjuan Li	Nanjing University of Posts and Telecommunications, China
Peng Liu	Institute of Telecommunications, China
Dana Petcu	West University of Timisoara, Romania
Xiaoling Qing	Nanjing University of Aeronautics and Astronautics, China
Wei Tan	IBM T.J. Watson Research Center, USA
Guillermo Taboada	University of A Coruna, Spain
Yang Xiang	Central Queensland University, Australia
Bin Xiao	Hong Kong Polytechnic University, Hong Kong, SAR China
Qingjun Xiao	Georgia State University, USA
Xianghan Zheng	Fuzhou University, China
Junwei Cao	Tsinghua University, China
Josef Noll	University of Oslo/UNIK, Norway
Nik Bessis	University of Derby, UK
Xiaolong Xu	Nanjing University of Posts and Telecommunications, China
Zhihui Du	Tsinghua University, China
Wei Xu	Tsinghua University, China
Xiangfei Meng	National Supercomputing Center in Tianjin, China
Chao-Tung Yang	Tunghai University, Taiwan, China
Yong Li	Beijing Jiaotong University, China

Keynote Speeches

Business Transformation with the Cloud and Big Data

Hui Lei

IBM T.J. Watson Research Center, New York, USA

Cloud computing and big data are transforming the business operations in many enterprises. Until recently, it was commonly recognized that there were two distinct classes of enterprise applications: namely Systems of Record and Systems of Engagement. While Systems of Record are traditional enterprise systems that maintain core business data such as those related to transactions, user demographics, and pricing, Systems of Engagement leverage new mobile and social networking capabilities to support loose and ephemeral interactions across employees and customers. With the advent of cloud computing, it is possible to integrate Systems of Record and Systems of Engagement at an unprecedented scale and speed. Key business insight can be generated from such integration, which can then be exploited to transform the way enterprises conduct their business. This has led to a new class of emerging, high-value enterprise applications called Systems of Insight. In this talk, I will discuss several examples of Systems of Insight developed at IBM Research, including novel business solutions that make use of social analytics and telematics analytics. I will also present a common platform for System-of-Insight as a Service, which addresses some key challenges in managing Systems of Insight throughout the lifecycle of development, deployment, and runtime.

Emerging Memories Advancing Big Data

Nong Xiao

National University of Defense Technology (NUDT), China

Big data challenges existing methods of data management and current computer design. As a foundation part of computer systems, storage plays a critical role in advancing big data applications, typically like data and knowledge analytics, computational scientific discovery and so on. However, compared to the fast development of computing technologies like multicore, multi-thread etc., the performance of storage evolves much more slowly and has become the bottleneck of the entire computer systems. With flash memories gaining popularity and other emerging non-volatile memories such as PCM, STT-RAM and RRAM etc. appearing, storage media has potentials to reshape systems for data storing and accessing, likely filling the gap between computing and storing.

In this talk, I will discuss typical emerging memories and expose their characteristics to storage system design. I will then highlight flash memories and provide some scenarios where they can be used regarding our research and experience of optimizing computer architecture and systems in the past. Additionally, since mersisters have been attracting much attention, I will introduce some work we have done based on them. At last, I will predict storage development with a focus on storage class memories that may potentially unify memory and storage.

Contents

Applications

Big Data and Social Network

Security and Privacy

Cloud Architecture

Guarding Fast Data Delivery in Cloud: An Effective Approach to Isolating Performance Bottleneck During Slow Data Delivery

Zhenyun Zhuang$^{(\boxtimes)}$, Haricharan Ramachandra, and Badri Sridharan

LinkedIn Corporation, Mountain View, CA 94043, USA
{zzhuang,hramachandra,bsridharan}@linkedin.com

Abstract. Cloud-based products heavily rely on the fast data delivery between data centers and remote users - when data delivery is slow, the products' performance is crippled. When slow data delivery occurs, engineers need to investigate and root cause the issue.

To facilitate the investigations, we propose an algorithm to automatically identify the performance bottleneck. The algorithm aggregates information from multiple layers of data sender and receiver. It helps to automatically isolate the problem type by identifying which component of sender/receiver/network is the bottleneck. After isolation, successive efforts can be taken to root cause the exact problem. We also build a prototype to demonstrate the effectiveness of the algorithm.

1   Introduction

Cloud-based products (e.g., Cloud storage productions such as Amazon S3) involve data transfer between cloud data centers and remote users. For cloud environments, fast data delivery is critical to ensure high performing data products, as it translates to higher data throughput and hence less response time as experienced by the users. However, despite many techniques aimed at optimizing the data delivery through Internet (e.g. web page optimization, network acceleration), a critical problem we have continuously experienced is slow data delivery. This problem has two forms: (1) data is not delivered; and (2) data delivery is slow (i.e., taking long time to complete the delivery). For example, a web client is browsing a web page, and needs to download a jpeg file. The file could either not be downloaded at all, or the downloading process takes too much time.

When slow data transfer happens, it cripples application performance, and engineers need to investigate and root cause the issue. There are many possible causes. For instance, it could be caused by the data sender being overloaded and hence not sending data; or the network channel is slow; or the data receiver is too busy to read data from network buffer. Based on our experiences, root causing the issues is tedious and requires lots of experiences and expertise. Largely, possible causes of such a problem can be classified into three types: (1) client application causes; (2) network side causes; and (3) server application causes.

© Springer International Publishing Switzerland 2015
W. Qiang et al. (Eds.): CloudCom-Asia 2015, LNCS 9106, pp. 3–15, 2015.
DOI: 10.1007/978-3-319-28430-9_1

4 Z. Zhuang et al.

Engineers need to quickly isolate different types of causes, so they can focus on the particular suspicious component and perform deeper analysis to root cause. It is a challenging task due to several reasons. First, the data delivery involves multiple network entities including two machines (i.e., the sender and receiver) and the networking routes, which is in sharp contrast to the usual performance issues where only a single machine is involved. Second, the diagnosis involves multiple layers of information including application layer and transport layer. To isolate the causes, engineers have to check various places including client log, server log, network statistics, cpu usage, etc. These types of checking take much time and efforts, and often times requires experiences and expertise from the performance engineers. To save time and efforts, performance engineers are eagerly seeking more intelligent tools to help them quickly isolate root causes.

In this work, we specifically focus on fulfilling such a request: quickly identify the component that is to blame, be it the sender, the receiver, or the network. This work presents an algorithm and a prototype to help performance engineers. The algorithm can automatically isolate the root cause when slow data delivery occurs. Once the blame part is figured out, successive investigations can be conducted to nail down the real problem. These will be outside of the scope of this work, and are part of our future work.

For the remainder of the writing, we present necessary background in Sect. 2, motivate the problems using three scenarios in Sect. 3. We then present the design in Sect. 4. Based on the proposed design principles, we describe the solutions in Sect. 5. We build a prototype and perform evaluation in Sect. 6. We present related works in Sect. 7 and in Sect. 8 we conclude the work.

2 Background and Scope

Background: *TCP Transport Protocol:* Transmission Control Protocol (TCP) is one of the transport-layer protocols that provides ordered and reliable delivery of streamed bytes. It is the most widely used transport protocol Today. TCP features flow control to avoid overloading the receiver. The receiver sets up a dedicated receiver buffer, and the sender sets up a corresponding send buffer. TCP also have congestion control, retransmission mechanism, etc. to ensure the reliable and fast data transfer.

Application Protocol: Application protocols such as HTTP are built upon lower protocols. When application performance suffers, the immediate symptoms typically are higher latency and lower throughput. Though these two symptoms are often related, we specifically focus on solving the slow data transfer (i.e. lower throughput) problem. Depending on application protocols and how applications are built, application layers may emit additional logs. For instance, data receiver may log the time stamps of sending data requests, receiving the first byte of requested data and receiving the last byte of the requested data. Similarly, the data sender may log the time stamps of receiving the data requests, sending the first byte of requested data and sending the last byte of the requested data.

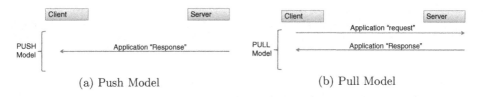

(a) Push Model   (b) Pull Model

Fig. 1. Push model and Pull model

Scope: To understand the causes of the low performing data delivery problem, we firstly need to understand the data delivery model which determines the flow of bytes. For simple presentation, we assume use "client" to denote the data receiver, while "server" to denote the data sender. This assumption is in line with today's web browsing paradigm where web browsers are data receivers and web servers are data senders.

Generally there are two models of data delivery: pull-based and push-based. In pull-based model, the client sends a request to the server, and the server sends back a response corresponding to the client request. In push-based model, the server pushes data to the client, without the client explicitly asking for that. Pull-based model is more complicated model, as the data flow of push-based model is only a subset of that of the pull-based model. In other words, the only difference between the two models is that pull-based model contains the additional "data request" phase, as shown in Fig. 1.

Though data downloading can be carried over with various protocols including TCP and UDP, most of today's data are transferred through TCP, given the dominance of Internet service. Hence, in this work we only focus on TCP-based data downloading. For easier presentation, we choose Linux platforms to present our designs and solutions due to Linux's popularity. However, the relevant designs and solutions will also apply to other platforms.

3   Problem Definition and Motivation Scenarios

Problem Definition: We use C to denote the client which receives data; S to denote the server that sends back the data. We firstly consider the pull-based delivery model, where for any data delivery, C sends a request Rq to S first; after receiving Rq, S will prepare the response data Rs and send back to C.

We assume the time taken in good data delivery scenarios is Tg, and the maximum of Tg is denoted as Tg_{max}. The delivery time is calculated as the time difference between when C's application sends out Rq and when that application receives the last byte of the Rs with read() call. The actual data delivery time is Ta. We say that the particular data delivery is slow when $Ta > Tg_{max}$.

Results of Three Motivation Scenarios: To further illustrate the nature of the problem we are addressing, we setup a simple experiment to demonstrate the application symptoms as well as the root causes of slow data downloading. In the experiment, a client (receiver) opens a TCP connection to download a bulk of

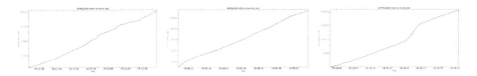

Fig. 2. Similar symptoms of slow data downloading, but caused by different types of bottlenecks: Server (a), Client (b) and Network (c)

data from a server. The environment are Gigbit LAN network, where the client machine and the server machine are directly connected. The RTT (round trip delay) is at sub-millisecond. Hence, in typical ("good") data transfer scenarios, the expected data transmission rate should not be less than 1 MB/s based on many practices.

For each scenario, we measure the data delivery rate on the client (receiver) side at application level. The three scenarios differ in where the performance bottleneck is, namely, the client (receiver), the server (sender) and the network. Each of these bottlenecks can cause slow data delivery. We plot the accumulated bytes received by the receiver (from application level) across the time line, which is in the format of hour:minute:second.

Receiver is the Bottleneck: In this scenario, the receiver application is not reading fast enough. The downloading progress as measured by the receiver is plotted in Fig. 2(a). The average throughput is about 3 KB/s.

Sender is the Bottleneck: In this scenario, the sender application is not reading fast enough. Average throughput is about 3 KB/s, as shown in Fig. 2(b).

Network is the Bottleneck: In this scenario, we introduce packet losses to the network, and the network protocol is not transmitting fast enough. The average throughput is about 5 KB/s, as shown in Fig. 2(c).

Summary: We have demonstrated that for all the three scenarios, the data transfer is slow - it takes up to one minute to download the 200 KB data. Though the exact throughput vary in three scenarios, the application symptoms are **similar**, regardless of the three different bottlenecks.

4   Design

We notice that various issues causing slow data transfer between machines are largely classified into 3 types based on where the problem is: (1) sender application; (2) receiver application; or (3) networking channel. We hope to quickly and automatically decide which type of problems is causing slow data downloading.

We also notice the pull-model is a super-set of push-model, and the push-model is exactly the response data delivery part of the pull-model. Hence we break down our solution into two parts. First, we focus on the push-model and derive the algorithm based on our observations with some experiments and our

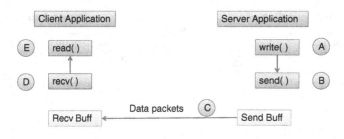

Fig. 3. The design for push model

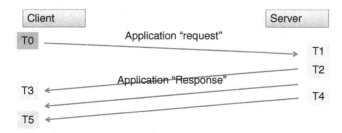

Fig. 4. Time line of pull model data transfer

analysis of network protocols. The solution relies on network-layer knowledge of network buffer queue size. Second, stepping on the solution of push-model, we extend the solution to solve the pull-model problem. The solution introduces a state-machine, and by incorporating the application-layer knowledge, it can deduct the bottleneck that causes the slow data transfer.

4.1 Design for the Push Model

Considering the data delivery process in push-model, a typical data flow in any TCP-based data transfer is illustrated in Fig. 3. Five steps with regard to system calls and network transmissions are as below: (1) at Step-A, the server application issues a write() system call and the application data is copied to socket send buffer. (2) at Step-B, the server's TCP layer issues send() call and sends some data to the network; The amount of data is subject to TCP's congestion control and flow control. (3) at Step-C, Network will route the data hop-by-hop to the receiver; IP routing protocol is in play in this part. (4) at Step-D, The client's TCP layer will receive the data with recv() system call. The data are put in receive buffer. (5) at Step-E, The client application issues read() call to receive the data and copy to user space.

During this course, there are could be 3 types of problems: (1) Server-application bottleneck. The server application may not have the data ready, this could be caused by multiple reasons. For instance, the server machine may be over-committed (e.g., running multiple applications), hence the particular application does not get chance to write(); the server application may have too many

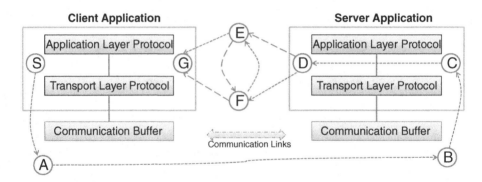

Fig. 5. The design for pull model

threads running and hence the write() thread is not being scheduled on cpus; the application's data-preparation business logic is slow, etc. (2) Client-application bottleneck. The client application may be too busy to issue read() call. Similar to server side, possible causes include machine over-commitment, and so on. (3) Network-side bottleneck. The network (including the TCP protocol) is unable to delivery data fast enough. Possible causes are poor network channel (e.g., high losses, low bandwidth) or poor TCP/IP configuration or tunings. Invariably, the symptom is the data are not being able to pushed through to the receiver side.

After careful analysis of the symptoms and the causes, we propose to isolate the problem into the above 3 types based on the knowledge of the TCP sockets, specifically, the queue lengths of the send and receive buffers.

In normal scenarios where data delivery is fast, the receiver's receive queue size should mostly be zero, indicating that the receiver is fast enough to consume the data received by copying the data from socket buffer to user space. The sender's send queue should be non-zero, indicating the fact that the data are produced fast enough to supply the consumption. On the other hand, when the receive queue is non-zero, that indicates client-side bottleneck; when the send queue is zero, that indicates the server-side bottleneck.

When data transfer is indeed slow, and neither the *server application* nor the *sender application* is the slowing down the data transfer, we can conclude that the data transmission (i.e., the network) between machines is slow. Such network-side issue may involve both ends of the transmission. Also note that this type of issue is not limited to networking or transport protocol themselves (e.g., TCP/IP), it may be caused by the slowdown of the sender/receiver OS (e.g., TCP/IP protocol processing).

4.2   Design for the Pull Model

The solution for pull-model is similar to the push-model solution, with the key difference of added phase of client-request. The server will not send back data until it receives data request. We illustrate the time line of a typical pull-data transfer in Fig. 4. When client needs to do a pull-based data delivery, it firstly sends out a request at T0; after the network delivered the request to the server at T1, the server then prepares for the data, followed by sending back the data at T2 by a sequence of write() calls. The sending completes at T4. After network transmission, the client begins receiving data at T3, and completes receiving at T5. Note that though other time stamps are strictly ordered, the ordering of T3 and T4 may vary depending on the exact scenarios. Specifically, for small data transfer, T4 may precede T3 since a single write() call may suffice. For large data transfer, T3 typically precede T4.

From the above process, we can see that if the data request failed reach the server, then the data delivery will not happen, hence not completed. Though the symptom of this scenario is different from the scenario where the delivery is slow, we decide to solve both scenarios since the data are not completely delivered. To provide a complete solution for pull-model problem, we need to distinguish between various possible failing components (e.g., client/network/server) for the client-request phase as well.

We propose a state-machine based algorithm in the following. Figure 5 is a diagram illustrating the states monitored in a state machine according to some embodiments. The state machine narrows the location of a problem in a data transfer operation to one of three realms: a sender realm that encompasses the sender or provider of the data (e.g., a server), a receiver realm that encompasses the recipient or receiver of the data (e.g., a client), and realm that encompasses the communication link(s) that convey the transferred data. In Fig. 5, the states are depicted with or near the component(s) whose action or actions cause a change from one state to another.

A state engine process is fed the necessary knowledge to monitor or identify the progress of a data transfer from start (i.e., at state S) for a pull-based transfer or at state C for a push-based transfer C to finish (at state G). This may require the OS of the two entities (e.g., client and server in Fig. 5) and the applications that use the data (e.g., client application, server application) to emit certain types of information at certain times. For instance, the receiver (e.g., the recipients OS and/or the application) logs events at one or more protocol layers, such as generation and dispatch of a data request, transmission of the request from the receiver machine, receipt of a first portion of data, and receipt of the last portion of the data. Similarly, the data provider (e.g., the providers OS and/or the application) logs events at one or more protocol layers, such as receipt of a data request, preparation of the data, dispatch of the first portion of the data, and dispatch of the final portion of the data.

5   Solution

Our proposed algorithm aggregates the information from the sender and the receiver nodes. For each node, the information from both the application layer and transport layer are collected and utilized. The heart of the algorithm is the Bottleneck Determination Engine (BDE). Internally, the algorithm maintains the current state of the data transfer and performs state transition when appropriate, which is handled by the State Transition Engine (STE). The state transition is based on the information collected by the Information Collection (IC), which collects and aggregates the information from both ends of the data transfer.

The algorithm we present has the following key design principles: (1) *Distributed information aggregation*. The solution utilizes both client and server knowledge in order to identify the causes; (2) *Cross-layer information aggregation*. The solution utilizes both application layer knowledge (e.g., the various types of application logs) and transport layer knowledge (e.g., the send queue and receive queue sizes); and (3) *State-machine based expert system*. The state transform is triggered by the aggregated knowledge from different machines and different layers.

5.1   Collecting Queue Size Information

In order for the algorithm to identify the bottleneck, it needs to gather information about queue sizes at transport layer on both sender and receiver sides. There are many ways to collect such information. In this work, we list two examples of these tools/utilities, namely *netstat* and *ss*. These two utilities can be issued on the hosts as commands. (a) *netstat* (network statistics) is a command-line tool that displays network connections (both incoming and outgoing) and network protocol statistics. For the purpose of this work, we utilize the sizes of the send queue and the receive queue for the TCP/IP sockets. This tool is available in many OS including Unix and Windows. (b) *ss* command is used to show socket statistics. It can display stats for TCP as well as other types of sockets. Similar to netstat, ss can also display the send and receive queue sizes, which can be used by our algorithm.

When utilizing these tools/utilities to collect information about transport level queue sizes, the collection process is desired to meet the following requirements. First, the queue size collection process should continuously output the queue sizes for interested TCP/IP sockets during the duration of the slow data transfer. If the collection tools (e.g., netstat and ss) can only display the instantaneously values of that time point, they need to be invoked repeatedly to gather multiple data points during the data transfer course. The collection process is desired to be long enough to contain multiple data points to avoid false alert and ensure certain degree of confidence about the collected information.

5.2   State Transitions

A pull-based data transfer begins in state S when client (e.g., application) issues a data request. When the client (e.g., application protocol, transport protocol) logs queuing of the request, the data transfer transitions from state S to state A (the request has been queued in the clients send buffer). When the clients send buffer is empty or there is some other indication that the request was transmitted from the client, the data transfer transitions from state A to state B (the request has been transmitted on communication link(s)). When receipt of the request is logged by server (e.g., application protocol, transport protocol), the transfer transitions from state B to state C (the server application has received the request). A push-based data transfer may be considered to start at state C.

After the server prepares and sends a first portion of the data to be transferred (e.g., the first byte, the first packet), the data transfer operation transitions to state D (the data response is underway). Progress of the data transfer may now depend on the amount of data to be transferred and the speed with which it is conveyed by communication link(s).

Note that it could be that there are more than a single bottleneck exist at the same time. For example, both client and server can be bottlenecks. As another example, at first half the data transfer, the client is the bottleneck, while at the second half the data transfer, the network is the bottleneck. Though it is possible to handle these complicated cases easily by splitting the time duration into smaller units and output the algorithm output for all these smaller units, for simpler presentation, we choose to focus on the scenarios where only a single bottleneck exists. However, the presented algorithm/solution can be easily converted to accommodate the above more complicated scenarios.

The determination regarding which component is the bottleneck when the state is stuck at States of D/E/F/G/H depends on the transport layer information. Specifically, below is the table illustrates all the three scenarios and the information needed to make the corresponding the decision. Specifically, if the Receive Queue on the client is not zero, then the client application is the bottleneck. If the Send Queue on the server is zero, then the server application is the bottleneck. If the client's Receive Queue is zero and the server's Send Queue is non-zero, then the network component is the bottleneck.

6   Evaluation

6.1   Prototype

We implemented the proposed algorithm with a prototype in Python. The prototype works in off-line mode, and it can be used to determine the bottleneck during a specific time duration where the data transfer is slow. The prototype includes a netstat-based information collection script (i.e., "netstat -Ttp") to repeatedly output the TCP connection information on both the sender and receiver hosts. A random delay period is injected into two succeeding netstat invocations, with the average delay value being 2 s. The output of each netstat invocation is prefixed by a time stamp in the granularity of 1 ms.

Table 1. Three scenarios in push model

Bottleneck	Recv. Que.	Send Que.	Notes
Client	Non-zero	Any	Blocked receiving
Server	Any	Zero	Blocked sending
Network	Zero	Non-zero	Blocked delivery

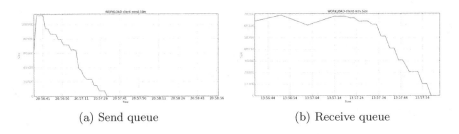

(a) Send queue   (b) Receive queue

Fig. 6. Client is the bottleneck

The prototype firstly takes the user input of a configuration file. The file defines the *directional data transfer connection*, the information source and the time duration. The directional data transfer connection is defined by tuple of (src host, src port, dst host, dst port). Note that unlike common definitions of TCP connections which are bidirectionally, the definition of connections here is directional, as the slow data transfer has the notion of *direction*. The information source is a local directory which contains the *netstat* output from both ends. The time duration is defined by the beginning time stamp and the ending time stamp (Table 1).

The prototype also takes user input of an "input" directory, where the *netstat* outputs are stored. Since data transfer is directional, the prototype treats each TCP connection as two *directional data transfer connections*. Thus, each extracted TCP data transfer is recorded in a separate csv file. Each csv file contains lines of tuples in the format of (time-stamp, recv-queue, send-queue). One special treatment about the processing is the handling of time zones. It is possible for the two hosts of the sender and receiver to be in different time zones. Because of this, the output of the netstat might be mis-aligned if not treated accordingly. To accommodate the possible differences in time zones, the configuration file allows the user to specify time zones for the two end hosts. Internally, the prototype will perform time zone conversion to align the csv data.

Based on the user input in the configuration file, the prototype only extracts the interested directional data transfer connections. Furthermore, only the interested time periods are extracted based on the user-input. Given a particular directional data transfer connection which needs to diagnose, the prototype determines the bottleneck based on the rules presented earlier.

To reliably determine the bottleneck, the prototype also filters out some spike values which might cause false alert. Specifically, the decision about whether a queue size is zero or non-zero has to persist for at least certain number of invocations and at least certain amount of time duration. For instance, in one debugging case, with the average netstat invocation of 2 s, in order for the prototype to conclude that the client side is the bottleneck, the receive queue size needs to be zero for at least 10 s.

6.2 Results

We evaluated the built off-line prototype. The usage scenario is as follows. After the user noticed slow data transfer during some time period in the form of beginning time stamp and ending time stamp, the user runs the prototype to diagnose the issue and determine the bottleneck.

We have used this prototype to identify bottlenecks in the following LinkedIn production investigations/issues: (1) Databus [1] bootstrapping; where a client receives bootstrapping events from the server using TCP/IP protocol; (2) Voldemort [7] cluster expansion; where data receivers fetch data from data senders using TCP/IP protocol. We have found that for the first scenario, the Databus server (i.e., the data sender) is the bottleneck; while in the second scenario, the networking part is the bottleneck.

To demonstrate all the three scenarios where each of the scenarios has a different type of bottleneck (e.g. sender/receiver/networking), we designed a set of experiments using the a custom-build workloads. The workload can mimic the three types of bottleneck based on user inputs. For each of the scenarios, a single separate TCP connection is created.

Client is Slowly Receiving: We force the receiver (i.e., the client) to slow down the data receiving by injecting delays between the calls to read() in application code, which represents a scenario where the client is the bottleneck. As shown in Fig. 6, the client side (receiver side) has certain RecvQ buildup, which is indicated by the non-zero values. These non-zero values last for a while, so that the algorithm can conclude that the client is the bottleneck.

Server is Slowly Sending: We force the sender to slow down the data sending. We inject delays between the calls to write() in application code, which represents a scenario where the sender is the bottleneck. As shown in Fig. 7, the server side has zero SendQ values. The lasting period is significant to allow the algorithm to conclude the sender bottleneck. Note that there is a spike about the SendQ. Since the spike does not persist for long enough period, it is internally filtered out by the algorithm.

Network is Slowly Transmitting: We created a scenario where TCP is slowly transmitting the data. Specifically, we inject delays and latencies to the network

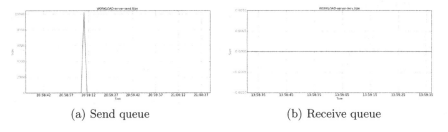

(a) Send queue   (b) Receive queue

Fig. 7. Server is the bottleneck

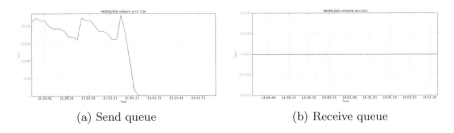

(a) Send queue   (b) Receive queue

Fig. 8. Network is the bottleneck

path, such that TCP can only transmit at a very low throughput. As shown in Fig. 8, the SendQ values are non-zero, while the RecvQ is zero. These values of typical of a fast data delivery. Since we see very low data transfer, the algorithm conclude that it must be networking problem.

7   Related Work

Networking protocol is a critical component of distributed system. Many protocols and algorithms [5, 6, 8] are proposed to ensure and improve high transmission rate in various scenarios. To estimate the available bandwidth of a link or path, various methods have been proposed, each with different tradeoffs between accuracy and intrusiveness [4].

Many works have been done to diagnose computer performance problem, but most of them focus on the performance of a specific component, for instance, system performance bottleneck about a particular OS (e.g., Linux kernel 2.6.18 [3]) or a particular protocol (TCP Reno [2], etc.). Though there are scattered knowledge/experiences about debugging the targeted slow-data-transfer problem, to our best knowledge, we have not seen any algorithm or prototype that is equivalent to our proposed algorithm/prototype. For the overall problem of slow data transfer, all sources we can find only rely on performance extertise/experience, and there is no single algorithm or resource to achieve the same as our algorithm does. Moreover, we did not find any automated prototype/tool for that purpose.

8   Conclusion

We proposed and implemented an algorithm to automatically determine the bottleneck component along the data transfer path. It reduces the efforts when diagnosing the problem and eventually root-causing the problem.

References

1. Das, S., Botev, C., et al.: All aboard the databus!: Linkedin's scalable consistent change data capture platform. In: SoCC 2012, New York, NY, USA (2012)
2. Grieco, L.A., Mascolo, S.: Performance evaluation and comparison of Westwood+, New Reno, and Vegas TCP congestion control. SIGCOMM Comput. Commun. Rev. **34**(2), 25–38 (2004)
3. https://bugzilla.redhat.com/show_bug.cgi?id=705989: Performance regression on tcp_stream throughput
4. Nam, S.Y., Kim, S.J., Lee, S., Kim, H.S.: Estimation of the available bandwidth ratio of a remote link or path segments. Comput. Netw. **57**(1), 61–77 (2013). http://dx.doi.org/10.1016/j.comnet.2012.08.015
5. Schulzrinne, H., Casner, S., Frederick, R., Jacobson, V.: RTP: a transport protocol for real-time applications (2003)
6. Sinha, P., Nandagopal, T., Venkitaraman, N., Sivakumar, R., Bharghavan, V.: WTCP: a reliable transport protocol for wireless wide-area networks. Wirel. Netw. **8**(2/3), 301–316 (2002)
7. Sumbaly, R., Kreps, J., Gao, L., Feinberg, A., Soman, C., Shah, S.: Serving large-scale batch computed data with project voldemort. In: Proceedings of the 10th USENIX Conference on File and Storage Technologies, FAST 2012, Berkeley, CA, USA, p. 18 (2012)
8. Zhu, J., Roy, S., Kim, J.H.: Performance modelling of TCP enhancements in terrestrial-satellite hybrid networks. IEEE/ACM Trans. Netw. **14**(4), 753–766 (2006)

Schedule Compaction and Deadline Constrained DAG Scheduling for IaaS Cloud

Fuhui Wu[✉], Qingbo Wu, Yusong Tan, Wei Wang, and Xiaoli Sun

College of Computer, National University of Defense Technology,
410073 Changsha, China
{fuhui.wu,qingbo.wu,yusong.tan,wei.wang,xiaoli.sun}@nudt.edu.cn

Abstract. Most cloud workflow scheduling algorithms assume that resources are charged under an ideal pay-as-you-go model, which may not be the case in real production cloud systems. Currently, most IaaS cloud providers charge users on billing cycle basis. If a resource is terminated before one billing cycle, the payment is still rounded up to one cycle. To address this problem, we firstly formalized it using bin-package method. Then, we propose a DAG schedule compaction algorithm of IC-⋆-SC, which compacts schedules generated by already exist algorithms to reduce resource requirement. Based on the compaction idea, we also propose a deadline constrained DAG scheduling algorithm of IC-SC. We compare our algorithms with state-of-the-art algorithms of IC-PCP and IC-PCPD2, and use 2 well-known scientific workflow applications for evaluation. Experimental results show that our algorithms reduce monetary cost drastically.

Keywords: Schedule compaction · IaaS cloud · Workflow scheduling · Deadline

1 Introduction

Many applications consist of a number of cooperative tasks which typically require more computing power beyond single machine capability. An easy and popular way is to describe complex applications using high-level representation in workflow method. As a single workflow can contain hundreds or thousands of tasks, the ever-growing data and computing requirements of workflows demand a higher-performance computing environment in order to execute the workflow in reasonable amount of time.

Traditional HPC systems are mostly dedicated and statically partitioned per administration policy. Owning a HPC system is inefficient in adapting to the surge of resource demand. With the development of cloud computing, workflow applications find a new way to solve this problem. Cloud computing targets at providing computing as service. With the IaaS cloud becoming mature, the long dream of providing computing service as the 5th utility [1] (after water, electricity, gas, and telephony) is gradually turning into reality.

© Springer International Publishing Switzerland 2015
W. Qiang et al. (Eds.): CloudCom-Asia 2015, LNCS 9106, pp. 16–28, 2015.
DOI: 10.1007/978-3-319-28430-9_2

In IaaS cloud environment, resources are served in a pay-per-use model and provisioned dynamically. Therefore, resources are provisioned on demand. To satisfy different QoS requirement of users, resource providers offer heterogeneous resources with various processing capabilities and prices. Usually, fast resource costs more money than that of slow resource. Thus, the cost/time trade-off problem becomes a hot topic in the literature. Ideally, users want to run their applications as fast as possible with minimum cost. If applications are not time-critical, a little delay can be tolerated for cost saving.

There are already algorithms concerning both time and monetary cost for cloud computing environment. However, most of them thought that leased resources can be completely utilized, and they are charged in an ideal pay-as-you-go model. This is not the case in real production system. As the Amazon EC2 cloud for example, they charge resources by hour. If a resource is terminated before one hour, the cost is still rounded up to one hour.

2   Related Works

The workflow scheduling problem has been studied extensively over the years. Kwok et al. [2] and Yu et al. [3] surveyed workflow scheduling on Multiprocessor system and distributed environments respectively. And Wieczorek et al. [4] focused on the taxonomies of the multi-criteria grid workflow scheduling algorithms. Most of them focus on decreasing the schedule length, which are classified as best-effort scheduling.

In contrast to best-effort scheduling, QoS constrained workflow scheduling is sometimes more close to that of real world application's requirement. In [5], Chen et al. studied the workflow scheduling problem with various QoS requirements and presented an Ant Colony Optimization (ACO) based approach. The target of their algorithm is to find a schedule that meets all QoS constraints and optimizes the user-preferred QoS objective.

Among all QoS constrained workflow scheduling problems in service oriented environment, deadline-constrained scheduling is one primary problem in the literature. DTL [6], DBL [7], and DET [8] are three heuristics that solve the deadline constrained workflow scheduling problem. Their basic idea is to distribute deadline over task partitions. Abrishami et al. [9,10] presented *Partial Critical Path* based scheduling algorithms, which distribute deadline in PCP-wise manner.

Motivation: To the best of our knowledge, there are few researches concerning about improving utilization of free time slots generated by task dependencies and billing cycle charged resources. In [11], they proposed a two-stage schedule compaction for duplication-based DAG scheduling. However, their target is to reduce resource number, which can not be directly applied to IaaS cloud system. In this paper, we address this highly unexplored aspect of DAG scheduling in IaaS cloud, and proposed a schedule compaction algorithm and a deadline constrained DAG scheduling algorithm in the following sections.

3   System Model and Definitions

3.1   System Model

A workflow is modeled by a directed acyclic graph (DAG): $\mathcal{G} = (\mathcal{T}, \mathcal{E})$, where $\mathcal{T}$ consists of a set of tasks $\{t_1, t_2, ..., t_{|\mathcal{T}|}\}$ and $\mathcal{E}$ consists of a set of directed edges $\{e_i^j | (t_i, t_j) \in \mathcal{E}\}$. Each task t_i is associated with a certain amount of computation workload wl_i. And each edge e_i^j carries d_i^j size of data send from task t_i to task t_j. Given a task graph, a task without any parent is called an *entry* task, and a task without any child is called an *exit* task. Without lose of generality, two dummy tasks t_{entry} and t_{exit} are added to the begin and end of $\mathcal{G}$. These dummy tasks have zero computation workload and zero-data dependencies to the actual *entry* or from the *exit* tasks.

There are theoretical unlimited resources provisioned by cloud providers and charged on a pay-as-you-go basis. The service provider offers several computation resource types $\mathcal{RT} = \{rt_1, rt_2, ..., rt_{|\mathcal{RT}|}\}$ to satisfy various QoS requirements of users. Each resource type rt_k is associated with a two-tuple information of $< ECU, price >$, where ECU denotes the processing capability, and $price$ denotes the monetary cost per charging unit. In general, scheduling task on faster resource costs more. In this paper, we assume a single cloud, where all resources are connected by a homogeneous network. The communication startup time (delay) L for each resource and communication bandwidth BW between any two resources are the same.

Hence, the execution time of a task on a resource is calculated as: $ET_{t_i}^{r_l} = ET_{t_i}^{rt(r_l)} = \frac{wl_i}{rt(r_l).ECU}$;[1] the data transmission time from task t_i to task t_j is calculated as: $TT_{t_i}^{t_j} = L + \frac{d_i^j}{BW}$. If two tasks are assigned to a same resource, the transmission time is zeroed. The monetary cost of network transmission is not considered in this paper. This assumption is reasonable as most popular commercial IaaS cloud providers charge users according to their leasing times for computing resources only.

3.2   Schedule Definition

Based on this system model, a schedule is defined as: $S = < \mathcal{R}, \mathcal{M}, makespan, cost >$. $\mathcal{R} = \{r_1, r_2, ..., r_{|\mathcal{R}|}\}$ is a set of used resources. $\mathcal{M}$ consists of all mappings of task to resource, with $m_{t_i}^{r_l} = < t_i, r_l, ST_{t_i}^{r_l}, FT_{t_i}^{r_l} >$, where $ST_{t_i}^{r_l}$ is the start time of task t_i on resource r_l and $FT_{t_i}^{r_l}$ is the finish time of task t_i on resource r_l. Then, the start time $st(r_l)$ and termination time $ft(r_l)$ of a resource r_l are calculated as: $st(r_l) = \min_{m_{t_i}^{r_l} \in \mathcal{M}} \{ST_{t_i}^{r_l}\}$, $ft(r_l) = \max_{m_{t_i}^{r_l} \in \mathcal{M}} \{FT_{t_i}^{r_l}\}$. And, the rounded up lease time $lt(r_l)$ for r_l is set with the minimal integer multiples of billing cycle length τ: $lt(r_l) = \lceil \frac{ft(r_l) - st(r_l)}{\tau} \rceil \times \tau$.

---

[1] $rt(r_l)$ is the resource type of resource r_l.

For a generated schedule S, the *makespan* is the overall schedule length. And the total cost C is the overall monetary payment. They are calculated as:

$$makespan = \max_{m_{t_i}^{r_l} \in \mathcal{M}} \{FT_{t_i}^{r_l}\}, \quad C = \sum_{l=1}^{l=|\mathcal{R}|} \frac{lt(r_l)}{\tau} \times rt(r_l).price.$$

3.3   Bin-Package Problem Formalization

In this paper, we take the DAG scheduling problem in IaaS cloud as a bin-package problem, where each resource is taken as a bin, and tasks of DAG are taken as materials to be packed. However, the times consumed by one task when assigned to various resource types are different, which makes it a non-traditional bin-package problem. In order to explain our algorithms, we first make some definitions as following:

Definition 1. *All tasks in $\mathcal{T}$ are divided into two parts of already assigned tasks $\mathcal{T}_a$ and unassigned tasks $\mathcal{T}_u$: $\mathcal{T} = \mathcal{T}_a \cup \mathcal{T}_u$.*

Definition 2. *$\mathcal{T}_{r_l}$: It contains all tasks that are assigned to resource r_l. An execution order is imposed on all tasks of $\mathcal{T}_{r_l}$, and a position $pos(t_i)$, starting from 0, is assigned for each task t_i.*

Definition 3. *$EST(t_i)$, $EFT(t_i)$: The earliest start time of task t_i, which is determined by its immediate parents and the pre-execution task t_k if exist.*

$$EST(t_i) = \begin{cases} \max\{\max_{t_j \in \mathcal{P}(t_i)}\{EFT(t_j) + TT_{t_j}^{t_i}\}, EFT(t_k)\}, & \exists t_k : pos(t_i) = pos(t_k) + 1 \\ \max_{t_j \in \mathcal{P}(t_i)}\{EFT(t_j) + TT_{t_j}^{t_i}\}, & otherwise \end{cases}$$
$$(1)$$

$$EFT(t_i) = EST(t_i) + ET(t_i) \tag{2}$$

$\mathcal{P}(t_i)$ contains all the immediate predecessors of task t_i. Assuming $\mathcal{RT}.fastest$ being the fastest resource type in $\mathcal{RT}$, then the ET is calculated as:

$$ET(t_j) = \begin{cases} ET_{t_j}^{r_l}, & \exists r_l : m_{t_j}^{r_l} \in \mathcal{M} \\ ET_{t_j}^{\mathcal{RT}.fastest}, & otherwise \end{cases} \tag{3}$$

Definition 4. *$ST(t_i), FT(t_i)$: The start time and finish time of already assigned task t_i. They are set with the earliest start time and earliest finish time on mapped resource r_l. Hence, they can be changed (usually be delayed) with the updating of EST and EFT.*

Definition 5. *$LFT(t_i), LST(t_i)$: The latest finish time of task t_i is determined by its immediate children and position if it is assigned to resource.*

$$LFT(t_i) = \begin{cases} \min\{\min_{t_j \in \mathcal{C}(t_i)}\{LST(t_j) - TT_{t_j}^{t_i}\}, LST(t_k)\}, & \exists t_k : pos(t_k) = pos(t_i) + 1 \\ \min\{\min_{t_j \in \mathcal{C}(t_i)}\{LST(t_j) - TT_{t_j}^{t_i}\}, st(r_l) + lt(r_l)\}, & pos(t_i) = |\mathcal{T}_{r_l}| - 1 \\ \min_{t_j \in \mathcal{C}(t_i)}\{LST(t_j) - TT_{t_j}^{t_i}\}, & otherwise \end{cases}$$
$$(4)$$

$$LST(t_i) = LFT(t_i) - ET(t_i) \tag{5}$$

$\mathcal{C}(t_i)$ contains all the immediate successors of t_i. To calculate the LFT of task t_i, the $\min\limits_{t_j \in \mathcal{C}(t_i)} \{LST(t_j) - TT_{t_j}^{t_i}\}$ part assures that its immediate children are able to start no late than the LST of them. If there is a task t_k assigned right behind it, it should finish no late than the LST of t_k. If it is the last task of assigned resource r_l, the $LFT(t_i)$ should not exceed $st(r_l) + lt(r_l)$. Otherwise, the monetary cost of r_l will increase.

Definition 6. *Insert Operation: A legal insert operation denotes a valid task-resource mapping for this special bin-package problem. To define a legal insert operation, we introduce a concept of slot first. A slot is a time fragment in a resource r_l. Assuming the position of slot in resource r_l is $pos(slot)$, the start time $st(slot)$ and finish time $ft(slot)$ are defined as:*

1. *If $\mathcal{T}_{r_l} = NULL$, meaning r_l is a new resource without any task assigned to it, then $st(slot) = 0$, $ft(slot) = +\infty$, and $pos(slot) = 0$;*
2. *else,*
   (a) if $pos(slot) = 0$, then $st(slot) = \max\{LFT(t_i) - lt(r_l), 0\}$, and $ft(slot) = LST(t_j)$, where $pos(t_i) = |\mathcal{T}_{r_l}| - 1$ and $pos(t_j) = pos(slot)$;
   (b) else if $pos(slot) > 0$ and $pos(slot) < |\mathcal{T}_{r_l}|$, then $st(slot) = EFT(t_i)$, and $ft(slot) = LST(t_j)$, where $pos(t_i) = pos(slot) - 1$ and $pos(t_j) = pos(slot)$;
   (c) else $pos(slot) = |\mathcal{T}_{r_l}|$, then $st(slot) = EFT(t_i)$ and $ft(slot) = EST(t_j) + lt(r_l)$, where $pos(t_i) = |\mathcal{T}_{r_l}| - 1$ and $pos(t_j) = 0$.

Then, a task t_i is able to be inserted into a *slot* in resource r_l, subjecting to: (i) $(\max\{EST(t_i), st(slot)\} + ET_{t_i}^{r_l}) \leq ft(slot)$; (ii)$EFT(t_i) \leq LFT(t_i)$.

Definition 7. *Append operation: It is an extension in case 2(c) of insert operation, except that the $ft(slot)$ is set $+\infty$. And, a legal append operation should satisfy the same conditions as insert operation.*

4   IC-⋆-SC: DAG Schedule Compaction

In this section, we propose a DAG schedule compaction algorithm named IC-⋆-SC. The IC-⋆-SC algorithm itself doesn't assign tasks to resources. It works on schedules generated by already exist algorithms. The symbol of ⋆ in IC-⋆-SC denotes the algorithm that generates schedules to be compacted.

4.1   The Basis of IC-⋆-SC

The basic idea of IC-⋆-SC algorithm is to compact tasks, already assigned to resources, into fewer resources. As the resources leased from IaaS cloud are charged in billing cycle unit, the monetary cost can be reduced only if all tasks that cover a billing cycle are compacted to other leased resources. Hence, we propose a split-wise DAG schedule compaction algorithm.

To introduce the conception of *split*, all resources in $\mathcal{R}$ are divided into two parts of *compacted resource* ($\mathcal{R}_c$) and *scheduled resource* ($\mathcal{R}_s$), where $\mathcal{R}_s$ is initialized as R, and $\mathcal{R}_c$ is empty at the beginning. Each *split* is chosen only from $\mathcal{R}_s$. After compaction, all leased resources are in $\mathcal{R}_c$ and $\mathcal{R}_s$ is empty.

A *split* is a set of tasks covering one billing cycle of a resource in $\mathcal{R}_s$. Each resource r_l in $\mathcal{R}_s$ is divided into $nb\_split(r_l)$ splits in (6). An $index(r_l)$ parameter is set for every resource r_l to record current *split* position. After compaction of one *split* in a resource, it is increased by one and point to the next *split*. The *split* start time and finish time are computed in (7) and (8).

$$nb\_split(r_l) = \left\lceil \frac{\max\limits_{\forall t_i : m_{t_i}^{r_l} \in \mathcal{M}} \{LFT(t_i)\} - \min\limits_{\forall t_i : m_{t_i}^{r_l} \in \mathcal{M}} \{EST(t_i)\}}{\tau} \right\rceil \tag{6}$$

$$st(split) = \max\limits_{\forall t_i : m_{t_i}^{r_l} \in \mathcal{M}} \{LFT(t_i)\} - \tau \times (nb\_split(r_l) - index(r_l)) \tag{7}$$

$$ft(split) = st(split) + \tau \tag{8}$$

For each compaction process, a *nextSplit()* procedure is called to select a next *split*. There may be several *split* candidates from different resources in $\mathcal{R}_s$. The one with the smallest split start time will be selected. And the more expensive one is chosen if ties. After a *split* has been determined, corresponding tasks should be added to *split*. For a task t_i on selected $r(split)$ (the resource where *split* from), it is added into *split* if the start time and finish time $[ST_{t_i}^{r(split)}, FT_{t_i}^{r(split)}]$ intersects with $(st(split), ft(split))$.

4.2   The IC-⋆-SC Algorithm

The IC-⋆-SC algorithm consists of two main procedures of *latest time updating* and *split compaction*. The first procedure determines the *LST*s and *LFT*s of tasks on resources they are mapped. The target is to increase intervals between neighbor tasks on the same resource, which produces larger free time *slots* for compaction. The second procedure achieves schedule compaction actually. Algorithm 1 describes the overview procedure of IC-⋆-SC algorithm.

For initialization, it sets the *LST* and *LFT* for the dummy exit task t_{exit}. The *LST* is set with a deadline, if required. Otherwise, it is set with the makespan of the original schedule. After that, the IC-⋆-SC iteratively calls *latest time updating* and *split compaction* procedures until all candidate *splits* are processed.

To compute the latest times of all tasks. It tries to move tasks as late as possible on mapped resources, which can be implemented using a up-wards breadth-first search method from the dummy exit task t_{exit}. The *LFT* of a task is calculated using (4–5).

4.3   Split Compaction Procedure

For a selected *split*, it iteratively tries to insert tasks in the *split* to other resources. A success insert operation is defined in Definition 6. The resources in $\mathcal{R}_c$ are selected firstly. If the insertion fails, resources in $\mathcal{R}_s$ are selected.

Algorithm 1. The IC-⋆-SC Algorithm

Input: S: the schedule to be compacted;
    $makespan$: the makespan of schedule S;
    D : the deadline, if required, for $\mathcal{G}$.
Output: S': a new schedule after compaction of S.
 1: **if** deadline is required **then**
 2:    $LST(t_{exit}) = D$
 3: **else**
 4:    $LST(t_{exit}) = makespan$
 5: **end if**
 6: update the LSTs and LFTs of all tasks in $\mathcal{T}$;
 7: **while** has an unprocessed split **do**
 8:    $split \leftarrow nextSplit()$;
 9:    call for $compact(split)$;
10: **end while**

After the insertion attempt of all tasks in selected $split$, there are 2 cases. If all insertions succeed, all tasks in $r(split)$ are separated into three parts according to their positions. Tasks in $split$ are inserted to other resources. Tasks before $split$ are kept in $r(split)$ and the resource $r(split)$ is moved to $\mathcal{R}_c$. Tasks after $split$ are assigned to a new resource with the same resource type as $r(split)$, and this new resource is added to R_s. However, if the insertion fails, it should roll the schedule back to the point before compaction of $split$.

Algorithm 2. The $compact$ Procedure

Input: $split$: the split to be processed.
Output: S_{tmp}: an intermediate schedule after compaction of $split$.
 1: **for all** $t_i \in split$ **do**
 2:    insert t_i to the first-fit resource in R;
 3:    **if** insertion succeeds **then**
 4:       iteratively update the ESTs and EFTs of $succ(t_i)$ and the immediate task
          following t_i on $r(split)$;
 5:    **else**
 6:       **break**;
 7:    **end if**
 8: **end for**
 9: **if** all tasks in $split$ are inserted into other resources **then**
10:    (i) schedule all tasks, if exist, after $split$ to a new resource r_k with the same type
       as $r(split)$, and add r_k to $\mathcal{R}_s$;
11:    (ii) move $r(split)$, if there are still tasks assigned to it, to $\mathcal{R}_c$;
12:    (iii) update the LSTs and LFTs of all tasks in $\mathcal{T}$;
13: **else**
14:    rollback S_{tmp} to the point before compaction of $split$;
15: **end if**

4.4   Time Complexity

The most time consuming part of the overall IC-⋆-SC algorithm is the *updateLatestTime* and *compact* procedures. To update the LFTs, IC-⋆-SC adopts breadth-first traverse method from t_{exit}. It costs a time complexity of $O(|\mathcal{T}| + |\mathcal{E}|)$, which approximately equals to $O(|\mathcal{T}|^2)$ for dense workflows with a edge between any two tasks. As the hole IC-⋆-SC algorithm compact all tasks in split-wise manner, and there are at most $|\mathcal{T}|$ splits, the *updateLatestTime* contributes $|\mathcal{T}|^3$ time complexity to the IC-⋆-SC algorithm.

The most time consuming part of the *compact* procedure is the insert operations for *split* tasks. To update the ESTs and EFTs for each insertion attempt at line 3 of Algorithm 2, the time complexity is $O(|\mathcal{T}| + |\mathcal{E}|)$, approximating to $O(|\mathcal{T}|^2)$, in the worst case. Overall, each task will be insert once. Hence, the time complexity for $\mathcal{T}$ tasks of all *compact* procedure is $O(|\mathcal{T}|^3)$.

At last, the time complexity of IC-⋆-SC is $O(|\mathcal{T}|^3)$.

5   IC-SC: Deadline Constrained DAG Scheduling

To take advantage of the bin-package idea in IC-⋆-SC algorithm. We propose a deadline constrained DAG scheduling algorithm that implements the bin-package mechanism at scheduling time.

5.1   IC-SC Algorithm

Deadline constrained DAG scheduling is a widely studied problem in the literature [6–10]. IC-PCP [9] and IC-PCPD2 [9] are state-of-the-art algorithms for billing cycle aware IaaS cloud environment. In order to introduce the IC-SC algorithm, we first make an review of the IC-PCP and IC-PCPD2 algorithms. The core conception of them is *partial critical path(PCP)* [9]. The details of *PCP* are not repeated here.

There are two main categories of workflow scheduling algorithms: *list* scheduling and *clustering* heuristics:

IC-PCP is a *clustering* heuristic. Two parts, that compose IC-PCP *clustering* heuristic, are:

1. *Clustering*. It determines how tasks are mapped to resources. IC-PCP maps all tasks in one *PCP* to the same resource to reduce data transmission times.
2. *Ordering*. It determines the execution order of all tasks on the same resource. The IC-PCP algorithm schedules all *PCP* tasks to a same resource, which is selected using a monetary cost optimization mechanism defined by them.

IC-PCPD2 is a *list* scheduling heuristic. It firstly assign subdeadlines to all tasks using the similar strategy to the IC-PCP algorithm. The actual scheduling is carried out in *list* scheduling manner, including two parts:

1. *Task selection*. It determines which task to schedule next. The IC-PCPD2 algorithm randomly selects a ready task.

2. *Resource selection.* It determines which resource the selected task is assigned to. The IC-PCPD2 algorithm always selects the cheapest resource, either an existing resource or a new one.

According to Abrishami et al., both IC-PCP and IC-PCPD2 have a promising performance, with IC-PCP performing better than IC-PCPD2 in most cases. However, the IC-PCP needs to schedule all tasks on PCP to the same resource. Moreover, the position of PCP task are limited on assigned resource. To utilize the bin-package idea of IC-$\star$-SC, we propose a deadline constrained DAG scheduling heuristic, named IC-SC. It also schedules workflow tasks in PCP manner. But the IC-SC algorithm doesn't need all PCP tasks to be assigned to a same resource. Moreover, the position for PCP task is more free.

The main structure of IC-SC is the same as IC-PCP. We don't repeatedly explain them here. The most obvious difference with IC-PCP lies in scheduling of PCP tasks. The IC-SC algorithm schedules each task in current PCP separately. It first tries to schedule PCP task to an already exist resource. It uses the insertion mechanism defined by Definition 6. The cheapest already exist resource that is able to insert the scheduling PCP task is chosen. And the scheduling PCP task is inserted to the first-fit slot in the chosen resource. If there is not an already exist resource that is able to insert the scheduling PCP task, two cases are considered. First, it tries to append the scheduling task to an already exist resource. The cheapest one that is able to append is assumed to be r_l. When implementing the append operation, we make a restriction that it should not increase a complete free billing cycle. As there are already 3 billing cycles after rounded up for example, the 4th billing cycle should be utilized at least partly by the appended task. Otherwise, this append operation is illegal. Second, it tries to lease a new resource, assuming to be r_l', and inserts the scheduling task to it. If r_l exists and r_l is cheaper than r_l', the scheduling task is appended to r_l. Otherwise, it is inserted to a new resource r_l'. The reason behind the append operation is to utilize the last slot of an already exist resource and reduce communication cost.

5.2   Time Complexity

The first part of the algorithm is to initialize the parameters of each workflow task, which requires the same time complexity of $O(|\mathcal{T}|^2)$ as IC-PCP algorithm. Then the second part of the algorithm is recursively scheduling workflow tasks to resources. We consider its overall actions instead of entering into details. Overall, the Parents Assigning procedure schedules each workflow task only once, and updates the parameters of its successors and predecessors. To schedule a task, there are three cases in all. First, there are at most $O(|\mathcal{T}|)$ slots in all the already exist resources for insertion. If insertion fails, there are also at most $O(|\mathcal{T}|)$ resources for appending. If a new resource has to be started, there are constant number of available resource types. Hence, the time complexity to schedule $|\mathcal{T}|$ workflow tasks is $O(|\mathcal{T}|^2)$. On the other hand, each task has at most $(|\mathcal{T}| - 1)$ successors and predecessors, the time complexity of the updating part for all workflow tasks is also $O(|\mathcal{T}|^2)$. Consequently, the overall time complexity of the

Parents Assigning procedure is $O(|\mathcal{T}|^2)$, which is also the time complexity of the IC-SC algorithm.

6 Performance Evaluation

6.1 The Initial Schedule Generating and Comparative Algorithms

We choose two state-of-the-art scheduling algorithms in the literature to generate initial schedules: IC-PCP and IC-PCPD2. Both of them are designed for billing cycle charged IaaS cloud. The schedules generated by them meet deadline constraint. To evaluate the performance of IC-SC algorithm, both IC-PCP and IC-PCPD2 are also taken as comparative algorithms.

6.2 Experimental Workflows

Figure 1(a) and (b) shows the approximate structures of selected workflows we choose from [12] to evaluate the performance of our algorithm. They are:

- Montage: astronomy;
- CyberShake: earthquake science.

They developed a workflow generator, which can create synthetic workflows. Using this workflow generator, they create different sizes for each workflow application in terms of total number of tasks. These workflows are available in DAX (Directed Acyclic Graph in XML) format from their website[2], from which we choose 6 sizes (50, 100, 200, 300, 400, 500) for each workflow.

6.3 Experimental Setup

In our experiments, we assume a IaaS cloud environment offering 9 different computation services, with different processing capability and prices as shown in Fig. 1(c). They satisfy that slower resource has higher performance/cost ratio. The average bandwidth between computation resources is set to 20MBps.

An important parameter of the experiments is the billing cycle. To evaluate the impact of short and long billing cycles on our algorithms, we consider two different billing cycles in the experiments: a long one equal to one hour, and a short one equal to five minutes.

All costs are normalized by the cost of the schedule generate by HEFT [13] as baseline C_{base}. We implement a modified HEFT algorithm for IaaS cloud environment. It always chooses the resource that finish selected task earliest. And then, the normalized cost of a schedule is defined as: $NC = \frac{C}{C_{base}}$.

Finally, to evaluate the impact of slackness degree of a schedule on our compaction and scheduling algorithms, we need to assign a deadline to each workflow. We firstly define the fastest schedule for a workflow be the one get by HEFT [13]. The makespan of this schedule is denoted by M_F. And then, the deadline of a workflow is set to be $\alpha \times M_F$, $\alpha \in \{2.5, 3, 3.5, 4, 4.5\}$.

---

[2] http://confluence.pegasus.isi.edu/display/pegasus/WorkflowGenerator.

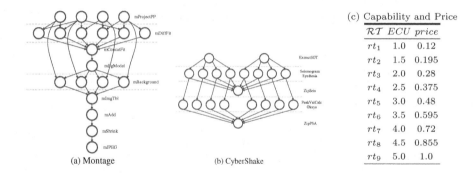

(a) Montage   (b) CyberShake

	$\mathcal{RT}$	ECU	price
(c) Capability and Price			
	rt_1	1.0	0.12
	rt_2	1.5	0.195
	rt_3	2.0	0.28
	rt_4	2.5	0.375
	rt_5	3.0	0.48
	rt_6	3.5	0.595
	rt_7	4.0	0.72
	rt_8	4.5	0.855
	rt_9	5.0	1.0

Fig. 1. The structures of benchmark scientific workflows, and information of available resource types.

6.4   Experimental Results

Figures 2 and 3 show the average cost of scheduling workflows with IC-PCP, IC-PCPD2, IC-SC algorithms, the average cost after compaction on schedules generated by IC-PCP, IC-PCPD2 algorithms. The IC-PCP-SC algorithm compacts schedules generated by IC-PCP algorithm, and the IC-PCPD2-SC algorithm compacts schedules generated by IC-PCPD2 algorithm correspondingly.

We can get that IC-⋆-SC is able to reduce cost dramatically after compaction over schedules generated by IC-PCP and IC-PCPD2 algorithms. They also show that the IC-SC algorithm is able to generate less monetary cost schedules than IC-PCP and IC-PCPD2 algorithms. With looser deadline constraint, the schedules generated by IC-SC cost even less than IC-PCP-SC and IC-PCPD2-SC.

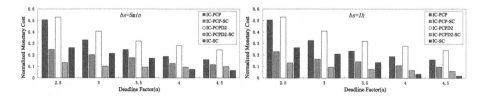

Fig. 2. Normalized costs of Montage application.

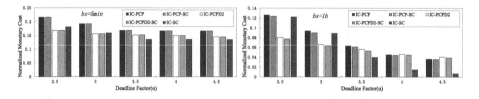

Fig. 3. Normalized costs of CyberShake application.

7 Conclusion

We developed a novel algorithm to reduce billing cycle charged resource requirement for schedules generated by any scheduling algorithm. When applied on a valid schedule, the proposed IC-⋆-SC algorithm compacts the schedule to fewer number of resources without increasing the schedule length or under a deadline constraint. We also proposed an efficient deadline constrained workflow scheduling algorithm for IaaS cloud. The time complexity is small, which makes it suitable for large scale workflow scheduling. Experiments on scientific workflows verified that IC-⋆-SC algorithm dramatically reduces the resource requirement of the schedules generated by IC-PCP and IC-PCPD2 algorithms. And, IC-SC performs better than the state-of-the-art algorithms, especial for workflows with loose deadline constraint.

Acknowledgments. This work is supported by project (2013AA01A212) from the National 863 Program of China, project (61202121) from the National Natural Science Foundation of China, Science and technology project (2013Y2-00043) in Guangzhou of China.

References

1. Buyya, R., Yeo, C.S., Venugopal, S., Broberg, J., Brandic, I.: Cloud computing and emerging IT platforms: vision, hype, and reality for delivering computing as the 5th utility. Future Gener. Comput. Syst. **25**(6), 599–616 (2009)
2. Kwok, Y.K., Ahmad, I.: Static scheduling algorithms for allocating directed task graphs to multiprocessors. ACM Comput. Surv. (CSUR) **31**(4), 406–471 (1999)
3. Yu, J., Buyya, R., Ramamohanarao, K.: Workflow scheduling algorithms for grid computing. In: Xhafa, F., Abraham, A. (eds.) Metaheuristics for Scheduling in Distributed Computing Environments, pp. 173–214. Springer, Heidelberg (2008)
4. Wieczorek, M., Hoheisel, A., Prodan, R.: Towards a general model of the multi-criteria workflow scheduling on the grid. Future Gener. Comput. Syst. **25**(3), 237–256 (2009)
5. Chen, W.-N., Zhang, J.: An ant colony optimization approach to a grid workflow scheduling problem with various QoS requirements. IEEE Trans. Syst. Man Cybern. Part C Appl. Rev. **39**(1), 29–43 (2009)
6. Yu, J., Buyya, R., Tham, C.K.: Cost-based scheduling of scientific workflow applications on utility grids. In: First International Conference on e-Science and Grid Computing, p. 8. IEEE (2005)
7. Yuan, Y., Li, X., Wang, Q., Zhang, Y.: Bottom level based heuristic for workflow scheduling in grids. Chin. J. Comput. **31**(2), 282 (2008)
8. Yuan, Y., Li, X., Wang, Q., Zhu, X.: Deadline division-based heuristic for cost optimization in workflow scheduling. Inf. Sci. **179**(15), 2562–2575 (2009)
9. Abrishami, S., Naghibzadeh, M., Epema, D.H.: Deadline-constrained workflow scheduling algorithms for infrastructure as a service clouds. Future Gener. Comput. Syst. **29**(1), 158–169 (2013)
10. Abrishami, S., Naghibzadeh, M., Epema, D.H.: Cost-driven scheduling of grid workflows using partial critical paths. IEEE Trans. Parallel Distrib. Syst. **23**(8), 1400–1414 (2012)

11. Bozdag, D., Ozguner, F., Catalyurek, U.V.: Compaction of schedules and a two-stage approach for duplication-based DAG scheduling. IEEE Trans. Parallel Distrib. Syst. **20**(6), 857–871 (2009)
12. Bharathi, S., Chervenak, A., Deelman, E., Mehta, G., Su, M. H., Vahi, K.: Characterization of scientific workflows. In: 2008 Third Workshop on Workflows in Support of Large-Scale Science, pp. 1–10. IEEE, November 2008
13. Topcuoglu, H., Hariri, S., Wu, M.Y.: Performance-effective and low-complexity task scheduling for heterogeneous computing. IEEE Trans. Parallel Distrib. Syst. **13**(3), 260–274 (2002)

Synthesizing Realistic CloudWorkload Traces for Studying Dynamic Resource System Management

Frederic Dumont[1]([✉]) and Jean-Marc Menaud[2]

[1] EASYVIRT, Nantes, France
frederic.dumont@easyvirt.com
[2] Mines Nantes/LINA/INRIA, Nantes, France
menaud@mines-nantes.fr

Abstract. Cloud Computing has become a new technical and economic model within companies. By virtualizing services, it allowed a more flexible management of datacenters capacities. However, its elasticity and its flexibility led to the explosion of virtual environments to manage. It's common for a system administrator to manage several hundreds or thousands virtual machines. Without appropriate tool, this administration task may be impossible to achieve. We purpose in this paper a decision support tool to detect virtual machines with atypical behavior. Virtual machines whose behavior is different from other VMs running in the data center are tagged as a typicals. This tool has been validated in production and being used by several companies.

Keywords: Cloud computing · Virtual machines · Datacenters · Clustering · Resource analysis

1 Introduction

In few years, Cloud computing has become a major key point for the societies development. Based on virtualization technologies, cloud computing allows to offer services on demand (with payment for use). These services can range from the provision of online applications (SaaS: Software As A Service) to virtual machines (IaaS). For the end user (which may be internal to the company), virtual machines (VM) are identical to physical machines. For IaaS service provider, the interest is both able to deploy VMs in a very short time (compared to the deployment of a physical server) and to share a same physical machine (a a host) between several VM. Thereby, virtualization optimize maintenance and investments costs. Today, IaaS Management systems (Vcenter, CloudStack, OpenStack) are mature and widely distributed inside data centers. Unfortunately, due to its ease of deployment, the number of virtual machines is growing exponentially. This virtual machines number explosion requires administrators to invest in some analysis tools for managing their hundreds to thousands of VMs. These analysis tools can identify VM with pre-defined behaviors. Typically,

© Springer International Publishing Switzerland 2015
W. Qiang et al. (Eds.): CloudCom-Asia 2015, LNCS 9106, pp. 29–41, 2015.
DOI: 10.1007/978-3-319-28430-9_3

a VM consuming any system resources for several months is a VM that can surely be stopped, freeing disk space and system resources to create a new VM. Tools such as VCOPS (VMware) and DCScope (Easyvirt) are based on limited predefined behaviors. Without powerful analysis tool, IaaS administrators can not cope with the ever increasing number of VMs. The main contribution of this paper is to propose a new VM analysis tool to exploite finely data center capacity. Evaluation of such tool is not so easy. In fact, to evaluate our tool we need an access to several real execution traces of VM. If some of them are available for some very large data centers, as those offered by Google, these traces are not significant in relation to the market reality. Indeed, cross-checking information in [1,2], we can determine that there are more than 3.8 million datacenters with 25 machines or less, against just over 140 000 datacenters with 200 machines or more. Thereby, we have collected real traces from several small datacenters. These reals traces allow to objectively analyze the datacenter activity (virtual machines and physical servers) during longs periods from one to six months. They were collected from different types of datacenters, from public to private cloud for small and large companies. These analysis tool was based on a specific partitioning algorithm which identifies VMs behaviors. To our knowledge, there are no tools allow to perform such analysis.

2   State of the Art

The objective of our work is to determine VM with predefined and atypical behaviors. The first one consists to analyse VM's activities (mainly processor and ram) and produce several sets of VM with fixed predefined behaviors. The second one identify VM whose resources consumption differ from others.

For predefined behaviors, severals solutions has already proposed (cite VCOPS, DCScope, VM turbo etc.). Theses solutions propose the same set of predefined behaviors. The first drawback of theses solutions still it remains difficult/impossible to change settings for each predefined behavior. For example, in VCops, a VM is classified idle if its cpu rate does not exceed 10 %. Define a single fixed percentage to analyze VM running in a heterogenous data center where several different physical server coexists is meaningless. The second drawback of these solutions is that they do not allow to filter the noise of the data collected. This is essential when you want to analyze hundreds of millions of data, which is typically our case study.

For atypical behaviors, no predefined filter based solution can not be used. In our work, a behavior is typical if it's shared by multiple VM. Our goal is to group by similarity VM. Atypical VM are those that can not be grouped with others.This classic approach of data analysis is based on a well-known technique: *Clustering*. Clustering involves to group a set of points, characterized by several dimensions into partitions (or clusters) of similar points. The similarity is expressed by the use of a distance measure between points. To partition a set of points, two clustering methods exist. The first is to associate a point to a single group. The second is to associate a point to several groups according to a membership criterion. Our aim is to determine VM with atypical behaviours, so we chose to study methods for associating a point to a single group. Partitioning methods can be classified into four classes defined according to the membership function: a distance, a density, a grid or a hierarchical approach.

The most widely used algorithm, K-MEANS [7], proposes to divide a set of points in k partitions to obtain a satisfactory similarity for all K partitions. This iterative approach seeks to determine K centroids, a point in space defining the center of a partition. The initial choice of K centroid is random. The algorithm puts points in the closest group (the nearest centroid) and recalculates the centroid of each group, redefining the position of all K centroids. The algorithm iterates by replacing points from these new centroids and ends when any of points doesn't change group. This algorithm has several problems. It's noise sensitive, necessarily not find the optimal configuration and it's necessary to define the number of K groups. These problems are blocking for our work insofar as we search atypical VM. To make better noise resistance, an improved K-means algorithm was proposed in PAM [6]. It consists to choose centroids from the set of points to analyze. The initial choice of k points (called medoid) is also random. The rest of the algorithm is similar to K-means excepted the new medoid must be a data point and not a space point. However, find new medoids is very expensive. Therefore, it's rarely used for large data sizes. This improvement requires any time to define in advance the number of partitions you want (the K number).

Algorithms based on density, such as DBSCAN [3], seek to identify partitions, based on point density. A point is considered neighbor from another if it's less than a fixed distance value. A point is considered dense if the number of neighbors exceeds a certain threshold. The algorithm seeks to merge dense points and thus consolidate partitions. Initially, the number of partitions is equal to the number of points. Two partitions merge if a point P of the first partition is connectable to a point of another partition, i.e. there exists a sequence of points between P and Q, the distance between them is less than the threshold. This iterative approach built partitions by merging them in order to obtain the largest. The advantage of this approach is that it doesn't require fixing the number of partitions in advance and it's relatively efficient in computation time. In contrast, the density threshold is difficult to determine.

Algorithms based on a grid such as BANG [8] seek to divide the space into cells then group the closest. This algorithm works similarly to algorithms based on density. BANG performs dense cells merging hierarchically, starting from the grid and successively merging neighbouring dense cells whose density difference doesn't exceed a threshold. As for density based algorithms, the threshold is difficult to determine.

Hierarchical algorithms seek either to group points until densest partitions (agglomeration approach) or start from a partition containing all points and then divide it into smaller partition (division approach). In the first case, initially each point represents a group. The algorithm searches the nearest two groups to combine them. It iterates until N (which can be equal to 1) groups (defined in advance) are found. The division approach is to divide partitions according to

an object dispersion criterion. The algorithm iterates until N groups are found
(N being the number of initial points). In both algorithms, the distance between
two groups can be calculated as: the minimum distance between all points of
two groups (single-link), the maximum distance between all points of the two
groups (complete-link), the average of distances between all points of the two
groups (average-link). As previous algorithms, the threshold is difficult to set.

3   Contributions

3.1   Collected Traces

As part of EasyVirt activities, we have been led to make numerous power consumption audits. These audits involved establishing software probes in client
datacenters and archiving datas in a MySQL database. These probes collect system resources consumption of the physical servers and virtual machines from
VMware® VCenter Server™. For confidentiality reasons, the companies in
which the data were collected may not be cited. Similarly, we can not distribute
collected raw data. In contrast, analysis can be distributed. EasyVirt has about
thirty traces. For this paper, we selected eight traces among them (some don't
contain enough informations).

Our prototype monitors datacenter activity using a software probe, developed in Java, connected to VMware® VCenter Server™. From a read-only
account, the solution recovers various information. Firstly, our solution build the
Datacenter/Cluster/ Server/VM architecture then retrieves static information of
physical servers. Thus, DCScope get data related to the hardware: server model,
CPU model, CPU frequency, cores number, RAM available, etc. Then, the solution recoveries virtual machine information related to their configuration: vCPU
(virtual CPU) number, allocated memory, reserved memory, VMware® Tools
status, VMDK size, etc. Following the recovery of these informations, DCScope
monitors physical servers and virtual machines in order to get their consumption
and virtual machines lifecycle. Various resources are monitored: processor, memory, network and disk. This monitoring is realized every 30 s, without impacting
VMware® VCenter performance. Data are stored in a classical relational database: MySQL. For performance reason, all data are not stored in a single table.
Each hour, raw data are stored in a specific table named consumption table.
In fact, database response times are dependent to the records number into the
table. Indeed, by performing a 'select' query in the table, MySQL browses all
records in order to check if each record meets the requirements ('where' clause).
Create indexes in the table reduces the computation time, but given the volume
of data processed, the indexing mechanism is not sufficient. Figure 1 represents
MySQL database size and consumption records number for the eight datacenters
of this study. With the aim to optimize disk space used by our database, we have
developed various scripts that automatically and periodically analyze, consolidate and purge collected data. Thus, information stored in each consumption
table are consolidated into a new abstract table. Data for each element (physical
servers and VM) are stored in only one record in the new table. This allows a

Companies	DC01	DC02	DC03	DC04
# Physical servers	10	8	7	8
# VM	123	122	83	151
BD size (in Mo)	9	25	18	61
# Consumption entries	1753743	4282260	2855559	9197709

Companies	DC05	DC06	DC07	DC08
# Physical servers	3	8	16	53
# VM	309	88	605	842
BD size (in Mo)	107	17	82	590
# Consumption entries	14325798	3892219	11633702	50803305

Fig. 1. Infrastructures

possible database purge by deleting consumption tables already summarized in order to free some disk space.

For this work, we focused on the RAM and CPU consumption. In fact, these two criteria are most relevant for our study and are the contention resource in a data center. Figure 2(a) to (h) show a graphical representation of VM behavior. In these graphics, each point represents the CPU and RAM average consumption for each VM. In Abscissa average RAM used and average CPU used in ordinate. Graphics are small, but we can observe for all collected traces, similarity between datacenters activities profiles. Indeed, we notice the presence of a very large set of VM close to the origin (dense masses). Around these dense masses, several VMs have a completely different activity. In third contribution, we will focus on these VMs called atypical VM.

3.2 Pre-determined Behaviour

DCScope solution developed by Easyvirt, can analyze in real time on a given time slot, pre-determined behavior. Behaviors are derived from VMware® specifications and recommendations as well as customer returns. Six behaviors are pre-defined:

- *idle* (without activity): if during a given time slot, 100 % of CPU and RAM consumption values are less than Y %. The activity rate (Y) being directly dependent on the CPU ability of the server where virtual machines are hosted.
- *lazy* (with unusual activity): if during a given time slot, a R resource (CPU or RAM) has consumption peaks (>30 %) during less than 10 % of audit time and idle during the remaining time.
- *undersized*: if during a given time slot, 100 % of CPU and RAM consumption values exceed 70 %.
- *busy*: Same as undersized except that the threshold is 90 %.
- *oversized*: if during a given time slot, 100 % of CPU and RAM consumption values are less than 30 %.
- *ghost* (off): if during previous and last analysed time slot, VM is off.

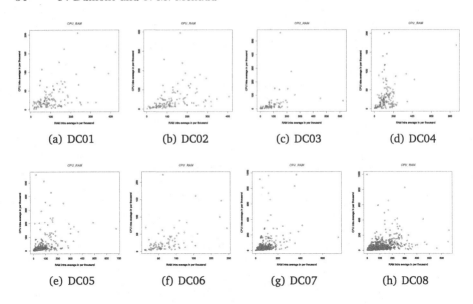

Fig. 2. Collected traces

However, in order to obtain a more robust VM detection than VMware, we have upgraded these pre-defined behaviors by adding a notion of noise. Indeed, in basic definitions, described previously, VM detection is strict. 100 % of points must be above or below a given threshold (according to the behavior). But, a machine always has an activity (peaks of activities) even if it is not used. This activity could result from the OS activity (updates) or an application with a temporary activity (anti-virus for example). Therefore, we modify predefined behaviors definitions in order to manage this particular activity that we call noise. Thus, for each pre-defined behavior, we vary the percentage of points following 4 thresholds: 100 %, 99.99 %, 99.9 % et 99 %. This factor of ten between our thresholds is related to classical high availability datacenter thresholds.

3.3   Atypical Behaviour

The state of the art has led us to develop our own multi-criterias, multi-resources and insensitive to noise partitioning algorithm. Our algorithm, based on a distance measure is multi-resources (CPU, RAM, Disk, Network etc.). It's also multi-criteria insofar each analyzed resource can itself be divided into criterias. Here, we define a criteria as a statistic. Available statistics are the first quartile (25 % of points are below this threshold), the median (as many points below and above this threshold), the average, the third quartile (75 % of points are below the threshold), the minimum value and the maximum value for each resource. Thus, it's possible to consolidate VM based on one or more criterias.

The distance used in our algorithm is calculated using a "rate of similarity", adjustable, allowing to define the minimum and maximum limits of statistical

---

Algorithm 1. searchGroup method

Input: listVM : VM list, listCriterias : criterias list, currentgroup : the current
      VM group

Result: The group of VM having the best VM number for selected criteria

1 c1 = first criteria into listCriterias;
2 bestgroup = a new empty VM group;

3 **while** *maximum interval value is not reached* **do**
4   tmpgroup = We search the best group of VM from listVM, having a close
        similarity for the criteria c1 –> call to searchGroup method;
5   **if** *tmpgroup.size > currentgroup.size* **then**
6   **if** *listCriterias isn't empty* **then**
7   tmpgroup = searchGroup(listVM, listCriteria, tmpgroup);
8   **if** *tmpgroup.size > bestgroup.size* **then**
9   bestgroup = tmpgroup;
10   **end**
11   **else**
12   bestgroup = tmpgroup;
13   **end**
14   **end**
15   interval is incremented;
16 **end**
17 **return** bestGroup;

---

values ranges. Thus, both VM have a strong similarity if for one or more given criterias, they belong to the same intervals. Our algorithm recursively searches the best group, comprising more VM depending on the similarity rate and criteria selected until all VMs belong to a group. A VM has an atypical behavior if it's alone in its group or the VM number in the group is less than or equal to an arbitrary value.

4   Evaluation

4.1   Pre-determined Behaviour

Our evaluations was realized on an Intel(R) Core(TM) i7-2760QM @2.40 GHz with 4 GB of RAM. Operating system used is Ubuntu 13.10 64 bits with kernel 3.11.0-26-generic, Java 1.7.0_45 (Oracle® version) and MySQL server 5.5.37. Computation time depend on records number into the database. For information, on a small database (DC01) the computation time is 2.40 s and 72.96 s on a large database (DC08).

Figures 3, 4 and 5 show percentages of VM with pre-defined behaviors for 8 data centers selected in this paper. Analysis are performed by one-hour time slot, within a time range of one week to twelve months. Figure 4 represents results obtained for busy, oversized and undersized behaviours. Busy VM investigation

Noise filtering (%)	100			99.99			99.9			99		
Inactivity threshold (%)	10	5	2	10	5	2	10	5	2	10	5	2
DC01	0.8	0	0	1.6	0.8	0	1.6	0.8	0	4.9	1.6	0
DC02	0	0	0	0	0	0	0.8	0	0	5.7	4.1	0
DC03	4.8	2.4	0	6.1	4.8	0	9.6	7.2	3.6	19.2	13.2	4.8
DC04	0	0	0	0	0	0	1.9	0.7	0	9.2	2.6	0
DC05	1.6	0.7	0.4	2.9	0.7	0.9	6.4	1.3	0.9	19.7	8.4	0.9
DC06	1.1	0	0	1.1	0	0	1.1	0	0	2.2	1.1	0
DC07	1.0	0.5	0.2	1.2	0.5	0.2	3.6	0.5	0.2	16.1	2.9	0.5
DC08	1.6	0.4	0	2.9	0.4	0	5.1	2.1	0	15.7	6.1	0.8

Fig. 3. Idle VM

	Buzy				Oversized				Undersized			
Noise filtering (%)	100	99.99	99.9	99	100	99.99	99.9	99	100	99.99	99.9	99
DC01	0	0	0	0	12.1	13.8	40.6	76.4	0	0	0	0
DC02	0	0	0	0	5.7	9.8	30.1	77.1	0	0	0	0
DC03	0	0	0	0	18.1	21.6	38.5	84.3	0	0	0	0
DC04	0	0	0	0	11.2	18.5	41.7	68.8	0.6	0.6	0.6	0.6
DC05	0	0	0	0	27.8	34.9	52.1	83.1	0.3	0.3	0.3	0.3
DC06	0	0	0	0	6.8	9.1	36.3	76.3	0	0	0	0
DC07	0	0	0	0	22.9	25.3	42.8	91.1	0	0	0.1	0.3
DC08	0.1	0.2	0.2	0.2	13.1	19.4	31.1	75.7	0.2	0.2	0.2	0.2

Fig. 4. Busy, oversized, undersized VM

Noise filtering (%)	100			99.99			99.9			99		
Inactivity threshold (%)	10	5	2	10	5	2	10	5	2	10	5	2
DC01	21.9	7.3	0	22.7	7.3	0	22.7	7.3	0	24.4	7.3	0
DC02	17.2	0.8	0	17.2	0.8	0	17.2	0.8	0	18.1	1.6	0
DC03	26.5	9.6	1.2	26.5	9.6	1.2	26.5	9.6	1.2	27.7	9.6	1.2
DC04	17.9	3.3	1.3	17.9	3.3	1.3	17.9	3.3	1.3	18.5	3.9	1.9
DC05	18.8	7.4	1.3	18.8	7.4	1.3	18.8	7.4	1.3	19.7	8.7	1.3
DC06	40.9	15.7	3.4	40.9	15.7	3.4	40.9	15.7	3.4	43.2	15.7	3.4
DC07	39.9	12.6	0.7	39.9	12.6	0.7	39.9	12.6	0.7	41.5	13.2	0.8
DC08	23.8	8.2	0.9	23.8	8.2	0.9	23.9	8.3	1.0	25.1	8.8	1.1

Fig. 5. Lazy VM

is important insofar as they are a very high activity. We don't know if this activity is normal or not but it permit to alert the administrator on the state of these VM (vigilance point). It is the same for undersized behavior. VM with this behavior are VM with high activity (superior to 70 %). Undersized VM can be overloaded, which will impact performance. In this way, undersized VM investigation informs the administrator before an eventual performance degradation and allocate additional resources. Conversely, looking for oversized VM allows the administrator to better allocate resources of virtual machines and potentially save system resources. Figure 3 shows results obtained for idle behaviour. As a reminder, idle VM is a virtual machine whitout activity. 100 % of points must be above a certain threshold (2 %, 5 % et 10 %). Trying to find idle VM permits to the administrator to shutdown or delete these VM with the aim to free system resources and optimize storage used. Figure 5 shows results obtained for lazy behaviour. It's a particular behaviour. Indeed, a virtual machine is tagged lazy if it is idle at least during 90 % of its running time and has peaks of activities superior to 30 % during 10 % of remaining time. Here we are looking for VM with a non-negligible activity during a short time (less than 10 % of running time). These VM look like idle VM but their profile need a more detailed analysis of the administrator.

By observing results of the various tables, we show the relevance of taking into consideration the noise in our analysis. For some behaviors as busy or undersized, the percentage of VM is not very different from 100 % to 99 %. This result is coherent insofar as it's rare to find one or more overloaded virtual machines in datacenters. In contrast, the result is different for oversized behaviour. The average percentage of VM obtained for 99 % is 5 times superior to the average percentage of VM obtained for 100 %. Similarly, the average percentage of VM for idle virtual machines with a threshold at 10 %, during 100 % is 1.36 % whereas it's 11.58 % during 99 % (multiplication by 8).

4.2   Atypical Behaviour

To validate our atypical VM algorithm, we analyzed partitioning methods results on real collected data by our prototype on thirty datacenters. For privacy reasons, companies may be cited and collected raw datas can't be distributed. For the sake of compactness, we have chosen to focus us on eight representative datacenters and compare our approach to K-MEANS partitioning algorithm. For this, we used R tool. R is a programming language for process, organize and rapidly represent large data sets. R also has a wide variety of clustering functions. In our experimentation, we used the 'boxplot' function (in order to obtain, from raw consumption data, the first quartile, the median, the average, the third quartile, the minimum value and the maximum value for each virtual machine) and the 'kmeans' function (in order to compare 'kmeans' function results and ours). In addition, all graphical representations presented in this paper were performed with R.

To compare K-means algorithm with our algorithm, we determined the groups number K by a first iteration of our algorithm. In these exemples, we

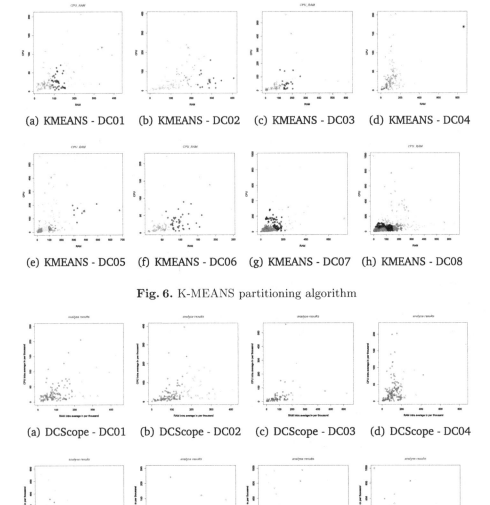

Fig. 6. K-MEANS partitioning algorithm

Fig. 7. Our partitioning algorithm (Color figure online)

take points on a two-dimensional space: CPU and RAM average for each VM. The horizontal axis is the average RAM usage and the vertical axis is the average CPU usage. In these graphs, each VM group is represented by a color and a symbol. Figure 7(a), (b), (c), (d), (e), (f), (g) and (h) describe the partitioning produced by our approach: a group with an important VM number and several small groups (1 to 3 VM). On Fig. 6(a), (b), (c), (d), (e), (f), (g) and (h) we can observe K-means algorithm. Each datacenter are partitioned into a large number of small groups, fairly consistent point number, little usable in our case.

To obtain these results, raw data stored in the database are initially exported in CSV then treated with R to calculate statistical values for each virtual machine. These statistical data are stored into files to be refilled and avoid multiple database access. Statistical informations of each VM are used by our algorithm to build the VM consumption model. Then, it calls the partitioning feature based on selected criteria. Groups obtained by our algorithm are graphically represented with R. Graphs obtained for K-MEANS are results of the 'kmeans' function implemented in R. Computation times for data process (excluding CSV export) and graphically represent our algorithm partitioning results and K-MEANS results are around the second (1.7 s on average).

For each case, we find 2 or 3 atypical VM. We recall that we define a VM with atypical behavior, a VM that doesn't have the same profile as others. Graphically, these VM are represented by separated points. To illustrate this fact, taking for examples Fig. 7(b), (c), (d), (e) and (h). We notice a highly isolated point. These are points with coordinates (165,395) for DC02, (152,561) for DC03, (866,169) for DC04, (677,160) for DC05 and (0,998) for DC08. Here, we have atypical VM, not belonging to any group. To compare with the K-means algorithm, let's look Fig. 6(b), (c), (d), (e) and (h). Regarding DC04, the K-means algorithm returns the same result that our algorithm. That is to say it has also isolated the same point (the atypical VM). In contrast, for DC02,DC03, DC05 and DC08, the K-means algorithm has included these virtual machines in groups. By observing these groups carefully, there are points (VM) having different consumption profiles. For example, in DC05, the virtual machine that we consider as atypical (coordinates 677,160) is in a group of VM whose CPU usage is extended from 300 to 700 permil. It's the same for DC03, the virtual machine that we consider as atypical (coordinates 152,561) is combined with another VM (coordinates 234,277). We also see this result for DC02. This shows that K-means algorithm is sensitive to noise, that is to say that it seeks to consolidate isolated points with others, although these points are distant. This sensitivity is due to the number k. Conversely, the K-means algorithm cut into smaller groups dense masses (blue and red masses on our results). This division is useless because VM in dense masses have similar profiles. These VM must be in the same group. This behavior is particularly visible on Figs. 6(b)–7(b) and 6(h)–7(h). The observations presented in this paper shows the effectiveness of our algorithm to look for atypical VM (separated points), unsensitive to noise, establishing groups of similar virtual machines.

5 Conclusion

This paper presents our work on cloud workload traces to exploite finely data center capacity. The three mains contributions of this paper are to collect real datacenter traces, to extend predefined VM behaviors, and to identify atypical virtual machines.

First, we have collected real traces from more than thirty datacenter and selected height. We have realized a simple analyses and show that VM in a real datacenter don't consume lot of resources. Many of them consume less than 5 % of the cpu.

The second contribution concerns the noise resistance addition in pre-determined behavior definitions. Behaviors are derived from VMware® specifications and recommendations. Six behaviors were defined:

- *idle* (no activity): 100 % of CPU and RAM consumption values are less than 10 %.
- *lazy* (with unusual activity): if a R resource (CPU or RAM) has consumption peaks (>30 %) during less than 10 % of audit time and idle during the remaining time.
- *undersized*: 100 % of CPU and RAM consumption values exceed 70 %.
- *busy*: Same as undersized except that the threshold is 90 %.
- *oversized*: 100 % of CPU and RAM consumption values are less than 30 %.
- *ghost* (off): if during previous and last analysed time slot, VM is off.

This noise resistance is importante because a machine always has an activity (peaks of activities) even if it is not used. This noise can be due to an anti-virus activity for example. So, we have decided to change above definitions in order to remove some activity peaks. For each pre-determined behavior, we vary the percentage of points following 4 thresholds: 100 %, 99.99 %, 99.9 % et 99 %. Results for eight datacenters shown the importance of taking into account the noise resistance in pre-determined behaviors.

Finally, we have developed a specific algorithm to identify atypical Virtual Machines. A virtual machine has an atypical behaviour if its consumption profile differs from others. We proposed in this paper a noise insensitive partitioning algorithm and shown the effectiveness of it on eight representative datacenter workloads by comparing our approach to K-MEANS partitioning algorithm. K-means algorithm is sensitive to noise. It seeks to consolidate isolated points with others, although these points are distant. Unlike it, our algorithm recursively searches the best group (the group with the higher VM number) until all VM belong to a group. So, for each datacenter profile, our algorithm has found 2 or 3 atypical VM.

References

1. Applied Computer Research Inc.: Defining the data center market and data center market size (2010)
2. Applied Computer Research Inc.: Identifying it markets and market size by number of servers (2011)
3. Ester, M., Kriegel, H.P., Sander, J., Xu, X.: A density-based algorithm for discovering clusters in large spatial databases with noise. In: Proceedings of 2nd International Conference on Knowledge Discovery and Data Mining, pp. 226–231 (1996)
4. Forestier, G.: Connaissances et clustering collaboratif d'objets complexes multi-sources. Ph.D. thesis, Universite de Strasbourg (2010)

5. Jain, A.K., Murty, M.N., Flynn, P.J.: Data clustering: a review. ACM Comput. Surv. **31**(3), 264–323 (1999)
6. Kaufman, L., Rousseeuw, P.J.: Finding Groups in Data: An Introduction to Cluster Analysis. Wiley, New York (1990)
7. MacQueen, J.B.: Some methods for classification and analysis of multivariate observations. In: Cam, L.M.L., Neyman, J. (eds.) Proceedings of the Fifth Berkeley Symposium on Mathematical Statistics and Probability, vol. 1, pp. 281–297. University of California Press (1967)
8. Schikuta, E., Erhart, M.: The BANG-clustering system: grid-based data analysis. In: Liu, X., Cohen, P., Berthold, M. (eds.) IDA 1997. LNCS, vol. 1280, pp. 513–524. Springer, Heidelberg (1997)

Energy-Efficient VM Placement Algorithms for Cloud Data Center

Xiuyan Lin, Zhanghui Liu, and Wenzhong Guo[✉]

Fujian Provincial Key Lab of the Network Computing
and Intelligent Information Processing, College of Mathematics
and Computer Science, Fuzhou University, Fuzhou 350116, China
Linxiuyan_lxy@sina.cn, 329717501@qq.com,
fzugwz@163.com

Abstract. Cloud is the computing paradigm which provides computing resource as a service through network. The client can use computing resource in a convenient and on-demand way, just like the water and the electricity we use daily. The mapping between virtual machine and physical machine is the key of the VM scheduling problem. Nowadays we advocate low-carbon life. It calls for the green cloud computing solutions whether protecting the environment or saving the cost of cloud suppliers. The proposed VM placement algorithm is energy-efficient, and considers the multi-dimentional resource constrains, such as CPU, memory, network bandwidth, and so on. The experimental results show that the proposed algorithms not only contribute a lot to energy saving, but also try best to meet the quality of service (QoS). Therefore, we make significant savings in operating cost and make full use of various resources in the cloud data center. The algorithm has promising prospect in application.

Keywords: Cloud resource scheduling · Energy-efficient · VM placement · Cloud data center · VM live migration

1 Introduction

Cloud computing is a new business computing paradigm and service model which is following after the parallel computing, distributed computing and the gird computing. In terms of the computing resource providing, cloud computing is a computing paradigm that provide the computing resource to the users as a service through the network. The client can use computing resource in a convenient and on-demand way, just like the water and the electricity we use daily. Cloud computing technology has become a hot issue in recent studies. Many large IT companies such as IBM, Amazon, Google, Microsoft and so on, all have made their own cloud computing system framework and provide the cloud service.

Cloud computing resource services providing is divided into three levels from the perspective of the specific application, that is Software as a service (SaaS), Platform as a Service (PaaS) and Infrastructure as a Service (IaaS). The three levels focus on different applications, but contain the same problem, namely resources, task scheduling problem. This paper focuses on the infrastructure as a service (IaaS).

© Springer International Publishing Switzerland 2015
W. Qiang et al. (Eds.): CloudCom-Asia 2015, LNCS 9106, pp. 42–54, 2015.
DOI: 10.1007/978-3-319-28430-9_4

Virtual machine scheduling problem is directly related to the stability, efficiency of resource, users' satisfaction and operating costs of cloud services. The data centers that share resources in the granularity of VM not only have high resource utilization, but also have better isolation of the application. In conclusion, the study of cloud computing data center virtual machine scheduling problems is very important.

With the expending scale of the cloud data centers, energy consumption has become an important part of the operating costs. For cloud providers, the full utilization of resources, maximizing profit is the main goal of their pursuits. It is critical to allocate quantitative resource efficiently according to the requirement of the application for the cloud system performance. That is the mapping of virtual machines and physical machines.

The contribution of this paper is a set of energy-efficient virtual machines placement algorithms for cloud data center, taking the multi-dimensional resource constraints into account, such as CPU, memory, network bandwidth, and so on. Two basic steps for virtual machine placement problem, the initial static allocation and dynamic migration algorithms are proposed. The experimental results show that our proposed algorithm can make full use of the resources of cloud data centers, while saving energy as much as possible and meeting the quality requirements of service (QoS).

The rest of this paper is organized as follows, the Sect. 2 of the paper is the discussion of the related study, and the analysis of relevant research work. The Sect. 3 describes the system model and the energy model used herein, laying the groundwork for the latter proposed algorithms. The Sect. 4 describes two specific algorithms and analyzes the relationship and the scheduling process between the two algorithms. In the Sect. 5, we conducted simulation experiments to demonstrate and analyze our results. And finally in Sect. 6, we summarize our research conclusions and discuss the future research.

2 Related Work

Research on the virtual machine placement has been the key of cloud resource scheduling, many scholars have carried out a lot of work, and put forward their own solutions for this problem. Here, we selected a small part of the solution as representatives, to make a brief summary and analysis.

Song et al. [4] proposed a virtual machines allocation mechanism based on two-layer on-demand resource for data center. The proposed local and global resource allocation algorithms with feedback optimize the allocation of resources. Li et al. [5] presented multiple objective genetic algorithms with the application service level objectives, to develop a framework for the virtual machine placement strategy.

In [6], they propose heuristic virtual machine placement algorithm. The loads on the physical machine balance the objectives of maximizing the number of served applications and minimizing the cost of migration. This work focuses on scalability, but in the case that all of the virtual machines can be placed, and the minimization of power and migration cost are the goals, there is no effect on the placement of virtual machines on a physical server.

Ardagna et al. [9] propose the VM placement solutions based on SLAs to maximize the profits of the cloud computing system. The issues they raised consider only the soft contract service level agreements.

The several solutions above are from the perspective of improving service, with emphasis on meeting the quality of service (QoS). As for the virtual machine placement algorithms, they use different strategies, basically focusing on the improving resource utilization without violating SLAs. However, these programs do not consider the energy consumption. Energy consumption has become an important part of operating costs, thus energy-efficient virtual machine placement algorithm is a critical research.

Cardosa et al. [11] studied the energy-aware virtual machine allocation method in virtualized enterprise computing environments, using a priori objective function. This method can effectively reduce energy consumption, but not for cloud computing data centers.

Verma et al. [12] studied the relevance between the CPU utilization and application, they put forward power and migration costs aware application layout in the virtualization system. The program only considers the CPU constraint, and the algorithm is complex and runs slowly.

Buyya et al. [13] put forward an energy-aware consolidation technique to reduce the total energy consumption of cloud computing systems. They describe a simple heuristic algorithm to integrate cloud computing systems processing work, but the algorithm can only be used for very small input.

Beloglazov et al. [8] investigate the principles of cloud computing architecture based on energy efficiency, and propose energy-efficient resource allocation strategy and scheduling algorithm. But they only consider one-dimensional limit, CPU resources, without considering the optimization of the network topology, nor optimizing the energy consumption of the refrigeration system and hot spots.

Li et al. [7] propose two algorithms respectively for the virtual machine placement and dynamic management initialization. With respect to literature [8], they take multiple dimensions resource limitations into account, but the quality of the service (QoS) has been neglected.

Katsaros et al. [10] have proposed the energy-aware virtual machine management framework, unlike the literature [7, 8] which use projected energy consumption model to estimate power consumption, they use a sensor device to measure the energy consumption. They measure the energy consumption more accurately, but introduce the additional hardware cost. So this solution is not practical.

3   The System Model

The resource cloud computing services provided is divided into three levels, namely software as a service (SaaS), Platform as a Service (PaaS) and Infrastructure as a Service (IaaS). This work focuses on infrastructure as a service (IaaS). In cloud infrastructure as a service (Infrastructure as a Service, IaaS), the cloud infrastructure providers allocate computing resources for Internet application. They sign and comply with the service level contract (Service Level Agreement, SLA). Service level contract contains one or more service level objectives. Service level objectives mean that

Internet application providers ensure application performance to network users. Cloud infrastructure providers should allocate appropriate resources to meet application service level objectives, improve resource utilization and reduce unnecessary overhead.

3.1 The IAAS System Model

In the cloud computing data center, infrastructure as a service (IaaS) model of the system is shown in Fig. 1. Users submit applications through their terminals, cloud service suppliers provide the required computing resources to users in the unit of virtual machine. Users rent these resources and pay the provider according to the amount of occupied resources and the length of occupied time. IAAS package the hardware equipment and other basic resources as a service to users. In IAAS environment, users can do almost anything they want to do, but they must consider how to make multiple machines work together. The biggest advantage of IAAS is that it allows users to dynamically apply or release nodes, bill according to usage. The number of running server in IAAS reaches as much as hundreds of thousands, so that the resources users can apply for is almost unlimited. Meanwhile IAAS is shared by the public, which has a higher efficiency of resource usage.

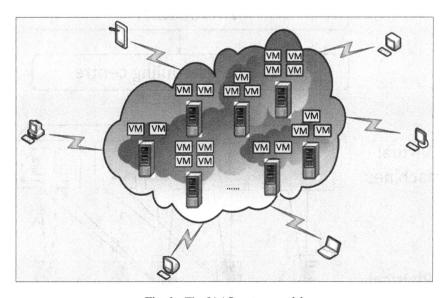

Fig. 1. The IAAS system model

3.2 System Model

The system model we use in the paper is shown in Fig. 2. The scheduling of virtual machine in the cloud data center is divided into two levels, the first one is the assignment of the virtual machine, we need to convert the users' demands into the resources demands, then allocate the computing resources to the applications. The second part is the placement of virtual machines, that is, we place the virtual machines onto the physical machines, to complete the mapping of VMs and PMs.

Clearly, the focus of our study is the second layer. We put the virtual machines to physical machines through resource dispatch center. It mainly consists of two steps, the initial placement of virtual machines and live migration of virtual machines. Our algorithm is for these two steps, the energy efficiency is an important goal of this article. We save the energy through turning more physical machines to be idle (or letting idle physical machine into hibernation). User submits application requests, we complete the mapping between virtual machines and physical machines according to the virtual machine that the application request for.

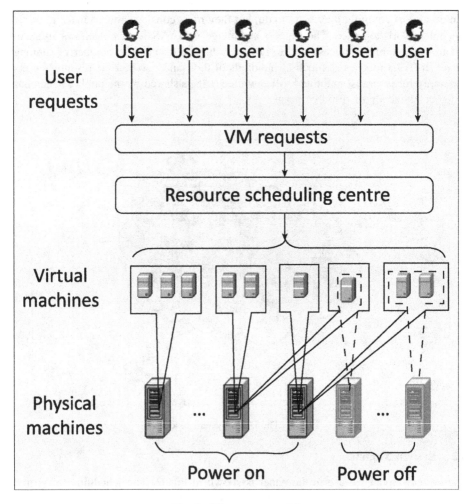

Fig. 2. The system model

3.3 Energy Model

There are many factors that affect energy consumption. Generally, the main cost of each node is decided by CPU, memory, hard disk capacity and network bandwidth. In these resources, CPU utilization has a top priority. CPU consumes a major portion of the energy. We assume that the physical machine of the cloud data center has three modes: idle mode, active mode and sleep mode. Numerous documents show that the average energy consumption of idle physical machine is 70 % of that running at full speed. Thus, in order to reduce the overall energy consumption, we need to convert the idle physical machine into sleep mode. The CPU utilization has a classical linear relationship with the workload of the entire system. That is to say, the CPU utilization is on behalf of energy efficiency to some extent. Based on this fact, we use the same energy model in the Ref. [8], as defined in Eq. (1) in the following.

$$P(u) = k \cdot P_{\max} + (1 - k) \cdot P_{\max} \cdot u \tag{1}$$

$P_{\max}$ represents the maximum power of the server with the peak workload, k denotes the proportion of the power the idle server consumed, then $1-k$ represents the consumption proportion server with workload, and u is the CPU utilization.

$$E = \int_{t0}^{t1} P(u(t))dt \tag{2}$$

The CPU utilization may change over time and load. Thus, the CPU utilization is a function of time, so we use u (t) to represent. Therefore, the total energy E of a physical node can be defined as a integral of the energy consumption function over a period of time, as shown in Eq. (2).

4 The Scheduling Algorithm

With the development of virtual technology, the key of the resource scheduling for cloud data center is the scheduling of virtual machines. We allocate computing resources reasonably depending on the application performance requirements. At the same time, we can integrate the virtual machine onto fewer physical machines. So we can switch the idle physical machine into hibernation, which can greatly reduce energy consumption. The VM placement mainly consists of two steps, the first one is the initial placement of virtual machines, to determine which physical machines the virtual machines were arranged on. The second is the live migration of virtual machines to reduce the number of active physical machines. The process includes when to migrate VMs, which virtual machines to migrate, and which physical machines to migrate.

Arranging the virtual machines to the appropriate physical machines is a fundamental step. The key is how to choose the appropriate physical machines. Our proposed algorithm is based on energy efficiency, and thus the energy consumption is one of our selection criteria. However, in addition to energy reduction, we also hope that our algorithm can meet the multi-objective optimization. Thus, we also consider the multi-dimensional resource constraints, while we introduce the balanced utilization as one of the selection criteria. Before presenting our algorithms, we need to define some of the parameters we used, as shown in Table 1.

Table 1. Parameter description

Parameters	Meaning
r_i^c	Computing capacity of a physical sever i
r_i^m	Memory capacity of a physical sever i
r_i^n	Network capacity of a physical sever i
r_{iR}^c	The remaining computing capacity of a physical sever i
r_{iR}^m	The remaining memory capacity of a physical sever i
r_{iR}^n	The remaining network capacity of a physical sever i
r_c	Computing capacity of the VM which is going to be placed
r_m	Memory capacity of the VM which is going to be placed
r_n	Network capacity of the VM which is going to be placed
CPU_i^u	Average CPU utilization of a physical sever i
MEM_i^u	Average CPU utilization of a physical sever i
NET_i^u	Average CPU utilization of a physical sever i

The overall average utilization of the physical machine i is defined as follows, as shown in formula (3), taking the average utilization of CPU, memory, network bandwidth.

$$IR_i^u = \frac{CPU_i^u + MEM_i^u + NET_i^u}{3} \tag{3}$$

The definition of unbalanced multi-dimension utilization of the physical machine i, we refer to Ref. [7], as the following Eq. (4) shown

$$IUV_i = \frac{(CPU_i^u - IR_i^u)^2 + (MEM_i^u - IR_i^u)^2 + (NET_i^u - IR_i^u)^2}{3} \tag{4}$$

The roughly steps we mentioned in the virtual machine placement selection algorithm are: (1) Select physical machines which can be set, i.e. CPU, memory and network bandwidth can meet the physical requirements of the virtual machine; (2) select the most appropriate physical machines can be set in accordance with the energy and standard balanced utilization. Therefore, the selection criteria is very important, we choose two factors here, the balanced utilization and the energy consumption to define the selection criteria, as shown in the Eq. (5).

$$SM_i = w_1 \cdot IUV_i + w_2 \cdot P_i(u) \tag{5}$$

In Eq. (5), w_1, w_2 represent the power imbalance rate and energy consumption values, the values must meet $w_1 + w_2 = 1$. For example, we set that w_1, w_2 equal to 0.5, representing the same weight of the unbalanced utilization and energy consumption. The specific algorithm is shown in the following Table 2:

Table 2. Energy-efficient virtual machine placement (EVMP)

Algorithm 1 energy-efficient virtual machine placement(EVMP)
Input: a new VM, a list of servers S-list in the cloud data center
Output: a VM-PM match
SM=max
for each server in the S-list
if the server i has enough resource for the VM then
calculate the selection criteria value of the server i in Eq(5)
if(SM_i <SM) then
SM= SM_i
Assign the VM to the server i

The EVMP algorithm is only the first part of the VM placement. Then we need to do the optimization of the current VM allocation. We should decide the VMs which to migrate and where to migrate them. To determine which VMs should be migrated, we need to introduce the node utilization threshold.

We set two thresholds, the upper and lower utilization thresholds for the servers. If the resource utilization of the server is higher than the upper thresholds, some VMs should be migrated from the server to reduce the utilization. Due to the fact that the migration of VM cost time and need additional resource to store the data, we should migrate the VMs as few as possible. If the utilization falls below the lower thresholds, all of the VMs on the server should be migrated. After all of the VMs migrated to other PMs, the server should be switched to the sleep mode to reduce the power consumption.

We propose a energy-efficient minimization migration algorithm for the optimization of the current VM allocation. The specific algorithm is shown in Table 3.

Table 3. Energy-efficient minimization migration algorithm (EMMA)

Algorithm 2   energy-efficient minimization migration algorithm (EMMA)
Input: a list of servers S-list in the cloud data center, the upper and lower thresholds vector
Output: a VM migration list
for each server in the S-list
if the utilization of server i exceeds the upper thresholds then
migrate the minimum number of VMs
if the utilization of server i falls behind the lower thresholds then
for each VM in the server i
migrate the VM through EVMP
turn the server to sleep mode

5   Simulation Result

In this section, we perform simulated experiments to evaluate our proposed algorithms described in last section. In our experiments, we use some metrics to evaluate the performance. The first metric is the total power consumption of the data center. The second metric is the imbalance utilization value of the resources in the cloud data center. For that the different types of applications require different resources, which inevitable results grate differences in the utilization of resources as the time went. The third metric is the SLA violation percentage, which is defined as the percentage of SLA violation events relatively to the total number of the processed time frames. The metric is directly related to the quality requirements of service (QoS). We assume that the cloud provider pays for the penalty in the case of SLA violation.

We have simulated a data center consisting of 200 heterogeneous physical machines. These PMs has three types, as shown in Table 4.

Table 4. Types of heterogeneous physical machines

Types	CPU/MIPS	Memory	Network/(Mbit $\cdot s-1$)
1	1000	8	50
2	2000	16	100
3	3000	24	150

In our simulated experiments, there are four different types of VMs. The parameters of the VMs are described in Table 5.

Table 5. Types of virtual machines

Types	CPU/MIPS	Memory	Network/(Mbit $\cdot s-1$)
1	100	1	10
2	200	1	20
3	300	2	10
4	400	4	40

In this paper, we conduct our experiments by comparing to the algorithms in [7, 8]. They are called E&E and M&M. The simulation results are presented in Fig. 3.

Figure 3a shows the relationship between the energy consumption and the IUV of resources. It shows that the energy consumption increases with the imbalance utilization value of the cloud data center. This is due to that the higher the imbalance utilization value of the cloud data center is, the more PMs will be operate to support the applications.

Figure 3b shows the relationship between the energy consumption and the IR of resources. We can see that an increase of the integrated resource utilization leads to a decrease of energy consumption. Therefore it is necessary to improve the integrated resource utilization.

In Fig. 4, we compare the power consumption of different algorithms. The results show that the M&M policy in [8] consume the minimum energy in the case of fewer VMs,

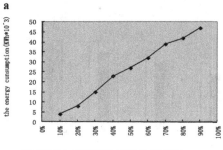

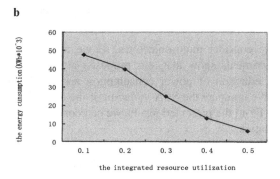

Fig. 3. (a) Relationship between the energy consumption and the IUV of resources. (b) Relationship between the energy consumption and the IR of resources

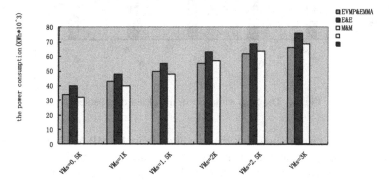

Fig. 4. The power consumption in the data center

our proposed algorithms, EVMP & EMMA save power in the case of bigger workload. This is because in [8], the power consumption is the only criteria to place the VMs.

Figure 5 shows the unbalance utilization value of the different policies. The results show that the E&E policy in [7] has the lowest unbalanced utilization value. This is due to that the unbalanced utilization value is the only criteria to place the VMs.

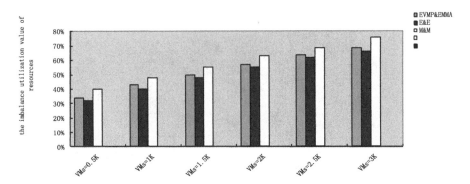

Fig. 5. The imbalance utilization value of the resources in the cloud data center

In this paper, we consider multi-objectives, so we make a trade-off among the objectives. The experiments show that, our proposed algorithms have good performance in both the reduction of power consumption and the balance utilization of different resources. We compare the SLA violation of the different policies, the results are shown in Fig. 6. From the presented results we can conclude that our policies have the lowest SLA violation.

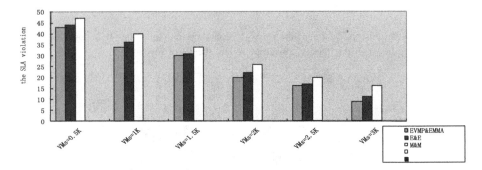

Fig. 6. The SLA violation in the cloud data center

The simulation results show that the two algorithms EVMP & EMMA have good performance in the three metrics, that is the total power consumption, the imbalance utilization value of the resources and the SLA violation percentage in the cloud data center.

6 Conclusion

In this paper, we propose VM placement algorithms, which are energy-efficient, and consider the multi-dimentional resource constrains, such as CPU, memory, network bandwidth, and so on. The experimental results show that the proposed algorithms not only contribute a lot to energy saving, but also try best to meet the quality of service (QoS). Therefore, we make significant savings in operating cost and make full use of various resources in the cloud data center. The algorithms have promising prospects in application.

However our proposed algorithms did not consider other part of the energy consumption, such as the cooling power consumption. The energy model in the paper does not capture the energy consumption accurately. Besides, we neglect the cost of migration, which may lead the additional pay to the consumers.

In the future research, we will make our best to overcome the shortcomings of the algorithms. We will improve the energy model to achieve higher accuracy in the measurement of the power consumption. Furthermore, we will conduct our experiments on the real cloud platform, and leverage our own technologies.

Acknowledgment. This work was supported in part by the Fujian province Education Scientific Research Project of Young and middle-aged teachers under Grant JA13356.

References

1. Bilal, K., Malik, S.U.R., Khalid, O., et al.: A taxonomy and survey on green data center networks. Future Gener. Comput. Syst. (2013)
2. Kim, N., Cho, J., Seo, E.: Energy-credit scheduler: an energy-aware virtual machine scheduler for cloud systems. Future Gener. Comput. Syst. **32**, 128–137 (2014)
3. Lucas-Simarro, J.L., Moreno-Vozmediano, R., Montero, R.S., et al.: Scheduling strategies for optimal service deployment across multiple clouds. Future Gener. Comput. Syst. **29**(6), 1431–1441 (2013)
4. Song, Y., Sun, Y., Shi, W.: A two-tiered on-demand resource allocation mechanism for VM-based data centers. IEEE Trans. Serv. Comput. **6**(1), 116–129 (2013)
5. Li, Q., Hao, Q., Xiao, L., Li, Z.: Adaptive management and multi-objective optimization of virtual machine placement in cloud computing. Chin. J. Comput. **34**(12), 2253–2264 (2011)
6. Tang, C., Steinder, M., Spreitzer, M., et al.: A scalable application placement controller for enterprise data centers. In: Proceedings of the 16th International Conference on World Wide Web, pp. 331–340. ACM (2007)
7. Li, H., Wang, J., Peng, J., Wang, J., Liu, T.: Energy-aware scheduling scheme using workload-aware consolidation technique in cloud data centres. China Commun. **10**, 114–124 (2013)
8. Beloglazov, A., Abawajy, J., Buyya, R.: Energy-aware resource allocation heuristics for efficient management of data centers for cloud computing. Future Gener. Comput. Syst. **28**(5), 755–768 (2012)
9. Ardagna, D., Panicucci, B., Trubian, M., et al.: Energy-aware autonomic resource allocation in multitier virtualized environments. IEEE Trans. Serv. Comput. **5**(1), 2–19 (2012)
10. Katsaros, G., Subirats, J., Fitó, J.O., et al.: A service framework for energy-aware monitoring and VM management in Clouds. Future Gener. Comput. Syst. **29**(8), 2077–2091 (2013)
11. Cardosa, M., Korupolu, M.R., Singh, A.: Shares and utilities based power consolidation in virtualized server environments. In: IFIP/IEEE International Symposium on Integrated Network Management (IM 2009), pp. 327–334. IEEE (2009)

12. Verma, A., Dasgupta, G., Nayak, T.K., et al.: Server workload analysis for power minimization using consolidation. In: Proceedings of the 2009 Conference on USENIX Annual Technical Conference, p. 28. USENIX Association (2009)
13. Buyya, R., Yeo, C.S., Venugopal, S.: Market-oriented cloud computing: vision, hype, and reality for delivering it services as computing utilities. In: 10th IEEE International Conference on High Performance Computing and Communications (HPCC 2008), pp. 5–13. IEEE (2008)
14. García, A.G., Espert, I.B., García, V.H.: SLA-driven dynamic cloud resource management. Future Gener. Comput. Syst. **31**, 1–11 (2014)
15. Tordsson, J., Montero, R.S., Moreno-Vozmediano, R., et al.: Cloud brokering mechanisms for optimized placement of virtual machines across multiple providers. Future Gener. Comput. Syst. **28**(2), 358–367 (2012)

A New Analytics Model for Large Scale Multidimensional Data Visualization

Jinson Zhang[1(✉)] and Mao Lin Huang[1,2]

[1] Faculty of Engineering and IT, School of Software,
University of Technology Sydney, Sydney, Australia
Jinson.Zhang@uts.edu.au
[2] School of Computer Software, Tianjin University, Tianjin, China

Abstract. With the rise of Big Data, the challenge for modern multidimensional data analysis and visualization is how it grows very quickly in size and complexity. In this paper, we first present a classification method called the *5Ws Dimensions* which classifies multidimensional data into the 5Ws definitions. The 5Ws Dimensions can be applied to multiple datasets such as text datasets, audio datasets and video datasets. Second, we establish a *Pair-Density* model to analyze the data patterns to compare the multidimensional data on the 5Ws patterns. Third, we created two additional parallel axes by using pair-density for visualization. The attributes has been shrunk to reduce data over-crowding in pair-density parallel coordinates. This has achieved more than 80 % clutter reduction without the loss of information. The experiment shows that our model can be efficiently used for Big Data analysis and visualization.

Keywords: Multidimensional data · Big Data · 5Ws dimension · Parallel coordinate · Pair-density · Shrunk attribute · Big Data visualization

1 Introduction

Big Data is considered to be structured or unstructured data that contains texts, images, audios, videos and other forms of data collected from multiple datasets, which grows rapidly in size and complexity. Big Data comes from everywhere in our life, and so is too big, too complex and moves too fast for us to analyze using traditional methods. For example, posting statuses or pictures on Facebook; uploading and watching videos on YouTube; sending and receiving messages through smart phones; broadcasting viruses over the Internet – all those activities collected by different datasets count as Big Data.

Based on Gartner's 3Vs definition [1], Big Data has three main characteristics: Volume, Velocity and Variety. The volume represents how datasets are extremely large and easily reach terabytes of information. The velocity describes how fast datasets are being produced. The variety illustrates the complexity of the datasets, including both structure and unstructured data which contains thousands of different attributes in multiple dimensions. Our approach establishes the analytic model for large scale multidimensional data.

Multidimensional data normally contains a large amount of noise data in different dimensions. Most current approaches try using different techniques to detach those

W. Qiang et al. (Eds.): CloudCom-Asia 2015, LNCS 9106, pp. 55–71, 2015.
DOI: 10.1007/978-3-319-28430-9_5

noise data, including data reduction, data integration and data clustering [5, 7]. Data reduction shrinks the data size to separate the noise data; data integration merges multiple data dimensions into coherent data attributes; and data clustering classifies the data into different groups which eliminates the noise data.

Data clustering plays a main role in multidimensional data analysis, which classifies the data dimensions into different groups, such as social media data clustering [8], airline flight data clustering [9], and petrol data clustering [10]. The cluster methods vary depending on the data structure, such as k-means cluster method [11], hierarchical cluster method [12] and density cluster method [13].

In this paper, we have further developed our previous works [6, 20] to classify the multidimensional data into the 5Ws dimensions based on their data behaviours, and then introduce 5Ws patterns crossing multiple datasets. Second, we establish pair-density patterns for analyzing the multidimensional data; four pair-density patterns are introduced to measure the different topics and patterns. Third, we created two additional parallel axes by using pair-density patterns in parallel coordinate visualization.

The paper is organized as follows; Sect. 2 illustrates the 5Ws dimensions and its patterns. Section 3 demonstrates the Pair-Density model. Section 4 shows the results of implementation. Section 5 describes related works, and Sect. 6 summarises our achievements and future works.

2   5Ws Dimensions

Multidimensional data contains texts, images, audios, videos and other forms of data, which occur every day in our lives. These include Facebook images, Twitter comments, YouTube videos or email contents. These multidimensional datasets grow very fast in size and complexity, which makes them hard to analyze using traditional database tools. Here, we analyze these data attributes and classify its behaviours into the 5Ws dimensions.

2.1   Multidimensional Data and Attributes

Assume that the first data incident, known as a data node, contains attributes

$$\{d_{11}, d_{12}, d_{13}, d_{1j}, \ldots, d_{1m}\},$$

where j indicates the jth dimension, an attribute d_{1j} illustrates the $1st$ data incident of the jth dimension. Therefore, the whole dataset can be illustrated as in (1) where $j = 1, 2, 3, \ldots m$ indicates the number of dimensions and $i = 1, 2, 3, \ldots n$ indicates the number of incidents. The total number of attributes $n \times m$ in the dataset can reach millions, even billions, in size.

$$D = \begin{Bmatrix} d_{11} & d_{12} & d_{13} & \cdots & d_{1j} & \cdots & d_{1m} \\ d_{21} & d_{22} & d_{23} & \cdots & d_{2j} & \cdots & d_{2m} \\ d_{31} & d_{32} & d_{33} & \cdots & d_{3j} & \cdots & d_{3m} \\ \vdots & \vdots & \vdots & & \vdots & & \vdots \\ d_{i1} & d_{i2} & d_{i3} & \cdots & d_{ij} & \cdots & d_{im} \\ \vdots & \vdots & \vdots & & \vdots & & \vdots \\ d_{n1} & d_{n2} & d_{n3} & \cdots & d_{nj} & \cdots & d_{nm} \end{Bmatrix} \qquad (1)$$

For example, during the 2014 FIFA World Cup Final between Germany and Argentina, there were 280 million Facebook interactions including posts, comments and likes across 88 million Facebook users [2]. Assume those interactions contained 5 dimensions, the total attributes in the entire dataset was $280 \times 5 = 1.4$ billion.

2.2 5Ws Behaviour Pattern

We classify the multidimensional data into the 5Ws dimensions based on its behaviours. The 5Ws dimensions are defined in this paper as; When the data occurred, Where the data came from, What the data contained, How the data was transferred, Why the data occurred, and Who received the data. Therefore, the dataset D can be demonstrated through the 5Ws pattern as

$$\textbf{When, Where, What, How, Why, Who}$$

$$D = \begin{Bmatrix} d_{1T} & d_{1P} & d_{1X} & d_{1Y} & d_{1Z} & d_{1Q} \\ d_{2T} & d_{2P} & d_{2X} & d_{2Y} & d_{2Z} & d_{2Q} \\ d_{3T} & d_{3P} & d_{3X} & d_{3Y} & d_{3Z} & d_{3Q} \\ \vdots & \vdots & \vdots & \vdots & \vdots & \vdots \\ d_{iT} & d_{iP} & d_{iX} & d_{iY} & d_{iZ} & d_{iQ} \\ \vdots & \vdots & \vdots & \vdots & \vdots & \vdots \\ d_{nT} & d_{nP} & d_{nX} & d_{nY} & d_{nZ} & d_{nQ} \end{Bmatrix} \qquad (2)$$

where $T = \{t_1, t_2, t_3, ...\}$ represents when the data occurred, $P = \{p_1, p_2, p_3, ...\}$ represents where the data came from, $X = \{x_1, x_2, x_3, ...\}$ represents what the data contained, $Y = \{y_1, y_2, y_3, ...\}$ represents how the data was transferred, $Z = \{z_1, z_2, z_3, ...\}$ represents why the data occurred and $Q = \{q_1, q_2, q_3, ...\}$ represents who received the data. A data incident d_i can be illustrated as a node using the 5Ws pattern as $d_i\{t, p, x, y, z, q\}$. The dataset D therefore can be defined as

$$D = \{d_1, d_2, d_3, \ldots, d_n\} \qquad (3)$$

We use a parallel axis to illustrate a dimension in the 5Ws behaviours pattern, in order to create the 5Ws parallel coordinates for visualization. Parallel coordinates are a popular information visualization tool for high-dimensional data introduced by Alfred Inselberg and Bernard Dimsdale [4]. Each parallel axis represents a dimensional data and polylines are drawn between independent axes at appropriate values. The data examined using the axes shows the data frequencies and the data relationships.

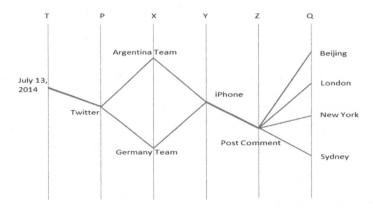

Fig. 1. Example of 5Ws parallel axes

Figure 1 shows an example using the 2014 FIFA World Cup Final between Germany and Argentina. Overall, Twitter users sent 618,725 messages per minute at the moment of Germany's victory [2]. Let us assume that the particular dataset contains the team names x_1 = *"Argentina Team"* and x_2 = *"Germany Team"*, which were posted through iPhone, and that countries which received the data were q_1 = *"Beijing"*, q_2 = *"London"*, q_3 = *"New York"* and q_4 = *"Sydney"*. These particular data incidents can be illustrated in the 5Ws parallel coordinates.

2.3   Dimension Clustering

The 5Ws dimensions can also be explored by clustering if necessary. For example, we want to explore the locations for who received the data by the countries and by the cities, Fig. 1 has, then be changed as Fig. 2 which shows clustering relationship between $Q1$ and $Q2$.

2.4   Shrunk Attributes

Each dimension contains hundreds, even thousands, of attributes, which can lead to the overcrowding of polylines in the pair-density parallel coordinates. To reduce this polyline cluttering, we create Shrunk Attributes (SA) to collect the attributes that are not displayed in each parallel axis, Fig. 3 shown the example given.

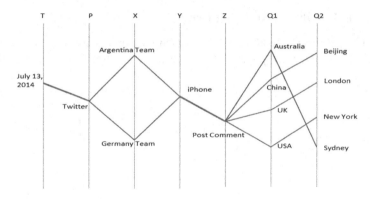

Fig. 2. Example of clustered 5Ws parallel axes

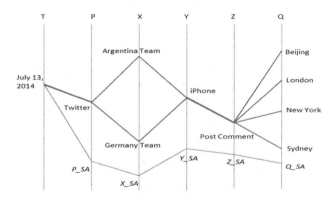

Fig. 3. Example of 5Ws parallel axes with SA

In Fig. 3, P_SA collects the attributes that are not displayed in the P axis, X_SA for X axis, Y_SA for Y axis, Z_SA for Z axis and Q_SA for Q axis.

Figures 1, 2 and 3 have all clearly illustrated the 5Ws behaviours patterns, but it has also raised an important issue: how do we compare between these patterns on What the data contained, How the data was transferred and Why the data occurred? To solve this issue, we established pair density to measure the 5Ws behaviours pattern.

3 Pair-Density Model

In this section, four Pair-Densities have been established, which are Sending Density via Receiving Density; Sending Density via Purpose Density; Sending Density via Transferring Density and Sending Density via Content Density. We will use Sending Density via Receiving Density to demonstrate the pair-density model.

3.1   Sending Density via Receiving Density

Based on (2) and (3), the sending pattern, which measures where the data came from for a particular attribute $d\{t, p, x, y, z\}$, is defined as a subset of $D(t, p, x, y, z)$

$$
D_{(t,p,x,y,z)} = \left\{
\begin{matrix}
d_t & d_p & d_x & d_y & d_z & d_{1Q} \\
d_t & d_p & d_x & d_y & d_z & d_{2Q} \\
d_t & d_p & d_x & d_y & d_z & d_{3Q} \\
\vdots & \vdots & \vdots & \vdots & \vdots & \vdots \\
d_t & d_p & d_x & d_y & d_z & d_{iQ} \\
\vdots & \vdots & \vdots & \vdots & \vdots & \vdots \\
d_t & d_p & d_x & d_y & d_z & d_{mQ}
\end{matrix}
\right\}
\tag{4}
$$
$$
= \{d \in D \mid d(t, p, x, y, z, Q)\}
$$

$Q = \{q_1, q_2, q_3, \dots q_m\}$ represents who received the particular attribute $d\{t, p, x, y, z\}$. The subset $D_{(t, p, x, y, z)}$ collects all data that has the same attribute, regardless of who received it. For example, Fig. 2 shows two sending patterns, $\{x = \text{``Argentina Team''}\}$ and $\{x = \text{``Germany Team''}\}$, regardless of which country or city receiving them.

The Sending Density (SD) measures the sender's pattern during data transferal. Based on (3) and (4) for particular attributes $\{t, p, x, y, z\}$, the Sending Density is defined as $SD_{(t, p, x, y, z)}$.

$$
SD_{(t,p,x,y,z)} = \frac{|D(t,p,x,y,z)|}{|D|} \times 100\%
\tag{5}
$$

The receiving pattern measures who received the data for particular attribute $d\{t, x, y, z, q\}$, which is defined as a subset of $D_{(t, x, y, z, q)}$

$$
D_{(t,x,y,z,q)} = \left\{
\begin{matrix}
d_t & d_{1P} & d_x & d_y & d_z & d_q \\
d_t & d_{2P} & d_x & d_y & d_z & d_q \\
d_t & d_{3P} & d_x & d_y & d_z & d_q \\
\vdots & \vdots & \vdots & \vdots & \vdots & \vdots \\
d_t & d_{iP} & d_x & d_y & d_z & d_q \\
\vdots & \vdots & \vdots & \vdots & \vdots & \vdots \\
d_t & d_{mP} & d_x & d_y & d_z & d_q
\end{matrix}
\right\}
\tag{6}
$$
$$
= \{d \in D \mid d(t, P, x, y, z, q)\}
$$

$P = \{p_1, p_2, p_3, \dots p_m\}$ represents where the particular attribute $d\{t, x, y, z, q\}$ came from. The subset $D_{(t, x, y, z, q)}$ collects all data that has the same attribute no matter where the data came from. For example, Fig. 2 shows eight receiving patterns;

{x = "Argentina Team", q = "Australia"}, {x = "Germany Team", q = "Australia"}, {x = "Argentina Team", q = "China"}, {x = "Germany Team", q = "China"}, {x = "Argentina Team", q = "UK"}, {x = "Germany Team", q = "UK"}, {x = "Argentina Team", q = "USA"}, {x = "Germany Team", q = "USA"}.

The Receiving Density (*RD*) measures the receiver's pattern during data transferal. Based on (3) and (4) for particular attributes {t, x, y, z. q}, the Receiving Density is defined as $RD_{(t, x, y, z, q)}$

$$RD_{(t,x,y,z,q)} = \frac{|D(t,x,y,z,q)|}{|D|} \times 100\% \tag{7}$$

The dataset D which illustrates the incidents summary represents the volume and velocity of Big Data. The 5Ws density $SD_{(t, p, x, y, z)}$ and $RD_{(t, x, y, z, q)}$ demonstrates the variety of Big Data utilising the patterns for sending and receiving.

3.2   Noise Data

The noise data is defined in this paper as the unknown or undefined attribute in the 5Ws density algorithm methods. We define the unknown attributes in P dimension as u_p; in X dimension as u_x; in Y dimension as u_y; in Z dimension as u_z; and in Q dimension as u_q. A subset for any unknown attribute is defined as

$$D_{(u)} = \{d \in D \mid d(t, p, x, y, z, q), p = u\_p \vee x = u\_x \vee y = u\_y \vee z = u\_z \vee q = u\_q\} \tag{8}$$

If the subset $D_{(u)}$ collects all the unknown attributes in the 5Ws pattern, then this improves the accuracy for the density algorithms. The $SD_{(t, p, x, y, z)}$ and $RD_{(t, x, y, z, q)}$ should then be re-defined as

$$SD_{(t,p,x,y,z)} = \frac{|D(t,p,x,y,z)|}{|D| - |D(u)|} \times 100\% \tag{9}$$

$$RD_{(t,x,y,z,q)} = \frac{|D(t,x,y,z,q)|}{|D| - |D(u)|} \times 100\% \tag{10}$$

$SD_{(t, p, x, y, z)}$ and $RD_{(t, x, y, z, q)}$ now represents the sender's and receiver's known patterns, which significantly improves the accuracy for Big Data analysis because both densities have avoided noise data.

3.3   Pair-Density Parallel Axes

Here, we create two additional axes by using $SD_{()}$ and $RD_{()}$. The value of each axis is arranged by alphabetical order, which ranges from 0 to 9, A to Z and a to z.

Five Shrunk Attributes, SAs, *p_sa, x_sa, y_sa, z_sa and q_sa,* collect the attributes that are not illustrated in each axis. The Sending Density and Receiving Density for SA are defined as

$$SD_{(SA)} = \frac{|D(t, p\_sa, x\_sa, y\_sa, z\_sa)|}{|D| - |D(u)|} \times 100\% \qquad (11)$$

$$RD_{(SA)} = \frac{|D(t, x\_sa, y\_sa, z\_sa, q\_sa)|}{|D| - |D(u)|} \times 100\% \qquad (12)$$

We will use Facebook interactions during the 2014 FIFA World Cup Final between Germany and Argentina [2], as our example to show the pair-density parallel coordinates. Let us assume that $SD_{(\text{"Facebook", "Germany Team", "iPad", "Like"})} = 40\%$, $SD_{(\text{"Facebook", "Argentina Team", "iPad", "Like"})} = 35\%$, $RD_{(\text{"Germany Team", "iPad", "Like", "Germany"})} = 20\%$, $RD_{(\text{"Argentina Team", "iPad", "Like", "Argentina"})} = 18\%$, $SD_{(Others)} = 25\%$ and $RD_{(Others)} = 62\%$. The pair-density parallel coordinate is shown in Fig. 4.

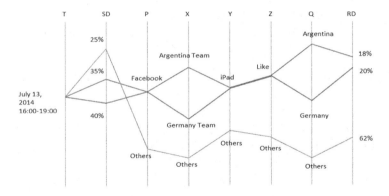

Fig. 4. Example of *SD-RD* parallel coordinates with SA

In Fig. 4, *"Others"* represents the SA that collect the other attributes for $p \neq$ *"Facebook"*, $x \neq$ *"Argentina Team"* or *"Germany Team"*, $y \neq$ *"iPad"*, $z \neq$ *"Like"*, $q \neq$ *"Argentina"* or *"Germany"*.

In Fig. 4, 40 % of Facebook senders supported the *"Germany Team"* compared to 35 % of senders who supported the *"Argentina Team"*. 20 % of Facebook receivers are located in *"Germany"* compared to 18 % of receivers in *"Argentina"*. 62 % of data goes to *"others"* countries and 25 % of data came from sources other than *"Facebook"*.

The axes $SD_{()}$ and $RD_{()}$, which were closest to axes *P* and *Q*, have demonstrated senders and receivers patterns which significantly improves measurement for multi-dimensional data. The pair-density parallel axes, combined with the alphabetical axes and numerical axes, provide the most analytical method for Big Data analysis and visualization. It also explores the particular data patterns that enable multidimensional data analysis and visualization to be very efficient since it can contract or expand as required.

3.4 Clustering in Pair-Density Parallel Axes

The clustering axis in pair-density parallel coordinates can be assigned as several data types or topics. It will lead the values of $SD_{()}$ and $RD_{()}$ to change because a dimension in the 5Ws subset has been added. For example, after adding dimension $P1$ as the clustered axis which contains attributes $P1\{text, image, video, etc.\}$, the subset has been changed to $\{T, P, P1, X, Y, Z, Q\}$. The value of $SD_{()}$ and $RD_{()}$ changes as well as a result. The clustered pair-density parallel coordinate is shown in Fig. 5.

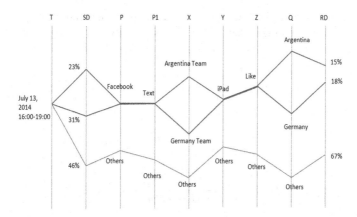

Fig. 5. Example of clustering in *SD-RD* parallel coordinates where *P1* is clustered axis *P*.

3.5 Other Pair-Densities

Sending Density via Purpose Density. Based on (9) and (10), Sending Density *(SD)* and Purpose Density *(PD)*, which measures where the data came from and why the data occurred, are defined as

$$SD_{(t,p,x,y,q)} = \frac{|D(t,p,x,y,q)|}{|D| - |D(u)|} \times 100\,\% \tag{13}$$

$$PD_{(t,x,y,z,q)} = \frac{|D(t,x,y,z,q)|}{|D| - |D(u)|} \times 100\,\% \tag{14}$$

$SD_{()}$ measures where the data came from for particular attribute $\{t, p, x, y, q\}$ regardless of why the data occurred. $PD_{()}$ measures why the data occurred for particular attribute $\{t, x, y, z, q\}$ regardless of where the data came from.

Sending Density via Transferring Density. Sending Density *(SD)* and Transferring Density *(TD)*, which measures where the data came from and how the data was transferred, are defined as

$$SD_{(t,p,x,z,q)} = \frac{|D(t,p,x,z,q)|}{|D| - |D(u)|} \times 100\,\% \tag{15}$$

$$TD_{(t,x,y,z,q)} = \frac{|D(t,x,y,z,q)|}{|D| - |D(u)|} \times 100\,\% \tag{16}$$

$SD_{()}$ measures where the data came from for particular attribute $\{t, p, x, z, q\}$ regardless of how the data was transferred. $TD_{()}$ measures how the data was transferred for particular attribute $\{t, x, y, z, q\}$ regardless of where the data came from.

Sending Density via Content Density. Sending Density *(SD)* and Content Density *(CD)*, which measures where the data came from and what the data contained, are defined as

$$SD_{(t,p,y,z,q)} = \frac{|D(t,p,y,z,q)|}{|D| - |D(u)|} \times 100\,\% \tag{17}$$

$$CD_{(t,x,y,z,q)} = \frac{|D(t,x,y,z,q)|}{|D| - |D(u)|} \times 100\,\% \tag{18}$$

$SD_{()}$ measures where the data came from for particular attribute $\{t, p, y, z, q\}$ regardless of what the data contained. $CD_{()}$ measures what the data contained for particular attribute $\{t, x, y, z, q\}$ regardless of where the data came from.

4   Implementation

We have tested our pair-density model by using six sample datasets from ISCX2012 network dataset [3], an example of Big Data with 20 data dimensions which contains 906,782 data incidents. The summary of these six sample datasets are shown in Table 1. Unknown traffics in those six datasets are traded as unknown nodes which are calculated and illustrated in the graph. We designed two stages to test our model for implementation. The first stage shows how 5Ws dimension works across 6 datasets. The second stage shows how the pair-density works with SA.

Table 1. Six sample datasets from ISCX2012

Dataset	Jun12	Jun13	Jun14	Jun15a	Jun15b	Jun15c
Network traffic node	133,193	275,528	171,380	101,482	94,911	130,288
Unknown TCP traffic	2	13,568	1,077	11,149	2	3
Unknown UDP traffic	254	414	6,172	36,149	20	36
Attacks	0	20,358	3,771	0	0	37,375
Source Ips	44	44	448	1,611	33	36
Destination Ips	2,610	2,645	7,959	15,067	2,164	1,656
Application names	21	85	95	69	19	19

4.1 5Ws Parallel Coordinate

The first test stage is shown in Fig. 6. The dimension P axis represents the source IPs which represented where the data came from. There were 1,948 attributes in the P axis. $P = "0.0.0.0"$ indicates that the source address was invalid.

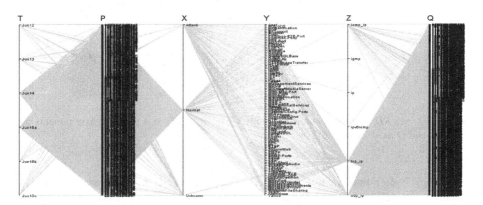

Fig. 6. 5Ws parallel coordinates without the pair-density algorithm where P axis contains 1,948 attributes, Y axis contains 105 attributes, and Q axis contains 24,372 attributes

The dimension X axis represents the data content, including *"Normal"* traffics, *"Attack"* traffics and *"Unknown"* traffics. The dimension Y axis represents the applications which describe how the data was transferred. There were 105 attributes in the Y axis. The dimension Z axis represents the protocol which illustrates why the data occurred. There were 6 attributes in the Z axis. The dimension Q axis represents the destination IPs which denotes who received the data. There were 24,374 attributes in the Q axis and Q = *"0.0.0.0"* means that the destination address was invalid. In total, 64,393 5Ws patterns are displayed in the graph from 906,782 data incidents. A lot of overlapping polylines and over-crowded attributes are shown in the P and Q axes as in Fig. 6.

4.2 The $SD_{()}$ via $RD_{()}$ Parallel Coordinate with SA

SA has been implemented on two axes; P axis and Q axis in order to reduce the attributes over-crowding in Fig. 7. We define SA for each subnet as *"00x.xxx.xxx.xxx"*, *"0xx.xxx.xxx.xxx"*, *"1xx.xxx.xxx.xxx"*, and *"2xx.xxx.xxx.xxx"* for where $SD_{(p)} < 1.0$ % or $RD_{(q)} < 1.0$ %. In another word, p or q = *"1xx.xxx.xxx.xxx"* including all IPs in the range of *{100–255. 1–255. 1–255. 1–255}* while $SD_{(p)} < 1.0$ % or $RD_{(q)} < 1.0$ %. For example, in P axis, if two attributes $SD_{(p = 111.111.111.111)} < 1.0$ % and $SD_{(p = 123.123.123.123)} < 1.0$ %, those two attributes will be shrunk into one attribute in the parallel coordinate as $SD_{(p = 1xx.xxx.xxx.xxx)} < 1.0$ %.

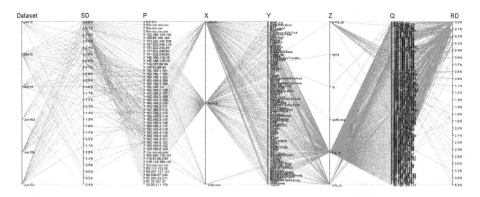

Fig. 7. $SD_{()}$ via $RD_{()}$ parallel coordinates with SA on P and Q axis where P axis remains 51 attributes and Q axis remains 200

In Fig. 7 after implementing SA, axis P contained 51 items, down from 1,948 attributes and axis Q had 24,372 attributes shrunk down to 200. The cluttering poly-lines and over-crowded attributes have been significantly reduced from 64,393 to 8,030. Cluttering has therefore been reduced by over 85 % without the loss of any information, which is a significant achievement. The attributes in each axis are represented as different topics such as *"Attack"* in X axis, or *"tcp_ip"* in Z axis. The data types can be extracted, such as *"http"* in Y axis. This provides comparisons between the different topics and types vital for business, government and organizational needs.

4.3 The $SD_{()}$ via $PD_{()}$ Parallel Coordinate with SA

SA has been applied on three axes; P axis, Y axis and Q axis, in $SD_{()}$ via $PD_{()}$ parallel coordinate shown in Fig. 8. Axis P contains 29 attributes shrunk from 1948. Axis Q has been reduced to 52 attributes from 24,374. To analysis "Attack" patterns, SA has been assigned as *"Other-Apps"* for attribute not containing *"Attack"*. The attributes in Y axis shrunk to 81 from 105.

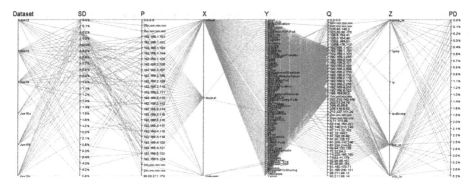

Fig. 8. $SD_{()}$ via $PD_{()}$ parallel coordinates with SA on P, Y and Q axis where P axis remains 29 attributes, Y axis has 81 attributes, and Q axis remains 52 attributes

In Fig. 8, the pattern *"Attack"* has been clearly illustrated between the X and Y axis. *"Normal"* attribute in X axis points to *"Other-Apps"* on Y axis as a result of using SA on the Y dimension. The cluttering polylines and over-crowded attributes have been significantly reduced from 64,393 to 3,404 after implementing SA.

4.4   Reduction of Polylines Cluttering

We have measured the polylines from the original 20 dimensions to our 5Ws dimensions in the parallel coordinates, and found out that the 5Ws parallel coordinates has significantly reduced the cluttered polylines and over-crowded attributes by more than 78 % shown in Fig. 9. This is a significant boost in the analysis of Big Data as it provides ease of access and clarity to our analysis.

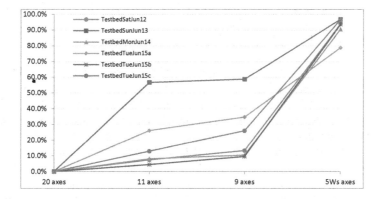

Fig. 9. Reduction for different axes between six datasets

Figure 10 shows the reduction of different SA between datasets. It has not only reduced polylines over-crowding in graphs, but also significantly reduced the data processing time for Big Data analysis and visualization – another major advantage of our model.

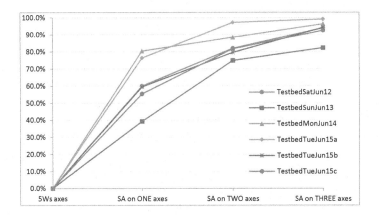

Fig. 10. Reduction for different SA between six datasets

5   Related Works

The multidimensional data analysis requires tools to explore the relationship between these dimensions. One powerful visual tool that explores multidimensional data is the parallel coordinate which is widely used for multidimensional data visualization. However, it has a problem while deals with large scale multidimensional data: the polylines clutter and over-crowd each other.

Xiaoru Yuan et al. [14] scattered points in parallel coordinates to combine the parallel coordinates and scatterplot scaling, which reduced data over-crowding. Matej Novotny and Helwig Hauser [15] grouped the data context into outliers, trends and focus, and set up three clustered parallel coordinates to reduce the data cluttering issues. Yi Chen et al. [23] used the parallel coordinates and enhanced ring (PCER) to explore the statistical results for students' scores to reveal any trend. Geoffrey Ellis and Alan Dix [16] developed three methods: raster algorithm, random algorithm and lines algorithm for measuring occlusion in parallel coordinates plots to provide tractable measurement of the clutter.

Most approaches for multidimensional data analysis and visualization are practiced on a single dataset such as text dataset, audio dataset, and image dataset. Xiaotong Liu et al. [17] developed a visual search engine based on CompactMap to stream the text data for visual analytics. Seunggwoo Jeon et al. [18] transformed unstructured email texts into a graph database and visualized them. Richard K. Lomotey and Ralph Deters [19] extracted the topics and terms from unstructured data by using the TouchR2 tool they created.

Researchers have tried to reduce multidimensional data in their visual approaches. Zhenwen Wang et al. [21] introduced ADraw for grouping the same attribute value nodes. Then they created virtual nodes to group the same attribute value nodes together. The different groups are separated by different colours in the visualization. Zhangye Wang et al. [8] clustered large-scale social data into users groups by using the information of user tag and user behaviour. The K-means algorithm has been deployed in their approach. Daniel Cheng et al. [22] proposed the Tile-Based Visual Analytics (TBVA) to explore one billion pieces of Twitter data. TBVA created tiled heat maps and tiled density strips for Big Data visualization. Quan Li et al. [24] proposed PatternTrack to detect visual patterns for multidimensional data, and mapped all dimension axes in concentric circles to integrate three level concentric groups: data values, patterns and gradient circles.

To the best of our knowledge, no previous work has used the 5Ws dimensions to classify the multidimensional data behaviours, nor has any work created two additional axes by using the pair-density in the parallel coordinate visualization. Common visualization methods trade each data as a node in visual graphics, and then find visual patterns to analyze the data. We have classified the data dimension first to obtain the 5Ws patterns, and then visualized those data patterns. Our method has significantly reduced the data processing time and the data cluttering for Big Data analysis and visualization.

6   Conclusions and Future Works

Pair-density model, a novel approach for multidimensional data analysis and visualization, has been introduced in this work. We have demonstrated the 5Ws patterns across multiple datasets, established the pair-density for Big Data analysis, and created two additional axes in pair-density parallel coordinates to reduce data over-crowding without the loss of any information. The shrunk attributes applied in each dimension axis enables the attributes to be contracted or expanded for better visualisation as necessary. The dimension clustering in pair-density parallel axes provides a clear view of visual structures and patterns for better understanding of Big Data.

The pair-sending not only measures multidimensional data patterns, but also provides comparisons for multiple datasets between different topics and data types. This provides more analytical features for Big Data analysis. Our model has reduced data over-crowding by at least 75 % in our testing. Even more, the pair-density pattern with shrunk attributes has reduced data cluttering by nearly 98 % based on our density algorithm. It has also significantly reduced the data processing time for Big Data analysis.

In the future, we plan to develop our pair-density model and deploy it in more areas and different datasets such as financial datasets and Facebook datasets. The combination of pair-density parallel coordinates and Treemaps is our next stage for Big Data analysis and visualization.

References

1. Stamford, Gartner Says Solving 'Big Data' Challenge Involves More Than Just Managing Volumes of Data. http://www.gartner.com/newsroom/id/1731916/. Accessed 27 June 2011
2. Lorenzetti, L.: World Cup scores big on Twitter and Facebook, Fortune. http://fortune.com/2014/07/14/world-cup-scores-big-on-twitter-and-facebook/. Accessed 14 July 2014
3. Shiravi, A., Shiravi, H., Tavallaee, M., Ghorbani, A.A.: Toward developing a systematic approach to generate benchmark datasets for intrusion detection. Comput. Secur. **31**(3), 357–374 (2012)
4. Inselberg, A., Dimnsdale, B.: Parallel coordinates: a tool for visualizing multi-dimensional geometry. In: Proceedings of First IEEE Conference on Visualization, pp. 361–378, October 1990
5. Qu, K., Lin, N., Lu, Y., Payan, D.G.: Multidimensional data integration and relationship inference. IEEE Intell. Syst. **17**(2), 21–27 (2002)
6. Zhang, J., Huang, M.L.: Density approach: a new model for BigData analysis and visualization. Concurrency and Computation: Practice and Experience. Wiley online Library, July 2014. DOI:10.1002/cpe.337
7. Yin, X.F.: Multidimensional data clustering based on fast kernel density estimation. In: Proceedings of 2013 International Conference on Machine Learning and Cybernetics, pp. 311–315, July 2013
8. Wang, Z., Zhou, J., Chen, W., Chen, C., Liao, J., Maciejewski, R.: A novel visual analytics approach for clustering large-scale social data. In: Proceedings of 2013 IEEE International Conference on Big Data (IEEE Big Data 2013), pp. 79–86, October 2013
9. Li, L., Gariel, M., Hansman, R.J., Palacios, R.: Anomaly detection in onboard-recored flight data using cluster analysis. In: Proceedings of 2011 IEEE/AIAA 30th Digital Avionics Systems Conference (DASC), pp. 4A4-1–4A4-11, October 2011
10. Nimmagadda, S.L., Dreher, H.: Petro-data cluster mining – knowledge building analysis of complex petroleum system. In: Proceedings of 2009 IEEE International Conference on Industrial Technology (ICIT 2009), pp. 1–8, February 2009
11. We, J., Yu, W.: Optimization and improvement based on k-means cluster algorithm. In: Proceedings of 2009 2nd International Symposium on Knowledge Acquisition and Modeling (KAM 2009), vol. 3, pp. 335–339, November 2009
12. Nie, B., Du, J., Liu, H., Xu, G., Wang, Z., He, Y., Li, B.: Crowds' classification using hierarchical cluster, rough sets, principal component analysis and its combination. In: Proceedings of 2009 International Forum on Computer Science-Technology and Application (IFCSTA 2009), pp. 287–290, December 2009
13. Zhang, J., Huang, M.L.: Detecting flood attacks through new density-pattern based approach. In: Proceedings of 2013 IEEE 15th International Conference on High Performance Computing and Communications (HPCC 2013), pp. 246–253, November 2013
14. Yuan, X., Guo, P., Xiao, H., Zhou, H., Qu, H.: Scattering points in parallel coordinates. IEEE Trans. Vis. Comput. Graph. **15**(6), 1001–1008 (2009)
15. Novotny, M., Hauser, H.: Outlier-preserving focus+context visualization in parallel coordinates. IEEE Trans. Vis. Comput. Graph. **12**(5), 893–900 (2006)
16. Ellis, G., Dix, A.: Enabling automatic clutter reduction in parallel coordinates plots. IEEE Trans. Vis. Comput. Graph. **12**(5), 717–724 (2006)
17. Liu, X., Hu, Y., North, S., Shen, H.W.: CompactMap: a mental map preserving visual interface for streaming text data. In: Proceedings of 2013 IEEE International Conference on Big Data (IEEE Big Data 2013), pp. 48–55, October 2013

18. Jeon, S., Khosiawan, Y., Hong, B.: Making a graph database from unstructured text. In: Proceedings of 2013 16th IEEE International Conference on Computational Science and Engineering (CSE), pp. 981–988, December 2013

19. Lomotey, R.K., Dters, R.: Topics and terms mining in unstructured data stores. In: Proceedings of 2013 16th IEEE International Conference on Computational Science and Engineering (CSE), pp. 854–861, December 2013

20. Zhang, J., Huang, M.L.: 5Ws model for BigData analysis and visualization. In: Proceedings of 2013 16th IEEE International Conference on Computational Science and Engineering (CSE), pp. 1021–1028, December 2013

21. Wang, Z., Xiao, W., Ge, B., Xu, H.: ADraw: a novel social network visualization tool with attribute-based layout and coloring. In: Proceedings of 2013 IEEE International Conference on Big Data (IEEE Big Data 2013), pp. 25–32, October 2013

22. Cheng, D., Schretlen, P., Kronenfeild, N., Bozowsky, N., Wright, W.: Tile based visual analytics for twitter big data exploratory analysis. In: Proceedings of 2013 IEEE International Conference on Big Data (IEEE Big Data 2013), pp. 2–4, October 2013

23. Chen, Y., Cheng, X., Chen, H.: A multidimensional data visualization method based on parallel coordinates and enhanced ring. In: Proceedings of 2011 International Conference on Computer Science and Network Technology (ICCSNT), vol. 4, pp. 2224–2229, December 2011

24. Li, Q., Chen, L., Liao, H., Yong, J.: PatternTrack: a visual pattern detection technique for multidimensional data. In: Proceedings of 2012 International Conference on Computer Science and Service System (CSSS), pp. 1360–1365, August 2012

AutoCSD: Automatic Cloud System Deployment in Data Centers

Tao Xie[1] and Haibao Chen[1,2]([⊠])

[1] Services Computing Technology and System Lab,
Cluster and Grid Computing Lab,
School of Computer Science and Technology,
Huazhong University of Science and Technology, Wuhan 430074, China
[2] School of Computer and Information Engineering,
Chuzhou University, Chuzhou 239000, China
chenhaibao@hust.edu.cn

Abstract. It is a huge challenge to deploy a cloud computing system in large-scale data centers. In order to help resolve this issue, we propose an automatic cloud system deployment approach with the characteristics of reliability, availability, and load balance. Specifically, we use workflow to deal with the dependencies among the automatic deployment processes of a cloud system. We also design a failover mechanism to avoid the single point failure of the deployment server. Besides, we adopt a load balancing algorithm to solve the bottleneck problem of deploying a cloud system.

We implement a prototype, and evaluate it with 16 physical nodes as well as a virtualized environment with 160 virtual machines. Experimental results show that the average deployment time under our approach is lower than that with traditional deployment methods. In addition, it achieves a cloud system deployment success ratio of up to 90 %, even in the high-concurrency environment.

Keywords: Data center · Cloud computing system · Deployment server · Reliability · Load balance

1 Introduction

Cloud computing, often referred to as simply "the cloud", is the delivery of on-demand computing resources over the Internet on a pay-for-use basis. It relies on the sharing of resources to achieve coherence and economies of scale, similar to a utility (like the electricity grid) over a network. With immense success and rapid growth within the past few years, cloud computing has been established as the dominant paradigm of the IT industry. To meet the increasing demand of computing and storage resources, cloud infrastructure providers are deploying planet-scale data centers across the world, consisting of hundreds of thousands, even millions of servers, which significantly increases the level of effort in deploying cloud computing management systems in these data centers. That is, it is

© Springer International Publishing Switzerland 2015
W. Qiang et al. (Eds.): CloudCom-Asia 2015, LNCS 9106, pp. 72–85, 2015.
DOI: 10.1007/978-3-319-28430-9_6

a huge challenge to deploy such system in large-scale data centers, because it is time-consuming and requires a lot of human labor.

Generally, the deployment of a cloud computing system involves hardware resources (e.g., physical hosts), software resources (e.g., operating systems, cloud computing middleware or databases), network resources, and storage resources. Much effort has been made to improve the efficiency of operating system deployment on physical nodes including, systemimager [1], Kickstart [2], etc. Currently, existing work on the deployment of the cloud system, e.g., DevStack [3], mainly focuses on system deployment automation. However, it ignores the reliability of system deployment, which results in a negative impact on the deployment progress of the cloud system.

In order to deploy the cloud system automatically in a data center, as well as guaranteeing the reliability of system deployment, we propose an approach of automatic cloud system deployment (AutoCSD) in data centers with the characteristics of reliability, availability, and load balance. Specifically, we use workflow to deal with the dependencies among the automatic deployment processes of the cloud system and we design a failover mechanism to avoid the single point failure of the deployment server which contains the installation files of the operating system and cloud middleware. Besides, we adopt a load balancing mechanism to solve the bottleneck problem of deploying the cloud system in a large-scale data center.

We implement a prototype, and evaluate it in a physical environment with 16 nodes as well as a virtualized environment with 160 virtual machines (VMs), respectively. Experimental results show that the average deployment time under our approach is lower than that with a traditional deployment method. In addition, it achieves a cloud system deployment success ratio of up to 90 %, even in the high-concurrency environment.

The contributions of this paper are two-fold:

- We propose an AutoCSD approach in data centers, which contains the workflow-based automatic deployment, failover mechanism, and load balancing mechanism.
- We implement a prototype, and evaluate it in a physical environment with 16 nodes as well as a virtualized environment with 160 VMs, respectively.

The rest of the paper is organized as follows. In Sect. 2, we present the background and motivation. In Sect. 3, we present our approach, including the overview, workflow, and major techniques of the cloud system automatic deployment. We evaluate our results in Sect. 4, and discuss some key issues and future work in Sect. 5. In Sect. 6, we discuss the related works. Finally, we provide concluding remarks in Sect. 7.

2 Background and Motivation

In this section, we first present the background to cloud computing and system deployment in a data center. Then, we discuss our motivations.

2.1 Background

With the rapid development of computing and storage technologies and the extreme success of the Internet, computing resources have become more powerful, cheaper, and more ubiquitously available than ever before. This technological shift has enabled the realization of a new computing paradigm called cloud computing. Technically, cloud computing is a large pool of easily accessible and readily usable virtualized resources, such as hardware, development platforms, and services that can be dynamically reconfigured to adjust to a variable load in terms of scalability, elasticity, and load balancing, and thus allow opportunities for optimal resource utilization.

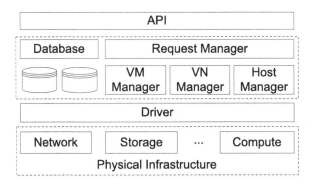

Fig. 1. The basic components of cloud platform

According to the National Institute of Standards and Technology's (NIST's) definition [4], the five essential characteristics of cloud computing are on-demand computing service, broad network access, resource pooling, rapid elasticity, and measured service. The architecture of cloud computing systems typically involves multiple cloud components communicating with each other over a loose coupling mechanism such as a messaging queue.

Typically, the main components of a cloud system are shown in Fig. 1, and are described as follows.

– Database Component. This contains all the data structures and information of the cloud platform.
– Request Manager Component. This exposes interfaces (e.g., XML-RPC), which can be used to call any components of the cloud platform.
– VM Manager Component. This component is responsible for managing VMs, e.g., creating, pausing, resuming, rebooting, shutting down, and destroying VMs.
– Virtual Network (VN) Manager Component. This component is responsible for creating and managing the VNs.
– Host Manager Component. This manages and monitors all physical hosts in the data center.

- Driver Component. A cloud system usually has a set of pluggable adaptors to interact with specific middleware, e.g., virtualization hypervisor. These adaptors are called drivers.
- Physical Resource Component. This mainly includes storage resources, network resources, and computing resources.

Due to the growing demands on the cloud computing service, a huge number of cloud computing platforms around the world are emerging, including public and commercial clouds (e.g., Amazon EC2 [5]) and private clouds which are built based on the open-source cloud systems, such as OpenStack [6] and Eucalyptus [7]. Actually, it is a huge challenge to deploy such systems in large-scale data centers, because it is time-consuming and requires much human labor.

2.2 Motivation

Cloud providers must be able to deploy cloud systems in large-scale data centers very quickly and very easily. Automatic deployment is an effective way of simplifying the complexity of such work in a data center and several projects aim to provide solutions for deploying cloud systems automatically.

For example, DevStack [3], which is an OpenStack community production to deploy OpenStack automatically, has evolved to support a large number of configuration options and alternative platforms and support services. This evolution of DevStack has grown well beyond the original intention and the majority of configuration combinations are rare. However, DevStack is not a general OpenStack installer and was never meant to be everything to everyone. Besides, it does not consider the reliability of cloud system deployment. Therefore, how to deploy the cloud system automatically in data centers as well as guaranteeing the reliability of system deployment has become a topic in current research work.

Motivated by these challenges, we aim to propose an automatic cloud system deployment approach, with the characteristics of reliability, availability, and load balance, which will be described in the following sections.

3 Automatic System Deployment

In this section, we present our AutoCSD approach in data centers. Specifically, in Sect. 3.1, we provide an overview of AutoCSD and then in Sect. 3.2 we show how AutoCSD works. Finally, in Sect. 3.3, we discuss the major techniques used in our approach.

3.1 The Overview

As discussed in Sect. 2.2, the existing approaches mainly focus on automating the deployment of the cloud system. In contrast, our AutoCSD approach considers the automation, reliability, availability, and load balancing of cloud system deployment at the same time.

As shown in Fig. 2, our AutoCSD consists of multiple loosely-coupled modules as follows:

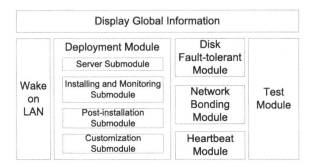

Fig. 2. The components of AutoCSD

- Wake-on-local area network (LAN) module. This is used to wake up the bare-metal nodes in the data center.
- Deployment module. This module automatically sets the configuration files and installs the system including the operating system and cloud middleware. Meanwhile, it is able to monitor the deployment progress of the cloud system. The functions of this module can be implemented by extending the open-source automatic installation software, e.g., Kickstart [2].
- Disk Fault-tolerant module. This module uses the RAID software technique [8] to turn the disks of the compute nodes into RAID, and mount them onto the compute nodes automatically.
- Network Bonding module. This provides a method for aggregating the multiple network interface cards (NICs) of the compute node into a single logical bonded NIC, and is able to switch the behavior of the single logical bonded NIC between load balancing (round-robin) mode and fault-tolerant (active-backup) mode. These functions can be implemented by extending the open-source bonding technique, e.g., Linux Bonding [9].
- Heartbeat module. This module automatically configures the heartbeat mechanism [10] of the nodes, which is used to realize the failover of the cloud system deployment server.
- Test module. This module is responsible for verifying the cloud system installation on the compute nodes, including the correctness of the configuration files, etc.
- Display Global Information module. This is used to demonstrate the relevant information of system deployment to the administrator of the data center.

3.2 The Workflow of System Deployment

The AutoCSD in the cloud data center can be divided into two parts: The first is to deploy the cloud system, including the operating system and cloud middleware, on the bare-metal nodes and the second is to test the cloud system deployed in the data center.

The workflow in cloud system deployment in a data center is shown in Fig. 3. When the bare-metal nodes to be installed and the storage devices are ready, the workflow wakes the nodes in the states of standby or shutdown. After that, these awakened nodes acquire the installation files of the operating system from the deployment server, then build an operating system and begin to configure the network, followed by the cloud middleware installation in the compute nodes. Finally, it starts the cloud system.

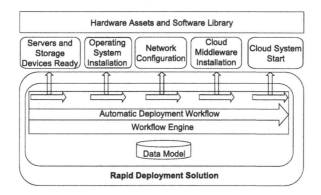

Fig. 3. The workflow of automatically deploying cloud system

Figure 4 gives the sequence chart of deploying the cloud system in the data center, which is described as follows:

- The node to be installed is woken up by the PXE (Pre-boot eXecution Environment) network card. Then it requests an IP from the LAN DHCP server.
- The DHCP server allocates an IP to this node and sends the IP information of the TFTP server as well as that of the cloud system deployment server to it.
- The node obtains the operating system installation files from the TFTP server, and starts an operating system on the disk.
- The operating system on the disk of the compute node obtains the cloud system installation packages from the deployment server.
- The cloud system is installed on the compute node. After the configuration files of the cloud system have been set, the cloud system is started.

When the cloud system is installed, our approach will test the cloud platform including the correctness of the configuration files, etc. As shown in Fig. 5, the test module will use APIs provided by the cloud system to check the cloud platform components of the database and request the manager, VM manager, virtual network manager, host manager, and driver.

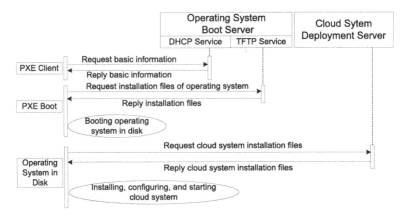

Fig. 4. The sequence chart of deploying cloud system, which contains the installation of operating system and cloud computing middleware on each compute node.

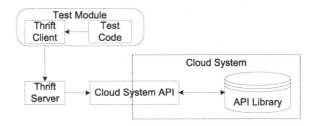

Fig. 5. The processes of test module

3.3 Major Techniques

Based on the workflow techniques and the open-source software, including Kickstart, software RAID, Linux Bonding, Heartbeat, etc., we implement a prototype of our AutoCSD.

In this section, we introduce the major techniques and strategies used in our approach. In order to improve the efficiency of cloud system deployment, we adopt automatic deployment workflow and Kickstart, which are described as follows.

- Workflow. This is the series of activities that are necessary to complete a task, i.e., in this paper, deploying a cloud system. As shown in Fig. 3, the deployment of a cloud system in a large-scale data center can be divided into five steps, each of which can be taken as an activity in workflow. Each step in a workflow has a specific step before it and a specific step after it, with the exception of the first step. In a linear workflow, the first step is usually initiated by an outside event. It is the core of our automatic cloud system deployment approach.
- Kickstart. In order to automatically install an operating system on machines, Red Hat created the Kickstart installation method. Using Kickstart, a sys-

tem administrator can create a single file containing the answers to all the questions that would normally be asked during a typical installation. Kickstart files can be kept on a single-server system and can be read by individual machines during the installation. This installation method can support the use of a single Kickstart file to install Red Hat Enterprise Linux on multiple machines.

Generally, the problem of load balancing occurs in the TFTP server, which stores the operating system installation files, and the deployment server, which contains the cloud system installation files. In order to solve the bottleneck problems of the TFTP server and deployment server in large-scale data centers, we use the strategy of multiple deployment servers and multiple TFTP servers, and adopt a load balancing mechanism (i.e., Domain Name System (DNS)-based load balancing) for these servers. Specifically, with the load balancing strategy, bare-metal nodes are allowed to obtain the boot configuration file, the kernel and the file system image from the multiple TFTP servers at the same time. In addition, it makes the nodes obtain files, e.g., Kickstart configuration files and the cloud system installation packages, from the multiple deployment servers simultaneously.

- Round-robin DNS-based load balancing is often used to load balance requests between a number of servers (i.e., in this paper, TFTP servers and deployment servers). Taking the TFTP server as an example, assuming that it has one domain name and three identical copies of the same service residing on three servers with three different IP addresses, when one node to be installed accesses the TFTP service, it will be sent to the first IP address; the second node, which accesses the TFTP service will be sent to the next IP address, and the third node will be sent to the third IP address. In each case, once the IP address is given out, it goes to the end of the list. The fourth node, therefore, will be sent to the first IP address, and so forth.

In order to keep the availability of the TFTP servers and the deployment servers, we implement a failover mechanism based on RAID, Linux network bonding, and Kickstart software. When a disk fails, the read request will be redirected to the mirror disk based on the technique of RAID1. When an NIC fails, another NIC will replace it, so that the network connection will not break. Similarly, when a server fails, another server will replace it, so that the service will not be disrupted.

- Software RAID. RAID, an acronym for Redundant Arrays of Independent Disks, is a way to virtualize multiple, independent hard disk drives into one or more arrays to improve performance, capacity and reliability (availability). Taking the pure RAID software as an example, RAID implementation is an application running on the host without any additional hardware. This type of RAID software uses hard disk drives which are attached to the host system via a built-in I/O interface or a host bus adapter (HBA). The RAID becomes active as soon as the operating system has loaded the RAID driver software. The primary advantage of this solution is its low cost.

- Linux network bonding. This is the creation of a single bonded interface by combining two or more Ethernet interfaces. This helps with the high availability of the network interface and offers performance improvement. The Linux bonding driver provides a method of aggregating multiple network interfaces into a single logical bonded interface. The behavior of the bonded interfaces depends upon the mode. Generally speaking, modes provide either hot standby or load-balancing services.
- Heartbeat. This is the main software product of the Linux-HA (High-Availability Linux) project. Specifically, it is a daemon that provides cluster infrastructure (communication and membership) services to its clients. This allows clients to know about the presence (or disappearance) of peer processes on other machines and to easily exchange messages with them.

4   Performance Evaluation

In this section we evaluate our AutoCSD in a physical environment with 16 nodes as well as a virtualized environment with 160 VMs, respectively. We first describe the experimental environment in Sect. 4.1, and then present our experiments and the results in Sect. 4.2.

4.1   Experimental Setup

Experimental platform. There are 18 identical physical nodes (note that two nodes act as the server to provide the system deployment service) in our platform, each of which has a quad-way six-core CPU, 16 GB memory, a SCSI disk, and a Gigabit Ethernet card.

Cloud system. A cloud system called CRANE is used in our experiments, which is the second generation of the support platform for ChinaGrid [11].

Operating system. The operating system used in our experiments is Red Hat Enterprise Linux (RHEL) 6. The size of the installation packages is about 3.7 GB.

Experimental classification. We design two experiments in this section, which are shown as follows.

- The acceleration of system deployment. In order to evaluate the performance of our AutoCSD in alleviating the bottleneck problem of the TFTP server and the deployment server, we take two servers (each one stores both the operating system installation files and the cloud system installation files) as an example to measure the acceleration of the system deployment.
- The success rate of the deployment system. In order to evaluate the success rate of system deployment on the compute nodes in a large-scale data center, due to the limitation of our physical platform, we carried out an experiment with 160 VMs.

4.2   Experimental Results

In this section, we carry out two experiments and present the results.

Acceleration of System Deployment. To evaluate the performance promotion resulting from the double servers, each of which is used to provide both the TFTP service and the cloud system deployment service, we conducted two groups of experiments. One group has two servers to provide the system deployment service, and the other has one server to provide this service. In each experiment, we gradually increased the number of nodes (i.e., the degree of concurrency), which will be installed in the cloud system at the same time, with a step of 2.

Figure 6 gives the experimental results. We can see that when the degree of concurrency is small (e.g., 2, 4, 6, and 8), neither has any effect on the deployment efficiency of the cloud system on bare-metal nodes.

With the degree of concurrency increasing, the system deployment time in stand-alone mode (i.e., single-server mode) increases. For example, the average time for system deployment under the stand-alone mode is 45 min when the degree of concurrency is 16, while that time is about 35 min when the degree of concurrency is 2.

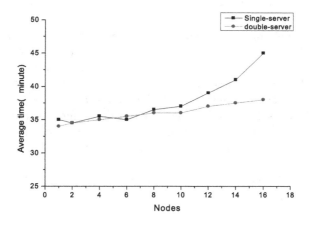

Fig. 6. The acceleration of cloud system deployment

It should be noted that the system deployment time under the double-server mode also increases, but has a slower growth rate. Besides, under the same degree of concurrency, the performance of the double-server mode outperforms that of the single-server mode. The main reason is that the double-server mode provides double the bandwidth of the single-server mode. Therefore, the average deployment time on the compute nodes under the single-server mode is significantly higher than that of double-server mode.

Success Rate of Deployment. The deployment success rate is an important metric in measuring the performance of the cloud system deployment approach.

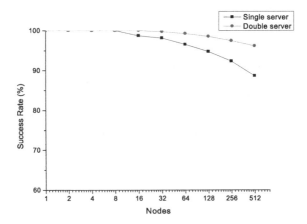

Fig. 7. The success rate of system deployment

Since there are insufficient physical nodes for use in our physical platform, we launched 160 VMs on 16 physical nodes for testing. In the experiment, we also gradually increased the number of VMs with a step of 2 to evaluate the success rates of the cloud system deployment on the VMs. Figure 7 shows the experimental results.

From Fig. 7, we can observe that the deployment success rate under both modes drops a little when the degree of concurrency is lower than a certain level (e.g., 8), because the server can afford to meet the requests of system deployment in time when the number of the deployment requests is small; when the degree of concurrency exceeds that level, the deployment success rate under the single-server mode declines rapidly. In contrast, the deployment success rate under the double-server mode is higher than that of the single-server mode, because the double-server mode can benefit from the fault-tolerant mechanism of our AutoCSD, which can mitigate the single point failure of the system deployment server and thus enhance the reliability of the deployment server and the availability of the system deployment service.

5   Discussion and Future Work

In this paper, we propose an AutoCSD approach in data centers with the characteristics of reliability, availability, and load balance. Under our approach, compute nodes in a LAN can be installed automatically. However, in many cases, the nodes of a cross-regional data center are connected by a wide area network (WAN). In the future, we will extend our approach to such cases.

In the section on performance evaluation, we took two servers as an example to test our approach. The experimental results show that the double-server mode outperforms the single-server mode in both system deployment acceleration and the system deployment success rate on bare-metal nodes. Intuitively, data centers with different numbers of nodes will need different numbers of servers. The design of a method to calculate the numbers of servers for different data centers will form our future work.

6   Related Work

Cloud computing, which refers to the concept of dynamically providing processing time and storage space from a ubiquitous "cloud" of computational resources, allows users to acquire and release the resources on demand and provide access to data from processing elements, while relegating the physical location and exact parameters of the resources. Cloud computing means scalability on demand and flexibility to meet business changes, and is easy to use and manage.

As the demands on the cloud computing service have increased, a large number of public cloud computing platforms have appeared, including Amazon EC2, Rackspace [12], and Microsoft Azure [13], etc. Google has also published the Google App Engine [14], allowing users to run web applications and other types of applications, such as scientific computing applications [15].

Besides, numerous private clouds have also been built based on open-source cloud systems, such as OpenStack, an open-source project launched in 2010. It was originally developed by Rackspace and National Aeronautics and Space Administration (NASA) and is currently maintained by the OpenStack Foundation with contributions from the major players in cloud computing. OpenStack provides an API and a dashboard to manage pools of computing, storage, and networking resources. Similarly, Eucalyptus [7] is another famous open-source project initiated in 2008. It is developed and maintained by Eucalyptus Systems. Eucalyptus allowed the building of the Amazon AWS-compatible private and hybrid clouds.

Automatic deployment is an effective way to simplify the complexity of such work in a data center and some approaches have been proposed, e.g., Ezilla Toolkit [16]. In addition, some open-source projects have also been presented for deploying cloud systems automatically. Taking OpenStack as an example, there have been many automatic deployment solutions, some of which are given below.

- Fuel [17] is now open source and contributed to OpenStack. It undertakes hardware discovery, network verification, OS provisioning and the deployment of OpenStack components. Fuel's distinct feature is its polished and easy to use Web User Interface (UI) that makes OpenStack installation seem simple.
- DevStack is an OpenStack community production to deploy OpenStack automatically. It has evolved to support a large number of configuration options and alternative platforms and support services.

However, these automated solutions do not consider the reliability of the cloud system deployment. Differently, the AutoCSD approach proposed in this paper considers the reliability, availability, and load balance of the system deployment in the data center.

7   Conclusion

With the increasing demands on the cloud computing service (e.g., Infrastructure as a Service, Platform as a Service, and Software as a Service), the scale of cloud data centers has become gradually bigger and bigger, significantly increasing the level of effort in deploying a cloud computing system. In this paper, we have proposed an AutoCSD approach with the characteristics of reliability, availability, and load balance.

Specifically, we use workflow to deal with the dependencies among the automatic deployment processes of the cloud system, and we design a failover mechanism to avoid the single point failure of the deployment server which contains the installation packages and configuration files of the operating system and cloud middleware. In addition, we adopt a load balancing algorithm to solve the bottleneck problem of deploying the cloud system in a large-scale data center. The experimental results show that the average deployment time under our approach is lower than that with traditional deployment methods. Our method also achieves a success ratio of cloud system deployment of up to 90 %, even in the high-concurrency environment.

Acknowledgement. The research is supported by National Science Foundation of China under grants No. 61232008, National 863 Hi-Tech Research & Development Program under grants No. 2014AA01A302 and No. 2015AA011402, Research Fund for the Doctoral Program of MOE under grant No. 20110142130005, Anhui Provincial Natural Science Foundation under grant No.1408085MF126, Youth Foundation of Chuzhou University under grant No. 2013RC006 and Scientific Research Foundation of Chuzhou University under grant No. 2014qd016.

References

1. Systemimager. http://sourceforge.net/projects/systemimager/?source=navbar. Accessed February 2015
2. Kickstart. https://access.redhat.com/documentation/en-US/Red_Hat_Enterprise_Linux/5/html/Installation_Guide/ch-kickstart2.html. Accessed February 2015
3. Devstack. http://docs.openstack.org/developer/devstack/overview.html. Accessed February 2015
4. Mell, P., Grance, T.: The nist definition of cloud computing. Nat. Inst. Stand. Technol. **53**(6), 50 (2009)
5. Amazon ec2. http://aws.amazon.com/ec2/. Accessed February 2015
6. Sefraoui, O., Aissaoui, M., Eleuldj, M.: Openstack: toward an open-source solution for cloud computing. Int. J. Comput. Appl. **55**(3), 38–42 (2012)
7. Nurmi, D., Wolski, R., Grzegorczyk, C., Obertelli, G., Soman, S., Youseff, L., Zagorodnov, D.: The eucalyptus open-source cloud-computing system. In Proccedings of 9th IEEE/ACM International Symposium on Cluster Computing and the Grid (CCGRID 2009), pp. 124–131. IEEE (2009)
8. Cortes, T.: Software raid and parallel file systems. In: High Performance Cluster Computing, pp. 463–496 (1999)

9. Linux channel bonding. http://sourceforge.net/projects/bonding/. Accessed February 2015
10. Heartbeat. http://linux-ha.org/wiki/Heartbeat. Accessed February 2015
11. Crane cloud system. http://www.chinagrid.edu.cn. Accessed February 2015
12. Rackspace cloud. http://www.rackspace.com/. Accessed February 2015
13. Microsoft azure. http://azure.microsoft.com/. Accessed February 2015
14. Google app engine. https://appengine.google.com/. Accessed February 2015
15. Prodan, R., Sperk, M.: Scientific computing with google app engine. Future Gener. Comput. Syst. **29**(7), 1851–1859 (2013)
16. Chen, H.-S., Wu, C.-H., Pan, Y.-L., Yu, H.-E., Chen, C.-M., Cheng, K.-Y.: Towards the automated fast deployment and clone of private cloud service: the ezilla toolkit. In: Proceedings of 2013 IEEE 5th International Conference on Cloud Computing Technology and Science (CloudCom), vol. 1, pp. 136–141. IEEE (2013)
17. Sobeslav, V., Komarek, A.: Opensource automation in cloud computing. In: Wong, W.E. (ed.) Proceedings of the 4th International Conference on Computer Engineering and Networks, pp. 805–812. Springer (2015)

An Architecture-Based Autonomous Engine for Services Configuration and Deployment in Hybrid Clouds

YanPing Liu[1,2], XingTu Lan[1,2], Xing Chen[1,2(✉)], Xin Feng[3], XiangHan Zheng[1,2], and WenZhong Guo[1,2]

[1] College of Mathematics and Computer Science, Fuzhou University,
Fuzhou 350108, China
`liuyanpingfzu@163.com, fzulanxingtu@gmail.com,`
`{chenxing,xianghan.zheng,guowenzhong}@fzu.edu.cn`
[2] Fujian Provincial Key Laboratory of Networking Computing and Intelligent Information Processing, Fuzhou University, Fuzhou 350108, China
[3] College of ZhiCheng, Fuzhou University, Fuzhou 350108, China
`fengxin8@21cn.com`

Abstract. As cloud computing offers scalability, extensibility, elasticity, flexibility and cost savings to the customers of cloud service providers, there is a growing trend towards migrating services to the cloud. Hybrid clouds, which comprise nodes both in the private cloud and in the public cloud, have emerged as a new model for service providers to deploy their services. However, to deploy services in hybrid clouds is a complex task as services are in essence distributed applications. What is more, there is heterogeneity among hybrid clouds. This article proposes an autonomous engine for services configuration and deployment in hybrid clouds. The automation is enabled by the definition of generic information model, which describes all the information relevant to the deployment and configuration of services with the same abstractions, including the required resources, service dependencies and business objectives. In addition, to shield the heterogeneity of hybrid clouds, we define mapping rules for model transformation. We also deploy a three-layer architecture application on Openstack and CloudStack to validate the correctness of our approach.

Keywords: Service deployment · Hybrid clouds · Cloud computing · Model · Architecture-based

1 Introduction

In recent years, there is a growing trend towards migrating services to the cloud. This trend is caused by the rise of cloud computing. Cloud computing offers scalability, extensibility, elasticity, flexibility, and cost savings to the customers of cloud service providers. After deploying services in the cloud, the running way of enterprise data centers is similar to Internet. Thus, the enterprise could allocate resources to required services immediately and pay the bill according to the actual resource usage. It greatly reduces the cost of supporting services.

© Springer International Publishing Switzerland 2015
W. Qiang et al. (Eds.): CloudCom-Asia 2015, LNCS 9106, pp. 86–98, 2015.
DOI: 10.1007/978-3-319-28430-9_7

Nevertheless, a single cloud cannot meet the requirements of service providers. First, the public cloud and the private cloud each have advantages. The public cloud has strong processing capability with low cost, whereas the private cloud has a good safety performance, which is more suitable for sensitive data. Second, some components of the service are limited by geographical factors. In this case, there must be multiple clouds distributed in different geographical position to support the service. Third, the different service components may have different requirements for cloud platforms. Thus, this paper presents a new deployment way that services are deployed in hybrid clouds. Hybrid clouds are structured and operated as a network of geographically distributed clouds for extended resource utilization, workload distribution, fault tolerance, cost effective (e.g., energy efficient), service/data placement and avoidance of "lock-in" to particular cloud providers.

However, it is not a simple task to automatically deploy services in hybrid clouds. First, as the service is a large-scale and distributed application system, how to specify the required resources (such as CPU, Memory, disk space, operating system and middle ware) and solve service dependencies (the relation between service components or others it depends on) are key problems. Second, how to automatically generate deployment plans remains to be solved. Third, there is heterogeneity among hybrid clouds. Due to the above reasons, it is difficult to deploy services in hybrid clouds in a generic way.

In our previous work [1], we have leveraged architecture-based runtime model for the management of diverse Cloud resources in a single cloud. An architecture-based runtime model is a causally connected self-representation of the associated system that emphasizes the structure, behavior, and goals of the system from a problem space perspective [2, 3]. It has been broadly adopted in the runtime management of software systems [4–6]. With the help of runtime models, administrators can obtain a better understanding of their systems and write model-level programs for management. We have developed a model-based runtime management tool called SM@RT (Supporting Model AT Run Time [7–9]), which provides the synchronization engine between a runtime model and its corresponding running system. SM@RT makes any state of the running system reflected to the runtime model, as well as any change to the runtime model applied to the running system in an on-the-fly fashion. And in our previous work [10], the application runtime environment can be managed.

In this paper, we present an autonomous engine based on architecture for services configuration and deployment in hybrid clouds. This engine automates the process of deployment and configuration from design to deployment. We use the demand information model to specify all the information relevant to the deployment and configuration of services with the same abstractions, including the required resources, service dependencies and business objectives. And the deployment information model is used to describe deployment plans. In addition, to shield the heterogeneity of hybrid clouds, we define mapping rules for model transformation. Thus the deployment model could be correctly transformed to cloud runtime models which abstract the resources of the cloud. And the changes of cloud runtime models would be applied to the running system.

The article is structured as follows. Section 2 describes the information models in detail. These information models are used to describe all the information relevant to the deployment and configuration of services with the same abstractions. Section 3 describes the mapping rules for model transformation. In Sect. 4, our approach is validated through a case study derived from a real scene. And then, some related works are introduced. Finally, the last section reflects on the explained aspects, focusing on the impact and potential extensions to this work.

2   Models for Services Deployment and Configuration

In general, services are large-scale and distributed applications which need multiple kinds of resources. And service dependencies are complexity. To shield the complexity of service deployment, the process from design to deployment should be automate. In this article, the automation is enabled by the definition of the generic information model, which describes all the information relevant to the deployment of the services with the same abstractions, including the required resources, service dependencies and business objectives.

2.1   Demand Information Model

The demand information model characterizes the services and deployable artifacts that provide them. It also provides a deployment and configuration perspective of the artifacts specified by the service provider. The presented abstractions keep the requirements of services that have been taken into account over the analysis, design, implementation and testing phases.

A *Demand Unit* aggregates functionality, such as services, libraries, web interfaces and business logic components, which will be available when the unit is deployed to the runtime environment. Figure 1 shows the main elements of the demand information model. The root element of the model is the *Demand Unit*, a subclass of *Resource* representing each indivisible artifact which provides functionality when provisioned to the environment. *Resource* type has two attributes, *name* and *expression*. As some resources are obtained from execution environment, we define *expression* attribute which could be composed by variables to describe the resource. (e.g. the ${ip} variable represents the IP address when the unit is deployed to the environment).

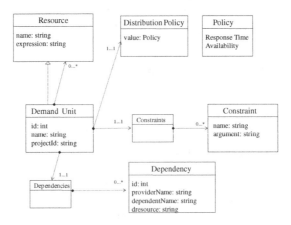

Fig. 1. The Demand Information Model

Constraints type specifies the required resources of the demand unit ranging from hardware resources to software resources. And the *value* attribute is the constraint of the resource. For example, we specify a resource named memory, and the value of this resource is 512 MB. It means that, when the unit is provisioned to the environment, the environment should allocate 512 MB memory to this unit.

Dependencies type specifies the dependency relationships of the services which would be deployed. The *providerName* attribute specifies which units it depends on, while the *dependentName* is the demand unit itself. And the *dresource* attribute specifies the resource contained on the *providerName* unit. Actually, the resource provided by the unit which name is *providerName* is the essential resource the demand unit depends on.

We use *Distribution Policy* to describe the character of the demand unit, as each unit has different requirement to the environment performance. Thus, when we deploy the unit to the environment, we could select the most suitable cloud platform as the environment. In this article, we define two types of *Distribution Policy*, *Response Time* and *Availability*.

2.2   Deployment Information Model

The deployment information model describes specific deployment plans to direct service deployment. In other words, it specifies how deployment units are deployed to cloud platforms. As shown in Fig. 2, the root element is the *Project* which is composed by clouds. And each *Project* has its own resources to support the service, such as images, volumes and servers.

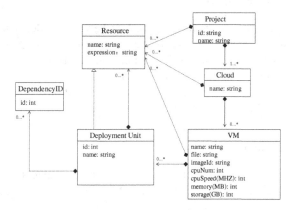

Fig. 2. The Deployment Information Model

A *VM* is the base execution platform for the services and modeled as *Deployment Unit*. The attributes in *VM* type are mainly used to specify the required resources to support the service. The *file* attribute specifies the required file for provisioning the functionality of the unit (such as WAR, JAR and EAR JEE deployable files and packaged SQL scripts). The *imageId* describes the required operating system and middle ware.

And it must be consistent with the corresponding information in cloud runtime models which abstract the resources of the cloud. As well as *cpuNum*, *cpuSpeed*, *memory* and *storage* specify the corresponding resources.

The *DependencyID* type presents the units which depend on the Deployment Unit. And the *id* attribute is consistent with the *id* of the *Dependency* type in the demand information model..

Obviously, the information of the demand information model is not enough to generate a deployment plan. Also the information of each cloud is needed, especially the information about the characteristics and residual resources of clouds. We use a XML file to describe the information of clouds mainly including the resources it contained, available images and some basic information.

3   Model Transformation

Up to this point, we have explained the main elements of the information model, the contained information and it relationships. However, these concepts are not sufficient for enabling an autonomic deployment, as there is heterogeneity among hybrid clouds. How to deploy deployment units to corresponding clouds according to the deployment information model remains to be solved. To solve the problem above, we define mapping rules to describe mapping relationships between the deployment information model and cloud runtime models which abstract the resources of the cloud. And the process of transformation is completed automatically according to the mapping rules.

3.1   Mapping Rules Definition

There are three types of basic mapping relationships between model elements. They are one-to-one mapping relationship, many-to-one mapping relationship and one-to-many relationship. Any other relationship can be demonstrated as a combination of them. One element in the information model is related to a certain element in the cloud model. Particularly, the attributes of the elements in the information model are also corresponding to the ones of related elements in cloud model.

3.2   Model Operation Transformation

Model operations are aimed to monitor some system parameters or execute some management tasks. There are five basic types of model operations, including "Get", "Set", "List", "Add" and "Remove".

In order to deploy deployment units according to deployment information model, operations on the deployment information model need to be transformed to ones on the cloud model. Also, for ensuring the correct synchronization between the runtime information model and the cloud model, operations on the runtime information model need to be transformed to ones on the cloud model. We have defined the model operation transformation rules, as shown in Table 1.

Table 1. Mapping Rules of Model Operation Transformation

	Mapping Rule
Example	A-> B A1.a1-> B1.b1
Get	Get A1.a1-> Get B1.b1
Set	Set A1.a1-> Set B1.b1
List	List*A-> List*B Get A.properties-> Get B.properties
Add	Add*A-> Add*B Set A.properties-> Set B.properties
Remove	Remove*A-> Remove*B

For instance, an element in the information model is mapped to the B element in the cloud model. Thus, the operation to add, remove or list the A elements is mapped to the same operation on the related B element. The operation to get or set the value of A's attribute is mapped to the same operation on the related attribute too.

In [1], we have implemented a prototype for integrate management of the hardware and software resources of virtual machines based on OpenStack and Hyperic to validate the feasibility and efficiency of our approach.

4 Case Study

We implement a three-layer architecture application named WebStore, and deploy it by our autonomous engine to CloudStack and OpenStack for validating the feasibility and efficiency of our approach. First, we use the demand information model to specify the relevant information of the application deployment. Second, the deployment information model is generated automatically according to the demand information model and cloud information. To achieve the deployment process, the deployment information model would be transformed to cloud runtime models.

4.1 WebStore Introduction

The WebStore service is composed by 9 components, as shown in Fig. 3. LoadBalancer is used to forward clients' request. Customer1 and Customer2 provide same service that customer could buy goods through the web. The Supplier1 and the Supplier2 are similar with the Customer1 and the Customer2. But they are used for suppliers to sell their goods through the web. LoadBalancer, Customer1, Customer2, Supplier1 and Supplier2 are web clients which should be deployed on a web server. And they have high requirements on response time.

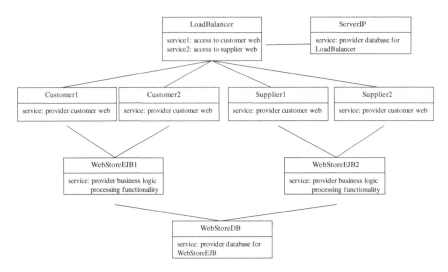

Fig. 3. The Components of WebStore

The requests from web client would be processed by WebStoreEJB1 and WebStor-eEJB2. They are EJB projects which should be deployed on an application server. As the functionality they provide with compute capability, we should specify more computing resources to them. In addition, ServerIP and WebStoreDB should be deployed on a database server, as they both provide the database service. The ServerIP mainly provide the information of servers for LoadBalancer. And the WebStoreDB is used to store the clients' and goods' information. Thus, we should specify more storage resource to them, as they are used for data storage. For the above components, the availability of the system is more important than the response time. Therefore, we deploy two groups of components on two type clouds. One cloud is OpenStack with the lowest response time, while the other is CloudStack which availability is the best.

The cloud runtime model of CloudStack and OpenStack are shown in Fig. 6. Given the architecture-based meta-models, we also need to identify the changes enabled by the models [11]. There are hundreds of management interfaces in Cloudstack and Openstack, so we can model them into the Access Model through specifying how to invoke the APIs to manipulate each type of elements in the models.

Based on meta-models and Access Models, the correct transformation between the deployment model and running systems and the correct synchronization between the runtime information models and running systems can be guaranteed by the SM@RT tool.

4.2   Construction of Demand Information Model

The demand information model mainly describes the basic information as well as specifies the required resources (such as osType, CPU, memory and storage). Not only that, the dependency relationship is described. Figure 4 shows an example of demand information model.

Figure 4 describes the demand information of LoadBalancer unit. This unit provides two resources when provisioned to the cloud. As IP and port information could be obtained only when the unit is deployed. We use variables to express first. After the unit is deployed, those variables would obtain the corresponding value from context.

LoadBalancer is a web client which has a high requirement on response time. Thus, we specify *Response Time* as the Distribution Policy. The demand unit has several constraints on the specified resources, including osType, middleware, cpuNum, cpuSpeed, memory and required file. Also, this unit depends on five other units. And the relationship is detailed description in Fig. 4.

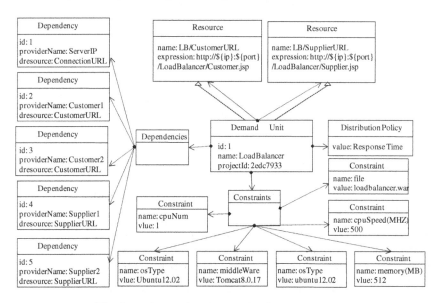

Fig. 4. An Example of Demand Information Model

4.3 Construction of Deployment Information Model

According to the cloud information in Fig. 5 and demand information model, the deployment information model is generated automatically, as shown in Fig. 6. In Fig. 6, the dependency relationships between components are presented by dashed arrow. Also, the dependent resource is shown on the line. For example, the Distribution Policy of LoadBalancer unit is *Response Time*. The OpenStack with lowest response time and enough residual resources is selected. In addition, LoadBalancer unit requires Ubuntu with version 12.02 as operating system and Tomcat with version 8.0.17 as middleware. The image belongs to OpenStack and project "2edc7933" which id is "14a95862" is selected as the operating system and middleware it contained matches requirement of LoadBalancer demand unit.

In order to deploy the deployment units to the corresponding clouds, we define mapping rules between them according to their mapping relationships. According to the mapping rules, the operations on the element in the deployment information model can be mapped to the operations on the related element in the cloud runtime models. Figure 7 shows an example of model operation transformation. According to the deployment information model, this deployment unit should be deployed in the cloud which name is CloudStack.

```
<cloud name="CloudStack" attribute="Availability" IPAdress="218.193.3.2">
    <ip>13/99</ip>
    <memory>2.63GB/11.38GB</memory>
    <cpuNum>2</cpuNum>
    <CPU>3.02GHZ/19.40GHZ</CPU>
    <storage>90.58GB/914.29GB</storage>
    <images >
        <image id="130e3ac4" projectId="2edc7933" osType="Ubuntu12.02" middleWare="JBoss7.11"></image>
        <image id="5bbb7191" projectId="2edc7933" oSType="Ubuntu12.02" middleWare="MySQL5.6"></image>
        <image id="14ac453e" projectId="2edc7933" osType="Ubuntu12.02" middleWare="JOnAs6.1.0"></image>
        <image id="342ab123" projectId="2edc7933" osType="Ubuntu12.01" middleWare="Tomcat8.0.17"></image>
        ...
    </images>
</cloud>
<cloud name="OpenStack" attribute="ResponseTime" IPAdress="218.193.3.2">
    <ip>4/51</ip>
    <memory>2.00GB/4.20GB</memory>
    <cpuNum>2</cpuNum>
    <CPU>2.50GHZ/11.60GHZ</CPU>
    <storage>38.20GB/456.00GB</storage>
    <images>
        <image id="14a95862" projectId="2edc7933" osType="Ubuntu12.02" middleWare="Tomcat8.0.17"></image>
        <image id="2f3dae49a" projectId="2edc7933" osType="Ubuntu12.02" middleWare="JBoss7.11"></image>
        <image id="4c1356cd" projectId="2edc7933" osType="Ubuntu12.02" middleWare="MySQL5.6"></image>
        <image id="5127ec49" projectId="2edc7933" osType="Ubuntu12.02" middleWare="JOnAs6.1.0"></image>
        ...
    </images>
</cloud>
```

Fig. 5. OpenStack Information (down) and CloudStack Information (up)

The CloudStack could be found by querying the cloud information XML file. And then execute operations on it, which are described as follows:

1. Query: Find the *VirtualMachines* element whose projectId is "2edc7933".
2. Add: Create a *VirtualMachine* element.
3. Set: Assign property values of the *VirtualMachine* element.

The original operation is mapped to the operation to create a *VirtualMachine* element in the Cloudstack runtime model. Model operation transformation is executed instruction-by-instruction. For instance, the action to query the *DeploymentUnits* node is mapped to the action to query the *VirtualMachines* node, whose *projectId* is "2edc7933". The action to add a *DeploymentUnit* node is mapped to the actions to add a *VirtualMachine* node. The actions to set the property values are mapped to the actions to set the values of related attributes too. According to the mapping relationships, the *templateId* attribute of related *VirtualMachine* element should be "130e3ac4", so there is an extra action to sign the property value. In the demand information model, we can specify any value of *cpuNum*, *cpuSpeed*, *memory* and *storage*. In the cloud runtime model, there is a customized scheme which allows users to specify a specific value. In this paper, the *ServiceOffering* configure *cpuNum*, *cpuSpeed* and *memory*. And the *serviceOfferingId* of customized scheme in CloudStack is "30b5690". The *DiskOffering* configure the storage size and the *diskOfferingId* of customized scheme is "6d174ab".

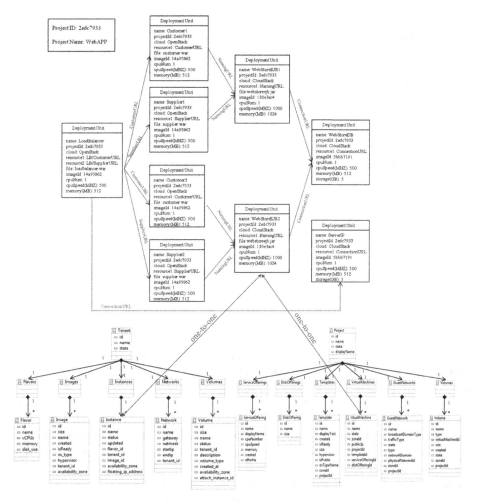

Fig. 6. Deployment Information Model of WebStore and Cloud Runtime Model of OpenStack (down-left), CloudStack (down-right)

The generated operation file is transferred to the Cloudstack runtime model. When the operation is executed, changes of the CloudStack runtime model will be applied on the running system.

5   Related Work

Following the IaaS model that has been dominant in the cloud service market, approaches in the literatures [12, 13] are geared towards deploying whole application stacks in virtual machines. While useful in their own right, the lack of separation between application and middleware components makes such approaches are not suitable for large-scale and distributed application. On the other hand, tools originating in [14, 15],

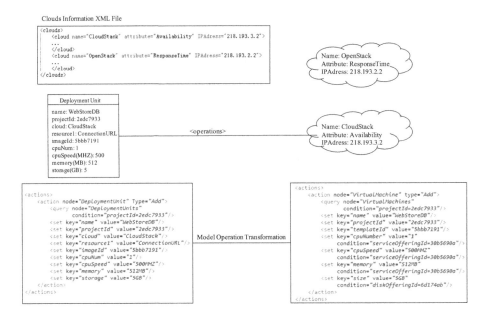

Fig. 7. An Example of Model Operation Transformation

allow for a finer degree of granularity in deploying both application and middleware components in a flexible manner. However, they require application developers to create their deployment plans.

In order to shield the complexity of services deployment, some researches provide the methods which could automate the process of deployment. The works [16, 17] that have applied model-driven to automate deployment of the enterprise application, they can be categorized as code generator by automatically generate code based on given some form of design model, whereas there are works [18] that proposed method automate from design to deployment. However, none of works mention above considers platform on cloud computing as target environment to deploy.

6   Conclusion and Future Work

We propose an autonomous engine for services configuration and deployment in hybrid clouds. We use information models to abstract the relevant information of service deployment. The demand information model is used to specify the required resources, service dependencies and business objectives. The deployment information model is used to describe the deployment plan which automatically generated according to the cloud information and demand information model. Not only that, to shield the heterogeneity between deployment information model and cloud runtime models, we define mapping rules to describe mapping relationships between them. In addition, the process of transformation is completed automatically according to the mapping rules.

In this article, we implement the autonomous engine for services configuration and deployment in hybrid clouds. However, to monitor and manage the services runtime is also an important problem. In future work, we would consider constructing a runtime information model which could be synchronized with cloud runtime models for services monitoring and management.

Acknowledgements. This work was supported by the National Natural Science Foundation of China under Grant No.61402111, the Fujian Province High School Science Fund for Distinguished Young Scholars under Grand No. JA12016, the Program for New Century Excellent Talents in Fujian Province University under Grant No. JA13021.

References

1. Chen, X., Zhang, Y., Huang, G., Zheng, X., Guo, W., Rong, C.: Architecture-based integrated management of diverse cloud resources. J. Cloud Comput. **3**, 11 (2014). Springer, Heidelberg
2. Bencomo, N., Blair, G.S., France, R.B.: Summary of the workshop Models@run.time at MoDELS 2006. In: Kühne, T. (ed.) MoDELS 2006. LNCS, vol. 4364, pp. 227–231. Springer, Heidelberg (2007)
3. Blair, G., Bencomo, N., France, R.: Models@ run.time. Computer **42**(10), 22–27 (2009)
4. Huang, G., Mei, H., Yang, F.: Runtime recovery and manipulation of software architecture of component-based systems. Automat. Softw. Eng. **13**(2), 257–281 (2006)
5. Occello, A., AM, D.P., Riveill, M.: A runtime model for monitoring software adaptation safety and its concretisation as a service. Models@ runtime **8**, 67–76 (2008)
6. Wu, Y., Huang, G., Song, H., Zhang, Y.: Model driven configuration of fault tolerance solutions for component-based software system. In: France, R.B., Kazmeier, J., Breu, R., Atkinson, C. (eds.) MODELS 2012. LNCS, vol. 7590, pp. 514–530. Springer, Heidelberg (2012)
7. Huang, G., Song, H., Mei, H.: SM@RT: applying architecture-based runtime management of internetware systems. Int. J. Softw. Inf. **3**(4), 439–464 (2009)
8. Song, H., Huang, G., Chauvel, F., Xiong, Y., Hu, Z., Sun, Y., Mei, H.: Supporting runtime software architecture: a bidirectional-transformation-based approach. J. Syst. Softw. **84**(5), 711–723 (2011)
9. Peking University: SM@RT: Supporting Models at Run-Time. http://code.google.com/p/smatrt/
10. Lan, X., Liu, Y., Chen, X., Huang, Y., Lin, B., Guo, W.: A model-based autonomous engine for application runtime environment configuration and deployment in PaaS Cloud. In: IEEE 6th International Conference on Cloud Computing Technology and Science on CloudCom (2014)
11. Huang, G., Chen, X., Zhang, Y., Zhang, X.: Towards architecture-based management of platforms in the cloud. Frontiers Comput. Sci **6**(4), 388–397 (2012)
12. Pawluk, P., Simmons, B., Smit, M., Litoiu, M., Mankovski, S.: Introducing STRATOS: A cloud broker service. In: 2012 IEEE 5th International Conference on CloudComputing (CLOUD), pp. 891–898 (2012)
13. Zaman, S., Grosu, D.: An online mechanism for dynamic VM provisioning and allocation in clouds. In: 2012 IEEE 5th International Conference on Cloud Computing (CLOUD), pp. 253–260 (2012)
14. Nelson-Smith, S.: Test-Driven Infrastructure with Chef. O'Reilly Media, Inc. (2011)

15. Juve, G., Deelman, E.: Automating application deployment in infrastructure clouds. In: 2011 IEEE Third International Conference on Cloud Computing Technology and Science (CloudCom), pp. 658–665 (2011)
16. Ruiz, J.L., Dueñas, J.C., Cuadrado, F.: Model-based context-aware deployment of distributed systems. IEEE Commun. Mag. **47**(6), 164–171 (2009)
17. Ceri, S., Fraternali, P., Bongio, A.: Web modeling language (WebML): a modeling language for designing web sites. Comput. Netw. **33**(1), 137–157 (2000)
18. De Sousa Saraiva, J., da Silva, A.R.: CMS-based web-application development using model-driven languages. In: Software Engineering Advances's 09 (ICSEA). IEEE (2009)

UNIO: A Unified I/O System Framework for Hybrid Scientific Workflow

Dan Huang$^{(\boxtimes)}$, Jiangling Yin, Jun Wang, Xuhong Zhang,
Junyao Zhang, and Jian Zhou

Department of Electrical and Computer Engineering,
University of Central Florida, Orlando, USA
{duang,jyin,jwang,xzhang,junyao,jzhou}@eecs.ucf.edu

Abstract. Recent years have seen an increasing number of Hybrid Scientific Applications. They often consist of one HPC simulation program along with its corresponding data analytics programs. Unfortunately, current computing platform settings do not accommodate this emerging workflow very well. This is mainly because HPC simulation programs store output data into a dedicated storage cluster equipped with Parallel File System (PFS). To perform analytics on data generated by simulation, data has to be migrated from storage cluster to compute cluster. This data migration could introduce severe delay which is especially true given an ever-increasing data size.

While the scale-up supercomputers equipped with dedicated PFS storage cluster still represent the mainstream HPC, ever increasing scale-out small-medium sized HPC clusters have been supplied to facilitate hybrid scientific workflow applications in fast-growing cloud computing infrastructures such as Amazon cluster compute instances. Different from traditional supercomputer setting, the limited network bandwidth in scale-out HPC clusters makes the data migration prohibitively expensive. To attack the problem, we develop a Unified I/O System Framework (UNIO) to avoid such migration overhead for scale-out small-medium sized HPC clusters. Our main idea is to enable both HPC simulation programs and analytics programs to run atop one unified file system, e.g. data-intensive file system (DIFS in brief). In UNIO, an I/O middleware component allows original HPC simulation programs to execute direct I/O operations over DIFS without any porting effort, while an I/O scheduler dynamically smoothes out both disk write and read traffic for both simulation and analysis programs. By experimenting with a real-world scientific workflow over a 46-node UNIO prototype, we found that UNIO is able to achieve comparable read/write I/O performance in small-medium sized HPC clusters equipped with parallel file system. More importantly, since UNIO completely avoids the most expensive data movement overhead, it achieves up to 3x speedups for hybrid scientific workflow applications compared with current solutions.

Keywords: HPC · MPI · Data-intensive · HDFS · Data migration · Scientific workflow

© Springer International Publishing Switzerland 2015
W. Qiang et al. (Eds.): CloudCom-Asia 2015, LNCS 9106, pp. 99–114, 2016.
DOI: 10.1007/978-3-319-28430-9_8

1   Introduction

HPC clusters are delivering increasingly finer resolution simulations and correspondingly generating larger datasets than ever before. Moreover, unlike previous scientific workflows, which are involved in only compute-intensive simulation, more recent simulations require multiple cycles of data-intensive analysis and/or visualization to fully understand the simulation results [8,10,11].

Here we formally define **Hybrid Scientific Workflow** as the specific type of workflow that cooperatively runs HPC simulations and HPC data analytics together. In this hybrid scientific HPC workflow, HPC simulations often generate massive amounts of data, and therefore are compute-intensive. While most analytic programs spend the majority of time reading, but performing minimal computational analysis on large data sets [23,24]. Lattice Quantum Chromodynamics(QCD) [9] + ADAT and FLASH [17] + ParaView [19] are two hybrid scientific workflow examples studied in this paper. QCD simulation allows us to understand the results of particle and nuclear physics experiments in terms of QCD, the theory of quarks and gluons, while ADAT is a quantum chromodynamics package used for carrying out post production data analysis; FLASH is a publicly available multiphysics multiscale simulation code capable of handling general compressible flow problems while ParaView is used to visualize data generated by FLASH.

This new workflow gives rise to a big challenge of transferring the ever increasingly heavy volumes of data across every level of network connection [2,14]. Existing HPC clusters usually consist of a dedicated compute cluster and a storage cluster as shown in Fig. 1(a). We define Fig. 1(a) as HPC-Traditional (HPC-Tra). Data migration is required for performing analytics on compute intensive resources against data stored on HPC Parallel File System. This data migration can result in prohibitive latency given a limited network bandwidth.

To deal with this data migration challenge, several researchers have developed successful solutions for HPC-Tra architectures from two aspects. The works in [12,16,22,33] develop in-situ techniques to reduce the amount of migrated data

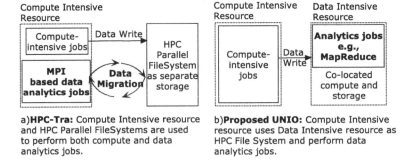

a)**HPC-Tra:** Compute Intensive resource and HPC Parallel FileSystems are used to perform both compute and data analytics jobs.

b)**Proposed UNIO:** Compute Intensive resource uses Data Intensive resource as HPC File System and perform data analytics jobs.

Fig. 1. Two Different configurations for running HPC data analytics applications (simulations + analytics)

by performing in-memory analytics before data reach the PFS storage. On the other hand, the works in [15,20,29,30] have made great efforts to improve overall I/O throughput such as ORNL's Jaguar and LANL's Roadrunner by adding I/O forwarding server. In summary, both methods can mitigate the data migration overhead rather than solve the problem from the root. This data migration issue will not lead to performance bottleneck in a supercomputer because of its sufficient network bandwidth ensuing from a dedicated high-speed interconnection network. However, in a small-to-medium sized HPC cluster setting, shared ethernet network bandwidth could impose data too heavy to move and therefore result in performance bottleneck. This is especially true when we run hybrid scientific workflow in a cloud-based HPC cluster.

In this paper, we develop a *framework-level* solution named UNIO (as shown in Fig. 1(b)) for hybrid scientific workflow applications that are not in the PetaFLOPS-scale or ExasFLOPS-scale and run in small to medium sized HPC clusters.

UNIO co-locates HPC analytics with Data-intensive File Systems (DIFS) as shared storage rather than with PFS as shared storage for two reasons. First, most PFS, such as PVFS [28] and Lustre [7], rely on hardware based reliability solutions like per I/O server RAID [31]. If one I/O node is disabled, the PFS storage system becomes inaccessible or very slow. In comparison, a DIFS DataNode failure will not affect storage performance and will in fact be completely transparent to the user. Second, DIFS exposes data location information to analytics program, therefore analytics task can be scheduled to access data locally, minimizing intra-cluster data transfer.

UNIO leverages the existing DIFS to support the semantics required for simulation I/O operations. The framework consists of two interlocked modules: (a) an I/O middle-ware, providing a generic I/O interoperable management interface for both HPC simulation and data analytics to access a unified DIFS; (b) an I/O scheduler, a leaky bucket read/write management scheme, avoiding I/O resource contention and balancing both compute and data intensive workloads on DIFS by preserving process locality.

Due to the budget issue, we conducted extensive experiments on our on-site cluster but configure with an identical system configuration as in Amazon's cluster compute instances (CCIs) [1,35], where physical nodes are allocated in a dedicated manner. Our experimental results show that UNIO is able to achieve comparable read/write I/O performance with small to medium sized HPC clusters equipped with parallel file system. Because UNIO completely avoids the expensive data movement overhead, it realizes up to 3x speedups in terms of workflow execution time for hybrid scientific workflow compared with current solutions.

The rest of this paper is organized as follow: In Sect. 2, we detail the UNIO design which contains an I/O middle-ware and an I/O scheduler. Section 3 presents evaluation methods, experimental results and corresponding analysis respectively. Section 4 illustrates the existing solutions and their limitations. Finally, Sect. 5 concludes the paper.

2   UNIO Design

To begin with, there are two problems to be solved— (a) I/O interface incompatibility: HPC simulation programs can not directly write to DIFS. (b) Resources Contention: HPC simulations are compute-intensive and normally write massive amounts of data, whereas HPC analytics spend most of the time in reading and performing minimal computations on large data sets. By running both simulations and analytics in a unified cluster, a potential problem of resources contention, especially for storage, is created between simulation writes and analytic reads.

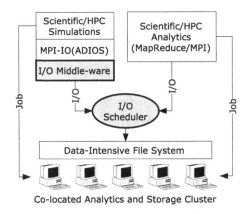

Fig. 2. High level system design

In this section, UNIO is presented to address these two problems and achieve the following goals:

- Compute-intensive cluster outputs its application results to the data-intensive resources.
- Analysis is performed on the data-intensive resource (There is no data migration, no matter how many times the analysis operations are performed.)

The overall design of UNIO is illustrated in Fig. 2, consisting of two major components: an I/O middle-ware and an I/O scheduler, each solves one of the problems mentioned above. They are briefly described below and will be detailed in the following subsections.

Component I: I/O Middle-ware. A middle-ware is developed to support various I/O methods adopted by simulation applications. It includes two subtasks: (a) providing a write interface for HPC simulations to write scientific data sets to data-intensive file system; (b) offering an interface to convert the exclusive $N - 1$ I/O into the concurrent $N - N$ I/O to enable parallel writing in DIFS.

Component II: I/O Scheduler. A scheduler is developed to balance the I/O consumption in the situation of co-existence of both simulations (writing result data sets) and analytics (reading/writing data sets).

2.1   Software Component I: I/O Middle-ware

Design Considerations. In order to demonstrate our approach, we employ the Hadoop Distributed file system (HDFS) [6] as our DIFS. We develop the I/O middle-ware as a virtual distributed HDFS software layer (VHDFS) between MPI-IO library and physical HDFS on the basis of Fuse-DFS [4], which is based on the Filesystem in Userspace project FUSE [5]. It supports basic I/O operations such as reads, writes and directory operations through Linux commands on HDFS. Fuse-DFS is chosen for the following reasons: (a) FUSE is capable of providing a standard file system interface without editing kernel code; (b) MPI applications are able to perform I/O operations against HDFS without any code modification; (c) current Fuse-DFS solutions already support basic I/O operations such as reads, write and directory operations through Linux commands, which reduce the whole cycle development effort; (d) Although FUSE introduces some overhead, our experimental results show that the benefit we gain by saving data movement can easily exceed the cost of the overhead produced even with one time run.

Random/Concurrent Write. Random write means that "processes can write to a random offset of a file at least one time" and concurrent write is that "multiple processes can write to a single file at the same time". Even with the support of Fuse-DFS, these two essential MPI/IO writing methods cannot be realized in HDFS. The main reason lays on the HDFS I/O characteristics, which can be explained in twofold: (a) HDFS uses append writes, each time the write is appended to the last saved files/offsets; (b) to simplify the consistency management, HDFS chooses exclusive writes/updates strategy which means any file can only be opened for write/append by one single process which obviously conflicts with concurrent write/updates. Fuse-DFS, as a translation layer, translates the applications I/O operations to HDFS I/O operations, however, it is not designed to incorporate the complicated parallel I/O operations into HDFS.

In light by PLFS [13,18], which utilized the log-structured file systems to improve the performance of checkpointing [27], we revised the original Fuse-DFS by adding an abstract layer VHDFS on top of HDFS as shown in Fig. 3 to support the two parallel writing functions (concurrent and random write). The basic idea of VHDFS is to create a virtual mapping layer that maps one logic file (exposed to the upper applications such as MPI simulations) to multiple physical files (in HDFS) and use index tables to manage and locate the specific physical files in HDFS.

VHDFS creates a folder on the underlying HDFS for every logical file to be written. From HDFS point of view, the basic structure of a folder consists of multiple physical files, transparent to calling applications. All data files are stored

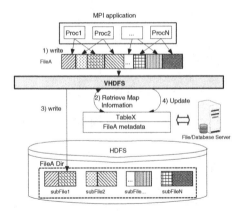

Fig. 3. VHDFS design architecture

in HDFS and spread across all the data nodes; meanwhile we adopt a table to store the index information of the mapping between the calling applications and the actual HDFS physical files. Since we only need to store metadata information in the tables which only requires basic database functions, a shared file based table or a database table is sufficient.

On one hand, when a process performs *random writes* to a virtual file, as shown in Fig. 3, the process will send a random write operation to VHDFS, which will search the metadata table to determine the actual data file and the actual write offset from the original random write offset. If the actual data file exists and the actual write offset is zero, the data file will be truncated to zero and then re-opened to write the new data; if the actual data file does exist but the actual write offset is non-zero, then the data will be appended to the actual data file; otherwise the data will be written to a newly created file. At last VHDFS will update the data extent records stored in the metadata table to reflect the new data mapping. Through this metadata recording strategy, a process is able to perform random write to a HDFS file. At the same time, when a process wants to read a virtual file, VHDFS will read the metadata table first, map the original read offset to the actual read offset of the actual data file, and then read the data files with actual read offsets from HDFS.

Another issue to address is the *concurrent write*. Suppose there are multiple processes trying to write to a same file concurrently, as process 1 through process N shown in Fig. 3 would write concurrently to FileA. When considering concurrent write, we must ensure the procedure of updating the same file are not interrupted by another process which means they should be independent as much as possible; otherwise there might be conflicts and the contents might be inconsistent. To tackle this problem, we ensure that every writing process is engaged to one single file in HDFS and it exclusively writes to that file with no conflicts with other processes. As illustrated in Fig. 3, when Proc1 wants to write some bytes to FileA, it would actually write to subFile1 in HDFS and record the mapping information between original writing information and actual writing information in the metadata table. With this one-one write mapping, UNIO avoids conflicts caused by multiple processes writing to one single file concurrently.

2.2   Software Component II: I/O System Scheduler

As specified above, the proposed I/O middle-ware creates the chance for different applications to access the DIFS concurrently. However, it introduces an extra problem that is the I/O resources contention. It is well-known that this workflow will not only constantly consume an amount of I/O bandwidth for read operations to satisfy the input of the analytics, but also periodically utilize I/O resources to perform a burst of simulation writes. As a result, continuous reads and periodical writes will exhaust I/O resources and give birth to contention. This situation is similar to network where the bandwidth is limited by burst traffic flows. In this section, we introduce the leaky bucket algorithm used in networks to balance the read/write requests from different sources.

Figure 4 is a block diagram of the proposed I/O scheduler, which consists of three major components: Write/Read pools, I/O Statistics and I/O Leaky bucket. Two buffer pools for write and read requests respectively are designed for storing input requests. The I/O Statistics collects the information to re-schedule the write and read requests, which means it takes charge of the write/read request statistics and determines the ratio of write/read requests for the input of the I/O leaky bucket. The leaky bucket is empty by default and it will pass the requests based on the ratio of write and read if it does not reach the watermark; otherwise the backpressure is generated to wait for the availability of the bucket. The balance between write and read requests is the key point of the scheduler, and the I/O leaky bucket actually executes the I/O requests in the bucket. Hence the problem becomes to correctly quantify the speed of read/write operations at all times. We develop an object function that aims to minimize the difference between the predefined read/write ratio (R_S) and actual read/write ratio (R_A) at certain time frame t so that the scheduler can dynamically adjust the write and read throughput accordingly. The equation is shown as follows.

$$\mu = |R_A - R_S| \tag{1}$$

Thus the goal is to minimize μ at each time slot. On one hand, R_S can be considered as a constant number which is able to be retrieved by collecting previous read/write ratios. On the other hand, R_A is obtained by the I/O monitor. On each data node, consider the "write" and "read" I/O operations as two independent queues at the beginning of the time frame. Denote the number of unfinished write operations in write queue as n_w, the number of unfinished read operation in read queue as n_r, the number of incoming write requests during the current time frame t as m_w, and the number of incoming read requests during the current time frame as m_r. Then the read/write ratio of current time frame can be calculated as:

$$R_A = \frac{n_w + m_w - v_w * t}{n_r + m_r - v_r * t} \tag{2}$$

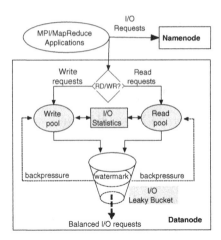

Fig. 4. Block diagram of I/O Scheduler in UNIO

where v_w and v_r are desired write and read speeds during the adjustment time period. The constraint is $v_w + v_r \leq V$ where V is the maximum I/O throughput for the data node. As a linear optimization problem, the object function is easy to solve and the computation overhead is low enough to be ignored in our experiments. That is

$$t = \frac{n_w + m_w - R_A * (m_r + n_r)}{v_w - R_A * v_r} \tag{3}$$

3 Experimental Results and Analysis

We test our proposed UNIO at UCF CASS cluster consisting of 46 nodes at two racks, one rack includes 15 compute nodes and one head node and the other rack contains 30 compute nodes, as shown in Table 1. As CASS's hardware is different with the supercomputers: each node has internal disks, we split the cluster into PFS storage part (15 nodes) and integrated data-intensive/compute-intensive cluster part (30 nodes) by racks. Data-intensive cluster is configured with Hadoop 0.20.203: one node for the NameNode and JobTracker, one node for the secondary NameNode with NFS v3.0 installed, and other 28 nodes as the DataNode/TaskTracker. In the same cluster as compute-intensive cluster, MPICH [1.4.1] is installed as compute framework. PVFS2 version [2.8.2] is installed as HPC storage system on the PFS storage part: one node as the metadata server for PVFS2, and other 14 nodes as the I/O servers. Most experiments are using this configuration unless indicated otherwise. Our proposed system is beneficial to a set of HPC applications with the following main characteristic: *i.e.* output data of compute-intensive jobs are large scale and analyzed afterward in a cluster with limited network resource.

Table 1. CASS cluster configuration

15 Compute nodes and 1 head node	
Make & Model	Dell PowerEdge 1950
CPU	2 Intel Xeon 5140, Dual Core, 2.33 GHz
RAM	4.0 GB DDR2, PC2-5300, 667 MHz
Internal HD	2 SATA 500GB (7200 RPM) or 2 SAS 147GB (15K RPM)
Network Connection	Intel Pro/1000 NIC
Operating System	Rocks 5.0 (Cent OS 5.1), Kernel:2.6.18-53.1.14.e15
30 Compute Nodes	
Make & Model	Sun V20z
CPU	2x AMD Opteron 242 @ 1.6 GHz
RAM	2 GB - registered DDR1/333 SDRAM
Internal HD	1x 146 GB Ultra320 SCSI HD
Network Connection	1x 10/100/1000 Ethernet connection
Operating System	Rocks 5.0 (Cent OS 5.1), Kernel:2.6.18-53.1.14.e15
Cluster Network	
Switch Make & Model	Nortel BayStack 5510-48T Gigabit Switch

We conduct two different sets of experiments, including one benchmark (IOR benchmark) and real applications—Flash with ParaView, in which Flash is typical HPC simulation applications running over the HPC clusters. To understand the simulation output, ParaView is usually launched.

FLASH [17] is a modular, parallel multiphysics simulation code capable of handling general compressible flow problems found in many astrophysical environments. It uses the MPI library for inter-processor communication. The HDF5 or Parallel-NetCDF library for parallel I/O is used to achieve portability and scalability on a variety of different parallel computers. In our experiments, HDF5 was used as the default parallel I/O library. The data FLASH generates will then be visualized by visualization software called **ParaView** [19]. Both FLASH and ParaView are developed using MPI. Thus this workflow is matched perfectly with the first configuration of current HPC data analytics architecture (shown in Fig. 1(a)). In this set of experiments, we compare the total execution time for the whole workflow between the traditional solution and our solution, as shown in follows.

$$T_{VHDFS} = T_{Flash} + T_{ParaView} \qquad (4)$$

$$T_{tra} = T'_{Flash} + T_{migration} + T'_{ParaView} \qquad (5)$$

where T_{VHDFS} and T_{tra} are the total execution time for our solution and traditional one, T_{Flash} and T'_{Flash} are Flash execution times; $T_{ParaView}$ and $T'_{ParaView}$ are ParaView execution times; $T_{migration}$ is the time of data movement from HPC parallel file system to compute resources.

3.1 Evaluating UNIO with Hybrid Workflow

In this section, we will firstly compare the performance of UNIO with/out I/O scheduler. Then we discuss the overhead of UNIO.

I/O Performance Test Using Mixed IOR and TeraSort/TestDFSIO Benchmarks. We evaluated the performance of UNIO and also studied how our I/O scheduler can help improve the I/O performance for hybrid scientific workflow. We scaled up the number of data nodes in the cluster and compared the write performance of two host file systems — PVFS2 and HDFS as illustrated in Fig. 5. During the experiments, we mount HDFS and PVFS2 as local file system at each data node if MPI processes are spawned on that node. For each node, three MPI processes are launched to handle I/O requests. We adopt a 64 MB block size and a 64 KB data transfer size. To mimic a hybrid scientific workflow, we run IOR as a representative parallel I/O write benchmark in parallel with two TeraSort and TestDFSIO read benchmarks as example analysis programs.

Figure 5 shows, when analysis programs are running simultaneously with IOR on HDFS, the IOR write performance without UNIO scheduler is significantly degraded in comparison to that of IOR on PVFS2. This result is consistent with our conjecture as discussed in Sect. 2.2. When both simulation and analysis programs run against the same file system (HDFS), some spiky write requests are significantly delayed in the outstanding queue by a large number of preceding read requests from analysis programs. With the assistance of UNIO I/O scheduler, we steal appropriate I/O bandwidth from preceding read requests by granting a burst of write requests higher priority. This results in ramping the write performance on HDFS up to a comparable level on PVFS2. As seen from the Fig. 6(a), (b), when the UNIO scheduling scheme is applied, the I/O read request is dynamically adjusted to a lower level (long response time) while a burst of incoming write requests is given the highest priority. Consequently, this achieves an average improvement of 30 % in terms of the write performance. The scheduler scales up well with an increasing number of nodes. With regards to the write bandwidth at a 45-node configuration, the average write throughput (30.1 MB/s) per node on HDFS is close to that of PVFS2 (35.8 MB/s).

To better understand how the UNIO I/O scheduler processes write and read requests effectively, we implemented a monitor placed at every data node (to intercept the DataInputStream in hdfs/server/datanode/DataXceiverServer.java & DataXceiver.java). The response time is calculated as a time interval between receiving InputStream request and acking InputStream close according to the system function System.currentTimeMillis(). As seen from Fig. 6(c), a majority of write operations on a single data node without UNIO scheduler have response

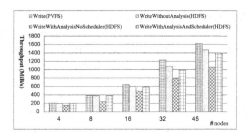

Fig. 5. IOR on VHDFS

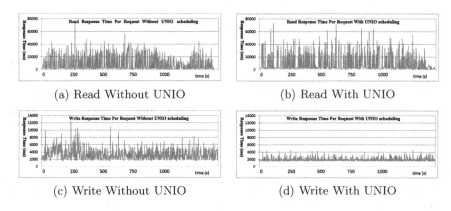

(a) Read Without UNIO

(b) Read With UNIO

(c) Write Without UNIO

(d) Write With UNIO

Fig. 6. Read/Write request response time

time ranging from 2.2 s to 4.7 s. With assistance of UNIO I/O scheduler, As illustrated in Fig. 6(d), the response times becomes much shorter, ranging from 1.5 s to 2.8 s. More importantly, the total program execution time for the two analysis programs are almost the same with or without the I/O scheduler. This could come from the following reasons. First, when a burst of write operations are processed together, these data are usually stored into many consecutive physical sectors on disk. Such a good locality preservation could potentially reduce the disk seek time for subsequent disk reads as illustrated in Fig. 6(b). Second, the Hadoop's task scheduler implements a good I/O load balancing mechanism. If one node is too long to respond, the map or reduce tasks are rescheduled to other nodes for execution.

3.2 Real Scientific Applications

In this section, we aim to mimic an in-situ data processing environment. Firstly, we run the applications over UNIO and traditional HPC architecture, and then the analytics programs multiple times over UNIO and traditional HPC settings.

FLASH and ParaView. We compared the performance of the workflow (Paraview after FLASH) in both the data-intensive cluster and the compute-intensive cluster. For the control experiments, we build this same configuration as Fig. 1(a). FLASH is launched in the compute part while PVFS2 is installed in storage part, in which one node is used for metaserver while the others are IO servers. In this configuration, FLASH writes its output and checkpoints into PVFS2 and ParaView is also reading data from PVFS2 for visualization. To test UNIO, HDFS is configured similarly on 30 nodes storage part. In this configuration, FLASH and ParaView are directly writing and reading from HDFS using our UNIO framework. Then we compare the completion time under different visualization repetitions. All the experiments were done three times and our results are based on the average. Specifically, when using UNIO, we launch MPI processes for FLASH and ParaView with the same order on the 30 nodes in order to distributively write/read data to/from HDFS.

Figure 7(a) shows the performance comparison of running FLAHS+ParaView over UNIO and HPC-Tra. The completion time is influenced by the number of analysis/visualization performed. Moreover, given the increasing number of repeated visualization operations, the performance gain of UNIO grows exponentially. For example, in an initial setting with one FLASH simulation and one ParaView visualization, the execution times of UNIO and HPC-Tra over PVFS2 are comparable. When we increase the visualization iteration by 64 times, UNIO finishes its execution three times faster than the traditional HPC architecture. As shown in Fig. 7(b), the time breaking down for paraview shows the reason: with the increasing number of analysis conducted, the time for data migration in the traditional HPC configuration has increased significantly when (between compute resource and PVFS2) more analysis repetition times are performed, which leads to a larger percentage of the whole completion time.

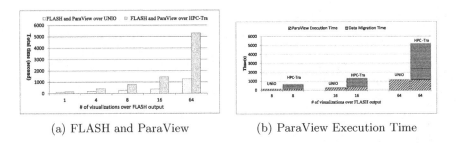

(a) FLASH and ParaView   (b) ParaView Execution Time

Fig. 7. ParaView execution time comparison on UNIO and traditional HPC

4 Related Work

With the increasing popularity of HPC data analytics, many researchers have been aware of these challenges. Specifically, the data movement overhead, incurred between simulation clusters and analysis clusters, becomes a major contributor to the performance bottleneck. In recent years, various solutions have

Table 2. Comparison among existing solutions

Working level	Representatives	Data migration	Scalability	Portability
Data	ISABELA-QA, Scala-H-Trace	Yes	N/A	N/A
File system	Shim layer	No	No	Yes
Framework	*Proposed UNIO*	No	Yes	Yes

been developed to reduce the data transfer overhead and they are summarized in different levels according to the modification level, as shown in Table 2.

The most straightforward way is to reduce the transfer overhead by decreasing the data size. Researchers proposed *data-level* approaches such as sampling, indexing and data compression. By using these approaches, the systems only transfer a small amount of the total data set per query, thus reducing the network transfer overhead. For example, random sampling reduces the data size by abstracting the statistical properties of the data. *Bitmap indexing* [3] is adopted by a number of scientific applications [26] because of better indexing compression rate and fast boolean query operations. The ISABELA-QA project compressed the data and then created indexing of the compressed data for query-driven analytics [22]. Wu *et al.* developed scalable performance data extraction techniques, and combined customization of metric selection with on-the-fly analysis to reduce the data volume in root cause analysis [34]. In summary, these approaches alleviate the network transfer overhead into a reasonable range for some specific applications. However, they cannot completely eliminate the overhead as our proposed framework-based solution.

Tantisiriroj *et al.* proposed a *file system-level* approach which implemented a PVFS shim layer that enabled the Hadoop applications to run on parallel virtual file system – PVFS2 [32]. By adding functions of preselecting, file layout and replication via a shim layer into PVFS2, this work allows both simulation and analytics programs to run on a single PVFS cluster without the any data movement. However, one major limitation of this shim layer is lack of appropriate interaction with runtime and other higher level layers in the framework. For example, experimental results of a recent work [25] show that the performance (in terms of throughput) degrades considerably when mixing the two workloads (periodically checkpoint write and consistently analytics read) together. It is because the two types of workloads are competing same I/O resources and the resource allocation problem cannot be well-addressed at the file-system level. To tackle this scheduling problem of contention, Hindman *et al.* presented Mesos, a hierarchical scheduling mechanism, to decide the resource allocation for different frameworks (such as Hadoop and MPI) running on the same cluster [21]. Our paper provides a *a framework level* solution to eliminate the data migration overhead from the root.

5   Conclusion

In this paper, we have proposed a Unified I/O Framework (UNIO) to address the data migration challenges imposed by **Hybrid Scientific Workflow** after examining the limitations of current HPC architectures and disadvantages of existing solutions. Our proposed UNIO consists of two major modules, the I/O middle-ware and the I/O scheduler. The former enables both HPC simulations and data analytics to run on one unified storage and the later dynamically balances read/write operations in the cluster to avoid resources contention and improve the overall performance. Extensive experiments are conducted to prove the effectiveness of both modules. By experimenting with a real-world scientific workflow over a 46-node UNIO prototype, we found that UNIO is able to achieve comparable read/write I/O performance with current small to medium HPC cluster equipped with parallel file system, and up to 3x speedups for hybrid scientific workflow compared with current solutions.

References

1. Amazon Inc.: High performance computing (hpc). https://aws.amazon.com/hpc/
2. Data-Intensive Computing: Finding the Right Program Models. http://hpdc2010.eecs.northwestern.edu/HPDC2010Bryant.pdf
3. Fastbit: An efficient compressed bitmap index technology. https://sdm.lbl.gov/fastbit/
4. Fuse-dfs. http://wiki.apache.org/hadoop/MountableHDFS
5. Fuse: Filesystem in userspace
6. The hadoop distributed file system. http://hadoop.apache.org/hdfs/
7. Lustre file system. http://www.lustre.org
8. Roadrunner open science. http://lanl.gov/roadrunner/rropenscience.shtml
9. US Lattice Quantum Chromodynamics. http://www.usqcd.org/usqcd-software/
10. Ahrens, J., Heitmann, K., Habib, S., Ankeny, L., McCormick, P., Inman, J., Armstrong, R., Ma, K.-L.: Quantitative and comparative visualization applied to cosmological simulations. **46**, 526–534 (2006)
11. Balaji, P., Chan, A., Gropp, W.D., Thakur, R., Lusk, E.R.: Non-data-communication overheads in MPI: analysis on blue gene/P. In: Lastovetsky, A., Kechadi, T., Dongarra, J. (eds.) EuroPVM/MPI 2008. LNCS, vol. 5205, pp. 13–22. Springer, Heidelberg (2008)
12. Bennett, J.C., Abbasi, H., Bremer, P.-T., Grout, R., Gyulassy, A., Jin, T., Klasky, S., Kolla, H., Parashar, M., Pascucci, V., Pebay, P., Thompson, D., Yu, H., Zhang, F., Chen, J.: Combining in-situ and in-transit processing to enable extreme-scale scientific analysis. In: Proceedings of the International Conference on High Performance Computing, Networking, Storage and Analysis, SC 2012, pp. 49:1–49:9. IEEE Computer Society Press, Los Alamitos (2012)
13. Bent, J., Gibson, G., Grider, G., McClelland, B., Nowoczynski, P., Nunez, J., Polte, M., Wingate, M.: PLFS: a checkpoint filesystem for parallel applications. In: 2009 ACM/IEEE Conference on Supercomputing, November 2009
14. Bryant, R.E.: Data-intensive supercomputing: the case for disc (2007)
15. Chen, Y., Chen, C., Sun, X.-H., Gropp, W.D., Thakur, R.: A decoupled execution paradigm for data-intensive high-end computing. In: 2012 IEEE International Conference on Cluster Computing (CLUSTER), pp. 200–208. IEEE (2012)

16. Klasky, S., et al.: In situ data processing for extreme-scale computing. In: SciDAC, Denver, CO, USA (2011)
17. FLASH Center for Computational Science. Flash user's guide
18. He, J., Bent, J., Torres, A., Grider, G., Gibson, G., Maltzahn, C., Sun, X.-H.: Discovering structure in unstructured i/o. In: PDSW (2012)
19. Henderson, A.: Paraview guide, a parallel visualization application
20. Hey, T., Tansley, S., Tolle, K. (eds.): The Fourth Paradigm: Data-Intensive Scientific Discovery. Microsoft Research, Redmond (2009)
21. Hindman, B., Konwinski, A., Zaharia, M., Ghodsi, A., Joseph, A.D., Katz, R., Shenker, S., Stoica, I.: Mesos: a platform for fine-grained resource sharing in the data center. In: Proceedings of the 8th USENIX Conference on NSDI, p. 22. USENIX Association, Berkeley (2011)
22. Lakshminarasimhan, S., Jenkins, J., Arkatkar, I., Gong, Z., Kolla, H., Ku, S.-H., Ethier, S., Chen, J., Chang, C.S., Klasky, S., Latham, R., Ross, R., Samatova, N.F.: ISABELA-QA: query-driven analytics with ISABELA-compressed extreme-scale scientific data. In: Proceedings of 2011 International Conference for High Performance Computing, Networking, Storage and Analysis, SC 2011, pp. 31:1–31:11. ACM, New York (2011)
23. Luo, Y., Guo, Z., Sun, Y., Plale, B., Qiu, J., Li, W.W.: A hierarchical framework for cross-domain mapreduce execution. In: Proceedings of the Second International Workshop on Emerging Computational Methods for the Life Sciences, pp. 15–22. ACM (2011)
24. Matsunaga, A., Tsugawa, M., Fortes, J.: Cloudblast: combining mapreduce and virtualization on distributed resources for bioinformatics applications. In: ESCIENCE, pp. 222–229. IEEE Computer Society, Washington, DC (2008)
25. Molina-Estolano, E., Gokhale, M., Maltzahn, C., May, J., Bent, J., Brandt, S.: Mixing hadoop and hpc workloads on parallel filesystems. In: Proceedings of the 4th Annual Workshop on Petascale Data Storage, PDSW 2009, pp. 1–5. ACM, New York (2009)
26. Rebel, O., Geddes, C.G.R., Cormier-Michel, E., Wu, K., Prabhat, Weber, G.H., Ushizima, D.M., Messmer, P., Hagen, H., Hamann, B., Bethel, W.: Automatic beam path analysis of laser wakefield particle acceleration data. Comput. Sci. Discov. **2**(1), 015005 (2009)
27. Rosenblum, M., Ousterhout, J.K.: The design and implementation of a log-structured file system. ACM Trans. Comput. Syst. **10**(1), 26–52 (1992)
28. Ross, R.B., Thakur, R., et al.: Pvfs: a parallel file system for linux clusters. In: Proceedings of the 4th Annual Linux Showcase and Conference, pp. 391–430 (2000)
29. Sun, X.-H., Byna, S., Chen, Y.: Server-based data push architecture for multiprocessor environments. J. Comput. Sci. Technol. **22**(5), 641–652 (2007)
30. Szalay, A.S., Kunszt, P.Z., Thakar, A., Gray, J., Slutz, D., Brunner, R.J.: Designing and mining multi-terabyte astronomy archives: the sloan digital sky survey. SIGMOD Rec. **29**(2), 451–462 (2000)
31. Tantisiriroj, W., Patil, S., Gibson, G.: Data-intensive file systems for internet services: a rose by any other name. Technical report (2008)
32. Tantisiriroj, W., Son, S.W., Patil, S., Lang, S.J., Gibson, G., Ross, R.B.: On the duality of data-intensive file system design: reconciling hdfs and pvfs. In: Proceedings of 2011 International Conference for High Performance Computing, Networking, Storage and Analysis, SC 2011, pp. 67:1–67:12. ACM, New York (2011)
33. Tiwari, D., Solihin, Y.: Mapreuse: reusing computation in an in-memory mapreduce system. In: IPDPS. IEEE Computer Society, Phoenix (2014)

34. Wu, X., Vijayakumar, K., Mueller, F., Ma, X., Roth, P.C.: Probabilistic communication and I/O tracing with deterministic replay at scale. In: ICPP, pp. 196–205. IEEE Computer Society, Washington, DC (2011)
35. Zhai, Y., Liu, M., Zhai, J., Ma, X., Chen, W.: Cloud versus in-house cluster: evaluating Amazon cluster compute instances for running MPI applications. In: State of the Practice Reports, SC 2011, pp. 11:1–11:10. ACM, New York (2011)

Survey on Software-Defined Networking

Jiangyong Chen[1,2(✉)], Xianghan Zheng[1,2(✉)], and Chunming Rong[3]

[1] College of Mathematics and Computer Science, Fuzhou University, Fuzhou 350108, China
29424176@qq.com
[2] Fujian Key Laboratory of Network Computing and Intelligent Information Processing,
Fuzhou 350108, China
[3] Department of Computer Science and Electronic Engineering, University of Stavanger,
Stavanger, Norway

Abstract. Recently, both the academia and industry have initiated research directed toward the integration of software-defined networking (SDN) technologies into the next generation of networking. In this paradigm, SDN transfers the control function from the traditional distributed network equipment to the controllable computing devices, which makes the underlying network infrastructure abstract to network services and applications. In this study, we survey OpenFlow-based SDN solutions that were recently proposed in both academia and industry. We consider technical issues, including SDN requirement, OpenFlow-based approach, challenges, and possible approaches. In addition, security breaches and possible solutions are described. Our survey is based on recent research publications.

Keywords: OpenFlow · SDN · Network virtualization · Security

1 Introduction

Traditional network architecture faces a few disadvantages [1]. First, protocols tend to be defined in isolation and solve a specific problem without the benefit of any fundamental abstractions. This condition has resulted in the primary limitation of traditional networks: complexity. Second, the complexity of traditional networks makes applying a consistent set of access, security, QoS, and other policies difficult for IT. Inconsistent policies cause an enterprise to become vulnerable to security breaches, non-compliance with regulations, and other negative consequences. Third, the network becomes complex with the addition of thousands of network devices that must be configured and managed, which makes the network unscalable. Finally, the vendors are dependent: carriers and enterprises seek to deploy new capabilities and services in rapid response to transforming business needs or user requirements. However, equipment product cycles of the vendors hinder their ability to respond, and these cycles can range from a period of three years or more.

As the next generation of network architecture, software-defined networking (SDN) technologies have many advantages. From the operator point of view, the core idea of SDN is the separation of the control plane and forwarding plane, which simplifies the network structure and layer, and reduces network construction and maintenance costs.

© Springer International Publishing Switzerland 2015
W. Qiang et al. (Eds.): CloudCom-Asia 2015, LNCS 9106, pp. 115–124, 2015.
DOI: 10.1007/978-3-319-28430-9_9

Centralizing the control of multi-vendor environments reduces complexity through automation and high innovation rate. Centralizing also increases network reliability and security, and granular network control.

SDN technologies have greatly improved in recent years. However, these technologies are still far from mature. In the promotion of Cisco and other manufacturers, IETF, IEEE, and other standards organizations removes links to SDN and OpenFlow, and retains the programmability. Thus, the generalized concept of SDN is extended, which refers to various basic network architectures based on programmable open-interface software, and will control forwarding logic separation. Although the data center deployment case has had many SDNs, SDN deployment in a large-scale network has not been considered for the future.

In the following sections, we conduct a survey on SDN. Our survey is mainly based on a few typical research projects and recent research. In Sect. 2, we introduce the SDN requirement. Section 3 specifies the key component of OpenFlow-based SDN. Application scenarios are described in Sect. 4. Section 5 introduces challenges and possible solutions. Conclusions and open issues are presented in Sect. 6.

2   Requirement Statement

In this section, we briefly describe the requirement of SDN systems.

1. Availability, stability, and efficiency. These concepts are the basic requirement at the system level.
2. Transport requirement. The designed protocol and interaction model should deliver reliable and efficient data.
3. Network programmable. The routing policy in the control function should be user definable.
4. Network virtualization. Network virtualization can not only help the administrator from every new access-domain physical-connection network virtual function, but can also effectively reduce waste allocation.
5. Centralized control and visualization. The controller must realize enterprise authentication and authorization, as well as completely isolate each virtual network. The SDN controller must be able to control the communication rate.
6. Interworking. Interoperability requires an appropriate protocol that both sustains the SDN communication interfaces and provides backward compatibility with existing IP routing and multiprotocol label switching control-plane technologies.
7. Security. The security requirement of SDN mainly focuses on the application layer and control layer; digestion includes application authorization, authentication, and isolation-policy conflict.
8. Extendibility. The system must also be able to mitigate the effects of broadcast overhead and growth of network flow table entries.

3 Software Defined Network

OpenFlow, as a prototype implementation method of SDN, represents the SDN control-architecture forward-separation technology. OpenFlow-based SDN is different from the traditional network-distributed architecture and overturns the traditional network operating mode.

3.1 System Architecture

SDN architecture is mainly divided into an infrastructure layer, control surface layer, and application layer, as shown in Fig. 1.

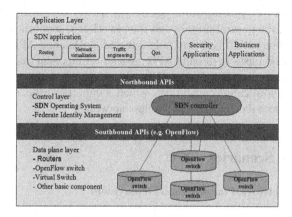

Fig. 1. Structure of OpenFlow switch.

The SDN controller is responsible for maintaining the global network view, and this controller updates the table information of switches in the flow through southbound APIs (e.g., OpenFlow protocol [2]) to realize centralized control of the network. The application layer interacts with the control layer through northbound APIs, which formulate relevant business rules (e.g., network configurations and application requirements) to realize network control and service programmability. In theory, SDN network routing considers a global view. Therefore, SDN has the natural advantage of in-routing decision-making.

3.2 Component Description

1. **OpenFlow Switch:** OpenFlow switch is responsible for the data forwarding function. Main technical details are composed of three parts [3]: flow table, security channel, and OpenFlow protocol.
2. **SDN Controller:** The control layer enforces all the policies to the underlying devices via southbound APIs using a network operating system (NOS). The crucial objective of NOS is to provide abstractions, essential services, and northbound APIs to developers. Communications between NOSs are realized through east–west APIs.

Early SDN technology has received considerable academic attention, such as the typical work of SDN of ForCES [4], 4D Architecture [5], RCP [6], SANE, [7] and Ethane [8].

3. **Southbound APIs:** Southbound APIs [9] are the crucial instrument for clearly separating control and data plane functionality. However, defining a switch-level negotiation framework that allows transparent translation among differentiated switches is necessary because of the interoperability and heterogeneity problem of existing protocols (e.g. OpenFlow from 1.0 to 1.4, etc.). Moreover, current TCAM storage (from 4K to 32K entries) is insufficient in a large-scale network (e.g., IDC).

4. **Northbound APIs:** The northbound interface is mostly a software ecosystem that translates application requirements into low-level service requests. Existing controllers, [10] such as Floodlight, Trema, Onix [18], and OpenDaylight, have proposed and defined their own specific northbound APIs. Therefore, the research includes an investigation of the general northbound framework that supports different types of currently northbound API and vertically oriented northbound API management.

5. **East–West APIs:** East–westbound APIs are a key component that includes import/export data among distributed controllers, algorithms for data consistency models, and monitoring/notification capabilities, which is important especially in large-scale networks, such as a data center and wireless networks.

4   Application Scenarios

4.1   Optimized Network Planning and Deployment

The WAN backbone network completely switches to OpenFlow network to plan a path of flow, which greatly optimizes network traffic. Network bandwidth utilization rate is greatly improved, the network becomes stable, management is simplified, and cost is reduced. Traffic engineering for flow control and route planning can clearly provide a clear picture of what happened inside the network. Traffic engineering will allow enterprises to control new service-need dynamic-service level protocol. Without the need for capacity expansion, the supplier can better control the quality of services because they can be completely based on the need to configure and remove network space.

4.2   Highly Extendable and Efficient Data Center

SDN and OpenFlow improve network manageability, utilization rate, and cost effectiveness of using programming. Since the beginning of 2012, all shaft connections of Google data centers have been using this architecture, and the network utilization rate has reached 95 %.

Google summarizes the advantages of SDN as follows: First, the network structure is unified, which simplifies the configuration, management, and optimization. Second, centralized traffic engineering achieves efficient use of cyber source. Third, the system can realize the rapid polymerization of cyber source and average distribution, and some network behaviors can be predicted.

OpenFlow protocol is at the early stage of development. Google research results show that the existing OpenFlow protocol is sufficient to support the development of many network applications. However, the router and the function of the controller for identification remains a topic under discussion, and function configuration is a problem that has yet to be addressed and resolved.

4.3 Good Virtualization and Security Control

To achieve communication between the monitoring of network virtual machine and network virtual isolation based on SDN, research on VxLAN network virtualization mechanism needs to be conducted based on multi-tunnel support, and efficient isolation of virtual machine communication network between multiple tenants needs to be achieved, as shown in Fig. 2.

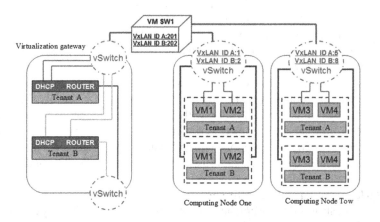

Fig. 2. Virtual machine communication monitoring.

On the basis of this observation, SDN realizes an integration network device monitor; network data-flow sampling statistics, and business user-flow monitoring technology by designing an SDN virtual machine network sensing system. The OpenFlow statistical method based on streaming technology facilitates the flow and priority of each data analysis in real-time virtual machine communication flow. The method can help to defend against network attacks. It can also be used to conduct network traffic monitoring of different tenants and applications, determine the current state of network fault diagnosis, address hidden dangers, and adjust the routing strategy in real time.

5   Challenges and Solutions

The existing hardware platform gives rise to a smooth evolution of virtual network compatibility and long-term coexistence of challenges. The old equipment determines how the new network can be fitted and ensures the smoothness of the function and performance, as well as provides support for business challenges.

5.1 Network Management Challenges

(1) **Forged traffic flows**
Network elements can be used to launch a DoS attack against OpenFlow switches (e.g., exhaust TCAMs) and controller resources from attackers.
Solution: Identifying abnormal flows by adopting intrusion detection systems with support for runtime root-cause analysis could be helpful.

(2) **Attack switch vulnerabilities**
One single switch could be used to slow down or drop network packets, clone or tamper network traffic, and even forge requests or inject traffic to overload neighboring switches or controllers.
Solution: These challenges can be addressed by using mechanisms of software attestation, such as autonomic trust management solutions for software components [11], to monitor and detect the behavior of network devices.

(3) **Attack control plane communications**
The network elements can be used to generate DoS attacks or for data theft. The TLS/SSL model is not enough to establish and ensure trust between controllers and switches. This lack of trust guarantees the creation of a virtual black hole network (e.g., using OpenFlow-based slicing techniques [12]), which allows data leakage during normal production traffic flows.
Solution: Securing communication with threshold cryptography across controller replicas [13], oligarchic trust models with multiple trust-anchor certification authorities or the use of dynamic, automated, and assured device association mechanisms may be considered to guarantee trust between the control plane and data plane devices.

(4) **Attack controller vulnerabilities**
Severe threats to SDN are crucial. A common intrusion detection system [14] may have difficulty in labeling a behavior as malicious through the exact combination of events that trigger a particular behavior.
Solution: Many techniques can be used in this situation, such as replication, employing diversity, and recovery.

(5) **Trust cannot be ensured between the management applications and controller**
The certification approach is the main difference from the referred threat. Techniques of certifying network devices are different from those used for applications.
Solution: Adopting mechanisms for autonomic trust management could guarantee that the application is trusted during its lifetime.

(6) **Attack administrative station vulnerabilities**
These administrative stations have been an exploitable target in the current network. Reprogramming the network from a single location is simplified.
Solution: The use of protocols requires double-credential verification and also using assured recovery mechanisms to guarantee a reliable state after reboot.

(7) **Lack of trusted resources for forensics and remediation**
The cause of a detected problem will be understood, which then initiates the process to a fast and secure recovery mode. The resources will be useful if their trustworthiness (authenticity and integrity) can be assured.

Solution: Logging and tracing are commonly used mechanisms and are needed both in the data and control planes. However, they should be indelible (a log that is guaranteed to be immutable and secure) to be effective. Furthermore, logs should be stored in remote and secure environments.

5.2   Trusted Association Establishment

Switch-Controller Association: A simple approach would be for controllers to keep authenticated white lists of known trusted devices. However, this approach lacks the edibility desired in SDN. Another way is to trust all switches until their trustworthiness dip goes below an accepted threshold. At that point, all devices and controllers would then automatically isolate the switch.

APP-Controller Association: A dynamic trust model [11] is required because software components exhibit changing behaviors from exhaustion, bugs, or attacks. In this study, the authors use a holistic notion of trust to allow a trustor to assess the trustworthiness of the trustee by observing its behavior and measuring this behavior based on quality attributes, such as availability, reliability, integrity, safety, maintainability, and confidentiality. The proposed model can also be applied to define, monitor, and ensure the trustworthiness of relationships among system entities.

Cloud infrastructure hardware trusted measurement based on TPM: The software execution platform process is initiated by using a TPM security module stored beforehand with software and hardware equipment boot measurement verification value based on the key step of starting process of cloud services platform equipment, such as integrity measurement. If the measure and verification value is consistent, then the device is reliable. Otherwise, further security updates are isolated. This method focuses on the underlying hardware process of the cloud infrastructure to ensure a credible verification process for hardware boot security of the cloud infrastructure (Fig. 3).

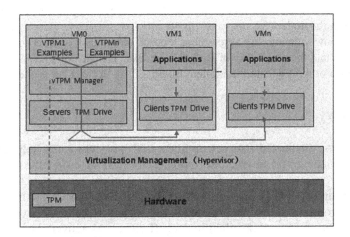

Fig. 3. TPM-based virtual machine management.

5.3   Detect, Correct, or Tolerate Faults

A switch that is dynamically associated with different controllers in a secure way could automatically tolerate faults. A switch increases control plane throughput and reduces control delay [36] by choosing the quickest-responding controller.

Increasing the data plane programmability would be helpful in this respect. One approach involves replacing some of the traditional functionalities of custom ASIC [37] by using common-purpose CPUs inside the switch. Another approach could act on behalf of the switch using a proxy element. With a common-purpose microcomputer, the element could simply attach to the switch by being deployed in a small black box.

Replication is one of the most important techniques to improve the dependability of the system. The application is replicated with some instances in the example. Controllers should be replicated as well.

Attackers discover vulnerable targets in the network using various scanning techniques. One method to defend against these attacks is the use of random virtual IP addresses using SDN. This technique involves managing a pool of virtual IP addresses in the Open-Flow controller, which is assigned to the network hosts, and hides real IP addresses from the outside network. Moving target defense is a form of adaptive cybersecurity.

5.4   Data Integrity and Confidentiality

A security technology [e.g., transport layer security (TLS)] can mitigate threats to mutual authentication between the controllers and their switches. Currently, Open-Flow specifications describe the use of TLS. However, TLS standard is not specified, and the security feature is optional. A full security specification must be defined to secure the connection and protect the data transmitted across the controller-switch interface. With a single controller controlling a set of network nodes, the necessary security may be provided through authentication with TLS. However, access control and authorization becomes complex in this way. The increase of potential unauthorized access could lead to the manipulation of the node configuration and malicious traffic through the node.

An OpenFlow vulnerability assessment focuses on the lack of TLS adoption from major vendors and the possibility of DoS [17] attacks. The lack of TLS use could lead to the insertion of modified and fraudulent rules.

6   Summary

In this study, we conducted a survey on SDN. SDN provides a new solution for the future development of the Internet. Currently, SDN must reconsider the real network deployment process and optimize performance, scalability, security, and distributed control requirements. Many open questions related to SDN have yet to be answered. First, a few proposals should be presented with regard to the SDN issue to test the real network and evaluate its availability and efficiency. Moreover, the interworking between SDN and conventional traditional network should be further studied.

Second, the inconsistent control logic and the extensibility of control plane of SDN remain to be discussed. Third, proposed SDN security reinforcement schemes are not comprehensive and are still theoretical.

References

1. Gude, N., Koponen, T., Pettit, J., Pfaff, B., Casado, M., McKeown, N., Shenker, S.: NOX: towards an operating system for networks. ACM SIGCOMM Comput. Commun. Rev. **38**(3), 105–110 (2008)
2. Jean, T., Puneet, S., Sujata, B., Justin, P.: Sdn and Openflow evolution: a standards perspective. Computer **47**, 22–29 (2014)
3. Mckeown, N., Anderson, T., Balakrishnan, H., Parulkar, G., Peterson, L., Rexford, J., Shenker, S., Turner, J.: OpenFlow: Enabling innovation in campus networks. ACM SIGCOMM Comput. Commun. Rev. **38**(2), 69–74 (2008)
4. Yang, L., Dantu, R., Anderson, T., Gopal, R.: Forwarding and control element separation (ForCES) framework. RFC 3746 (2004)
5. Greenberg, A., Hjalmtysson, G., Maltz, D.A., Myers, A., Rexford, J., Xie, G., Yan, H., Zhan, J., Zhang, H.: A clean slate 4D approach to network control and management. ACM SIGCOMM Comput. Commun. Rev. **35**(5), 41–54 (2005)
6. Caesar, M., Caldwell, D., Feamster, N., Rexford, J., Shaikh, A., Merwe, J.: Design and implementation of a routing control platform. In: Proceedings of the 2nd USENIX Symposium on Networked Systems Design and Implementation (NSDI), pp. 15–28. USENIX Association, Boston (2005)
7. Casado, M., Garfinkel, T., Akella, A., Freedman, MJ., Boneh, D., Mckeown, N., Shenker, S.: SANE: a protection architecture for enterprise networks. In: Proceedings of the 15th Conference on USENIX Security Symposium, pp. 137–151. USENIX Association, Vancouver (2006)
8. Casado, M., Freedman, M.J., Pettit, J., Luo, J., Mckeown, N., Shenker, S.: Ethane: taking control of the enterprise. In: Proceedings of the SIGCOMM 2007, pp. 1–12. ACM Press, Kyoto (2007)
9. Tootoonchian, A., Ganjali, Y.: HyperFlow: a distributed control plane for OpenFlow. In: Proceedings of the 2010 Internet Network Management Workshop/Workshop on Research on Enterprise Networking (INM/WREN) USENIX Association, San Jose (2010)
10. Koponen, T., Casado, M., Gude, N., Stribling, J., Poutievski, L., Zhu, M., Ramanathan, R., Iwata, Y., Inoue, H., Hama, T., Shenker, S.: Onix: A distributed control platform for large-scale production networks. SIGCOMM Comput. Commun. Rev. **38**(3), 105–110 (2008). In: Proceedings of the 9th USENIX Conference on Operating Systems Design and Implementation (OSDI). USENIX Association, Vancouver (2010)
11. Yan, Z., Prehofer, C.: Autonomic trust management for a component- based software system. IEEE Trans. Dep. Sec. Comput. **8**(6), 810–823 (2011)
12. Sherwood, R., et al.: FlowVisor: a network virtualization layer. Technical report, Deutsche Telekom Inc. R&D LabStanford University, Nicira Networks (2009)
13. Desmedt, Y.G.: Threshold cryptography. Eur. Trans. Telecommun. **5**(4), 449–458 (1994)
14. Heller, B., Sherwood, R., McKeown, N.: The controller placement problem. In: HotSDN. ACM (2012)
15. Kreutz, D., Ramos, F., Verissimo, P.: Towards secure and dependable software-defined networks. In: Proceedings of the Second ACM SIGCOMM Workshop on Hot Topics in Software Defined Networking. ACM, pp. 55–60 (2013)

16. Braga, R., Mota, E., Passito, A.: Lightweight DDoS flooding attack detection using NOX/OpenFlow. In: IEEE 35th Conference on Local Computer Networks (LCN). IEEE, pp. 408–415 (2010)
17. Fundation O N. Software-defined networking: the new norm for networks. ONF White Paper (2012)
18. Koponen, T., Casado, M., Gude, N., Stribling, J., Poutievski, L., Zhu, M., Ramanathan, R., Iwata, Y., Inoue, T. Hama, H., Shenker, S.: Onix: a distributed control platform for large-scale production networks. OSDI 2010 (2010)

Dynamic Load Sharing to Maximize Resource Utilization Within Cloud Federation

Md. S.Q. Zulkar Nine, Md. Abul Kalam Azad, Saad Abdullah,
and Nova Ahmed[✉]

ECE Department, North South University, Dhaka, Bangladesh
{zulkarnine,saad}@eecs.northsouth.edu,
ak.azad@live.com, nova@northsouth.edu

Abstract. It is evident in recent years that cloud has resource constraints. Client requests are at the highest priority of cloud services. Denial of client services will not only hamper profits but also tarnish reputation of the provider. In these cases an effort can be made to provision resources from the other cloud providers so that they can serve the request using their unused resources. In this way the idea of cloud federation has emerged. The idea is, if a cloud saturates its computational and storage resources, or it is requested to use resources in a geography where it has no footprint, it would still be able to satisfy such requests for service allocations sent from its clients. Our work contributes by offering a model for enacting the cloud federation. More precisely, we have introduced a cloud broker which decides whether a job sent to a particular cloud provider should be served there or routed to another provider. Our proposed model selects the best option to outsource the request without violating Service Layer Agreement (SLA).

Keywords: Cloud computing · Cloud federation · Dynamic load sharing · Cloud broker service · Analytic hierarchy process

1 Introduction

Open source operating system distribution provides great promises for Cloud computing is the most recent abstraction based distributed system that offers computing services as pay-as-you-go basis [1]. A wide range of services such as - Software as a service (SaaS), Infrastructure as a Service (IaaS), and Platform as a service (PaaS) are introduced to a diverse user group from another cloud, organization to single common users.

Currently, the interoperability among the heterogeneous cloud computing operators becomes an essential strategy as it directly involved with the profitability of the whole community. Moreover, due to the unpredictable nature of the distribution of user requests and user location, it is very difficult to meet user requirements and agreed Quality of Service (QoS) in time of load spike. As a result the resource sharing is inevitable in the evolution of the cloud ecosystem. Now, a newer concept 'Cloud of clouds' has emerged which is known as Cloud Confederation. Buyya et al. [1] explored various challenges that should be addressed to auto-scale the cloud resources in the

© Springer International Publishing Switzerland 2015
W. Qiang et al. (Eds.): CloudCom-Asia 2015, LNCS 9106, pp. 125–137, 2015.
DOI: 10.1007/978-3-319-28430-9_10

federated cloud. They introduced an architecture that is mediate by a cloud exchange based on competitive economic model. Such models are always prune to single point of failure and have poor scalability. Goiri et al. [2] provided several equations and functions (with parameters) to characterize the providers' federation in the cloud. The characterization includes outsourcing jobs to other providers, insourcing free resources to other providers and shutting down nodes when not in operation. Various protocols for the interoperability of the clouds are exploited by Bernstein et al. [3]. A well-constructed definition of cloud federation was offered by Kurze et al. [4] where they proposed a reference architecture of cloud federation that allows horizontal and vertical integration of new service models into the federation. In addition, they also address two economic problems like vendor lock-in, hold up and underinvestment. A Reservoir model is proposed by Rochwerger et al. [5] where independent providers make decision based on its own local preferences. In Reservoir model infrastructure providers own reservoir sites which make the reservoir cloud federation. Villegas et al. [6] exploited the layered model for inter-cloud federation which is mediated by a broker service. Nai-Wei et al. [7] exploited the computing resource allocation which is basically the management of processing power, network transmission and storage resource. They also defined a manager frame for cloud and cross-cloud federation architecture based on inter-trust context. Xiaoyu et al. [8] proposed an architecture for the Real-time Online Interactive Applications (ROIA).Cloud federation is defined as a business layer with increased security features and on-demand resource provisioning facilities. A co-operative game theoretic approaches for horizontal dynamic cloud federation is proposed in the work of Hassan et al. [9]. Lucas et al. [10] proposed a cloud broker tool that can discover resources and use service constraints, security policy and compliance policies to allocate the best one that matches the requirements. Yangui et al. [11] proposed CompatibleOne, an open source broker service for efficient resource provisioning from a federated cloud. They used CompitableOne Resource Description System to describe IaaS and PaaS resources in the federation of clouds. Kaiyang et al. [12] formulated the resource reservation problem as nonlinear integer programming model. To overcome the prohibitive complexity of the model a heuristic algorithm is introduced which can obtain near optimal solution.

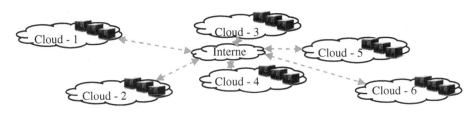

Fig. 1. Global cloud organization

We proposed an architecture to select efficiently the most appropriate federation member to outsource the user requests. This happens when two or more cloud providers interconnect their cloud computing environments for accommodating spikes on demand and load balancing. Taking the cloud federation into account, our proposed architecture includes a cloud broker for each cloud provider to decentralize the decision making. The cloud broker performs a vital operation of assessing the load of its server when there is a pending user request. In a situation when server has sufficient idle resources to entertain the user request, then the broker simply routes the request to its local datacenter. However, when the cloud provider is heavily loaded and cannot serve further requests, the broker selects the most appropriate cloud provider that will best serve the client request at hand while maintaining a good profit for itself. The analytic hierarchy process is an excellent structured technique that organizes and analyzes complex decisions based on some given criterion. In our model we consider four parameters, namely, quality of service, availability, security, and cost. Ensuring the most balanced tradeoff among these characteristics, the AHP algorithm makes the decision of selecting the proper cloud provider for the job.

The remaining paper is organized as follows: In Sect. 2 we elaborate the concept of cloud federation. Section 3 introduces an overview of the AHP. Our novel approach is explained in Sect. 4. In Sect. 5, we describe the simulation and experimentation to evaluate our proposed model.

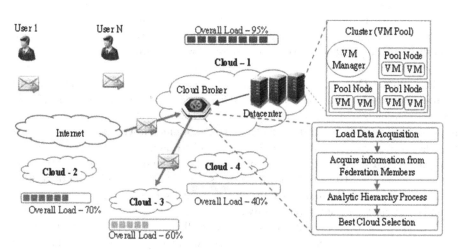

Fig. 2. Proposed cloud federation architecture

2 Cloud Federation

The concept of cloud federation arises from the concept of 'Cloud of Cloud' [1]. Cloud federation is the practice of interconnecting the cloud computing environments of two or more service providers for the purpose of load balancing traffic and accommodating spikes in demand. Cloud federation requires one provider to wholesale or rent computing resources to another cloud provider. Those resources become a temporary or permanent extension of the buyer's cloud computing environment, depending on the specific federation agreement between providers. Cloud federation offers two substantial benefits to cloud providers. First, it helps to utilize the underutilized resources of the clouds which can maximize the profit of the cloud providers. Second, it is essential to expand the geographic footprints as showed in Fig. 1 and to accommodate unpredictable user spikes in peak hours without acquiring any physical hardware.

In a typical cloud system normally two types of VMs are provided such as – (1) On-demand VMs – which are based on strict Service Layer Agreement (SLA) where requirements must be meet. (2) Spot VMs which a dedicated based on bidding to the user with low price and provider can terminate those VMs if they need to accommodate on-demand VMs. Cloud federation could be an efficient model to manage and provision resources to those On-demand and Spot VMs.

3 Background

In Bangladesh, computers are still a luxury for the masses but the trend is changing at a fast pace with time. Nowadays computers are everywhere from a mobile recharge shop to corporate world. People are buying different kinds of configuration of hardware as per their requirements and budget but one piece of the computer is common on every configuration of hardware and that is the software that runs the computer, which is Windows. It doesn't matter whether its XP, 7 or 8, it rules the PC market. In Bangladesh it is a notion that computer means Windows and Microsoft Office. The same situation is with smart phones. People using android OS does not what to move to windows phone OS or any other Mobile OS that has different look and feel than android. But people are using smart phones that are android like OS such as Firefox OS and other variants of Linux based mobile OS's. This indicates that with an windows like interface we can use other computer OS that are much lower on resources and can open new opportunities in third world countries like Bangladesh. Now the things have changed in the Linux world it has become very user friendly and usable thanks to the massive development of open source products. This paper discusses the opportunity and the acceptability of Linux OS at the all level of the society as well as the possibility to replace propitiatory and pirated software's.

Windows has been the star of software's of all time since the beginning. But in Bangladesh people use pirated copies of windows that are buggy and vulnerable to viruses which leads to regular re-installation of the system and which hampers the productivity and ease of use. In Linux there is less chance of virus attacks thus making it robust and secured. The people who uses computer for typing jobs, printing, browsing and other minimal stuff does not required a heavy system like windows to operate. Therefore in this paper we are trying to show the adaptability of Linux for specific tasks that lower level people works on. Our work aims to replace windows with Linux with minimal windows like desktop environment and study the acceptability of a new system and see if the minimal UI can increase the ease of use and interaction.

4 Analytical Hierarchy Process

The analytic hierarchy process (AHP) algorithm is developed by Saaty [13] in 1988 and is still one of the most popular decision making algorithms and widely used in decision making under multi-criteria situation. The AHP formulates the decision problem as a structure of hierarchy, for example goal, criteria, and sub-criteria. A prioritization procedure is then applied to assign the priority to each element of the hierarchical structure through a pairwise comparison. AHP uses a standardized comparison scale having nine levels of importance shown in Table 1 to make the pair-wise comparison. Let C = {C j/j = 1, 2,..., n} be the set of criteria. After pair-wise comparison on n criteria, the result can be preserved in a (n × n) evaluation matrix A where each element a_{ij} is the quotient of weights as shown in Table 1. Right eigenvector (w) corresponding to the largest eigenvector (λ_{max}) is used to calculate relative priorities (AW) as:

$$AW = \lambda_{max} \times w \tag{1}$$

In case of consistent pair-wise comparison, the matrix A has rank 1 and $\lambda_{max} = n$. The rows or columns are then normalized to obtain weights. This procedure is iterated for all the elements of the hierarchy. In order to synthesize the various priority vectors, these vectors are weighted with the global priority of the parent criteria and synthesized. This process starts at the top of the hierarchy. In this way, the local priorities are converted into the overall priorities. As the alternatives are located at the lowest level of the hierarchy, the newly calculated priorities represent their contributions to the top. In this way, the problem itself is placed on top of the hierarchy.

In AHP methodology, inconsistency in pair-wise comparison may occur due to subjective human judgment. Finally, the consistency ratio (CR) [5] is calculated as the ratio of the error. Therefore, it is important to check the consistency in response through a consistency index (CI) by using the following equation.

$$CI = \frac{\lambda_{max} - n}{n - 1} \tag{2}$$

CI and the random consistency index (RCI), which is shown in Table 1. If the value of CR is 10 % or less, the inconsistency is acceptable. If greater than 10 %, we need to revise the subjective judgment i.e. the respondents were not consistent while answering.

Table 1. Random consistency index

Matrix rank	1	2	3	4	5	6	7	8
CR	0	0	0.58	0.9	1.1	1.2	1.3	1.4

5 Proposed Architecture

In our proposed architecture we introduce a cloud broker service to each cloud system in the cloud federation. Due to the decentralized nature of the service it is highly scalable and free from single point failure. We adopt a simple cloud federation to model our proposed architecture as shown in Fig. 2.

Each cloud maintains two cloud broker daemon processes that actually decide whether the request will be served in the local cloud or the remote one. Cloud broker has two distinct responsibilities (1) assess and provision computing resources of the most viable cloud member. (2) Response to any incoming outsourcing requests. Two independent daemon processes are responsible to execute those tasks.

5.1 Information Acquisition

In the phase, upon receiving a request from the user, the cloud broker instance that is responsible for resource provisioning checks its local datacenter for computing resources. As we know local data center is serving spot VM, on-demand VM. In an ideal situation when datacenter has enough resources to provision, the broker just route the request to the local datacenter to serve. However, in a situation when the datacenter is heavily loaded and resource provisioning is quite impossible without dropping some spot VMs, then the cloud broker start to assess the possibility of provisioning the resources from a suitable cloud federation member. The cloud broker immediately dispatches a request message to the fellow cloud members asking whether they have enough resource to provision. Using the replies from the other cloud members, the broker makes a shortlist of the underutilized clouds who are willing the provision resources. In case when no suitable cloud is found (i.e. all the federation members are heavily loaded) the cloud broker can initiate measures to stop some spot VMs to provision more profitable on-demand VM.

5.2 Service Layer Agreement (SLA) Analysis

Service Layer Agreement is used by the parties who engaged in the electronic business where the minimum expectation and obligations are recorded including business parties, pricing policy, and properties of the resources required to process the service [3]. Even though the pricing seems most dominating feature in SLA, many other concerns like availability, security, quality of service might become crucial in SLA. So the most viable candidate must reflect all those requirements.

5.3 Multi-criteria Decision Making

When a cloud becomes heavily loaded it can route the request to a suitable cloud which is moderately loaded with requests. The decision is critical as it has to consider the SLA requirements of the users to select the most suitable cloud. It is a Multi-Criteria Decision Making (MCDM) problem. We proposed the Analytic Hierarchy Process

(AHP) to address the problem. It uses criteria such as price, security, quality of service and availability. Quality of service might be consist of average response time, reliability, elasticity. The decision is based on relative comparisons among those criteria. The shortlist created by the cloud broker is then ranked using AHP.

5.4   Signing Agreement

Cloud broker then start contacting with the high ranked clouds asking whether they are interested to utilize their unused resources and send a detailed resource and SLA requirements. In the receiving cloud the broker instance that is responsible for assessing incoming outsourcing request checks its current datacenter status and respond with an approval message. Then the request initiating cloud collects all the approval message and send confirmation to the most viable cloud and denial message to others. In this way a three way handshaking is used to confirm resource provisioning agreement.

5.5   Resource Provisioning

The confirmation request from the initiator cloud is used to provision the resources from the member cloud. When the member cloud has unutilized resources it just starts provisioning the requested resources. In case of partial resource availability the member cloud might drop some spot VMs to release resources.

Figure 2 shows the whole diagram of resource sharing inside federated cloud. Here each individual cloud system maintains a broker demon process which actually involved making decision about the most suitable cloud. Users are distributed in the whole world with regional distribution. In our simulated model we combine the requests of each region using the Userbase which is responsible for handling those requests through the internet. All those user requests are sent to the appropriate cloud system for acquire VMs. User requests are handled by the cloud broker in cloud end.

To evaluate our proposed model we have used a simulation environment. Large scale internet application simulation is very difficult to perform with hardware implementation. Small test-bed might be manageable; however, it is hard to accommodate huge unpredictable internet traffic. We have used CloudSim [14–16] to simulate our experiment.

Simulation Tools. Cloud Analyst is a robust large scale cloud simulator which facilitates realistic cloud environment with a large scale internet load. The main advantage of the simulator is its flexibility to test and design new ideas. It also ensures the repeatability of the experiment. It means simulator can generate similar results if it could be run with same parameters. However, we had to modify the simulator to implement our algorithms on the top of its basic structure.

Simulation Setup. For simulation purpose we have used 6 cloud providers in six geographic regions the architecture and operating cost of the clouds are provided in Table 2 below.

Table 2. Cost per VM per hour of the cloud federation members

Name	C1	C2	C3	C4	C5	C6
Cost per VM $/Hr	0.1	0.15	0.10	0.15	0.08	0.12

The resource capacity such as – number of VMs, image size, memory, bandwidth of each cloud is provided in Table 3 below.

Table 3. VM properties

Cloud	# VMs	Image size	Memory	BW
C1-C6	50	10000	512	1000

The processor capacity used in this simulation is 10000 MIPS. Here a detail illustration of Analytic hierarchy process is explained. It uses Quality of Service, availability, Price, and Security as criteria and the cloud systems in the federation as alternatives. It basically makes pair-wise comparison between criteria to extract the precedence of the criteria over one another.

Table 4. Criteria – criteria comparison matrix

	QoS	Security	Availability	Cost
QoS	1	3	3	1/5
Security	1/3	1	5	1/7
Availability	1/3	1/5	1	1/9
Cost	5	7	9	1

Table 4 shows that Quality of Service has moderate precedence over the Security, however, Quality of Service has inverse priority over the Cost by 1/5, means cost has strong priority over Quality of Service.

We can create pair-wise comparison between alternatives for each criteria which is given below (Tables 5, 6, 7 and 8).

Table 5. Pair-wise comparison among clouds for QoS

	C1	C2	C3	C4	C5	C6
C1	1	3	5	1	1	3
C2	1/3	1	3	1	5	1
C3	1/5	1/3	1	1	9	5
C4	1	1	1	1	3	7
C5	1	1/5	1/9	1/3	1	3
C6	1/3	1	1/5	1/7	1/3	1

Table 6. Pair-wise comparison among clouds for security

	C1	C2	C3	C4	C5	C6
C1	1	9	5	5	5	5
C2	1/9	1	9	5	5	5
C3	1/5	1/9	1	5	5	5
C4	1/5	1/5	1/5	1	5	5
C5	1/5	1/5	1/5	1/5	1	5
C6	1/5	1/5	1/5	1/5	1/5	1

Table 7. Pair-wise comparison among clouds for availability

	C1	C2	C3	C4	C5	C6
C1	1	3	5	7	7	3
C2	1/3	1	7	9	7	5
C3	1/5	1/7	1	3	3	7
C4	1/7	1/9	1/3	1	9	9
C5	1/7	1/7	1/3	1/9	1	9
C6	1/3	1/5	1/7	1/9	1/9	1

Table 8. Pair-wise comparison among clouds for cost

	C1	C2	C3	C4	C5	C6
C1	1	5	3	6	9	5
C2	1/5	1	5	3	9	3
C3	1/3	1/6	1	2	5	9
C4	1/6	1/3	1/2	1	9	9
C5	1/9	1/9	1/5	1/9	1	9
C6	1/5	1/3	1/9	1/9	1/9	1

The final ranking of the six candidate clouds are computed as shown is Table 9.

Table 9. The final ranking of the clouds

C1	C2	C3	C4	C5	C6
0.4345	0.2389	0.1506	0.0954	0.0500	0.0307

Here C1 gets the highest ranking as it is the local cloud and it gets the advantage of bandwidth, latency and cost. We choose the next highest cloud C2 to route the user request.

Non-federated Totally In-house. To evaluate our system, we have used the in-house request provisioning policy. In this case, no requests are forwarded to the cloud federation. All the requests are managed within the cloud itself. However, it initiate commands to drop the spot requests when it needs to accommodate the on-demand VMs.

Performance Matrices. We applied the following metrics to analyze the impact our proposed model:

Utilization. Utilization is the fraction of time when the VM is used.

$$Utilization, (\Delta u) = \frac{\sum_{i=1}^{vm} runtime(vm_i)}{vm_{max} \times total\ time} \tag{3}$$

where, vm is the total number of VMs which includes on-demand, spot VMs and vm_{max} is the maximum number of VMs that a provider can run simultaneously in its data center. Runtime (vm_i) shows the corresponding runtime for each VM.

Profit. Profit is the difference between the achieved revenue and operating cost during a period of time. For simplicity we have ignored the other costs involved.

$$Profit, P(\Delta t) = revenue\ (\Delta t) - Cost\ (\Delta t) \tag{4}$$

where revenue is the revenue obtained during Δt including on-demand, spot, contributed to the federation, and outsourced requests, whereas Cost (Δt) is the cost of the outsourcing VMs at the same period.

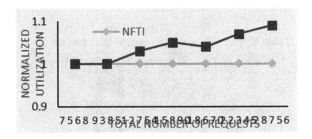

Fig. 3. Impact of the traffic on utilization

Number of Rejected On-demand VMs. This metric shows the number of on-demand VMs rejected. It considers only on-demand requests as they are the primary source of revenue and reputation.

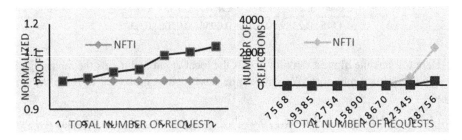

Fig. 4. (a) Impact of traffic on profit. (b) Impact of traffic on the number of on-demand VM rejections.

Increase of Utilization. We have used the formula (3) to compute the utilization of the cloud system and normalized it using NFTI utilization as base line value. As we have mentioned earlier that the NFTI policy does not explore the situation of exploiting other cloud federation members, the use of normalized values can illustrate the efficiency of the federation. It can be seen from the Fig. 3 that, for limited number of requests both approaches have same utilization as they can accommodate them in-house. However, with the gradual increase of the number of requests the NFTI can accommodate the on-demand requests by dropping the spot requests where our proposed model can increase the utilization further by using the idle capacities of the member clouds of the federation. The simulation reveals that our proposed model provides 11 % more utilization than the NFTI policy and shows an incremental trend as the number of requests increases.

Impact on Profit. The profit can be computed by using Eq. (4). In Fig. 4(a), it can be seen that profits are merely same for both model when the number of requests are low. It is very obvious, as the local datacenter is capable to process the VM requests. However, when the number of requests increases our model provides increased profit due to its outsourcing capability. SLA agreements are also sustained as we choose the most appropriate cloud for resource provisioning. Simulation shows that profit has an increasing trend with the increase of client requests, because our system choose the resources with optimal costing. This minimal cost helps to increase the profit of the cloud provider. For 28756 client requests our model ensures 12 % more profit that the NFTI policy.

Decrease in On-demand VM Rejections. The last and most important performance matric is the number of on-demand VMs that are rejected. It can be seen from the Fig. 4 (b) that both model can sustain without rejecting the on-demand VMs for a certain level of user traffic. In such situation NFTI start to drop the spot VMs and accommodate the on-demand VMs. However, after that it starts rejecting the on-demand requests with a merely exponential rate where our proposed model is highly resilient due to its federated resource provisioning capability. In the simulated environment, our proposed model reduces the number of VM rejection by a factor of 5.

6 Conclusion

Cloud architecture comes with a high elasticity of resource provisioning and utilization. However, the instant on-demand computing facilities might be a challenging with the increase of client requests. Due to the unpredicted nature of the load, it become difficult to acquire hardware and other resources to accommodate the increased load. To address such situation Cloud federation is a useful paradigm for resource sharing in cloud. In our work, we proposed a decentralized broker based approach that independently decide about the incoming client requests and route to the most viable cloud member in the federation. We model the satiation as a Multi-criteria Decision Making (MCDM) problem and investigated the parameters that are involved to select the best possible option to route the requests. Our study provides a novel solution to the select the most suitable cloud where user request can be routed in time of resource shortage due to heavy user load. For the performance analysis we simulate our model and compare in with NFTI policy for three individual performance matrices – utilization, profit, and number of VM rejections. Our proposed architecture outperforms the NFTI in all three cases. It increases utilization 10 %, profit 12 % while confirms a 5 fold reduction in number of the VM rejections. We plan to investigate other decision strategies and parameters that can be used to improve the architecture.

References

1. Buyya, R., Ranjan, R., Calheiros, R.N.: InterCloud: utility-oriented federation of cloud computing environments for scaling of application services. In: Hsu, C.-H., Yang, L.T., Park, J.H., Yeo, S.-S. (eds.) ICA3PP 2010, Part I. LNCS, vol. 6081, pp. 13–31. Springer, Heidelberg (2010)
2. Goiri, I., et al.: Characterizing cloud federation for enhancing providers' profit. In: 2010 IEEE 3rd International Conference on Cloud Computing (CLOUD). IEEE (2010)
3. Bernstein, D., et al.: Blueprint for the intercloud-protocols and formats for cloud computing interoperability. In: 2009 Internet and Web Applications and Services, ICIW 2009 (2009). Kurze, T., et al.: Cloud federation. In: Cloud Computing 2011, The Second International Conference on Cloud Computing, GRIDs, and Virtualization (2011)
4. Rochwerger, B., et al.: The reservoir model and architecture for open federated cloud computing. IBM J. Res. Dev. 53(4), 1–4 (2009)
5. Villegas, D., et al.: Cloud federation in a layered service model. J. Comput. Syst. Sci. 78(5), 1330–1344 (2012)
6. Lo, N.-W., et al.: An efficient resource allocation scheme for cross-cloud federation. In: 2012 International Conference on Anti-Counterfeiting, Security and Identification (ASID), pp. 1–4, 24–26 August 2012
7. Yang, X., et al.: A business-oriented cloud federation model for real-time applications. Future Gener. Comput. Syst. 28(8), 1158–1167 (2012)
8. Hassan, M.M., et al.: Cooperative game-based distributed resource allocation in horizontal dynamic cloud federation platform. Inf. Syst. Front. 1–20 (2012)
9. Lucas-Simarro, J.L., et al.: A cloud broker architecture for multicloud environments. Large Scale Netw.-Centric Distrib. Syst. 359–376 (2013). doi:10.1002/9781118640708.ch15
10. Yangui, S., et al.: CompatibleOne: the open source cloud broker. J. Grid Comput. 12(1), 93–109 (2014)
11. Liu, K., et al.: Dynamic resource reservation via broker federation in cloud service: a fine-grained heuristic-based approach. In: 2014 IEEE Global Communications Conference (GLOBECOM), pp. 2338–2343, 8–12 December 2014
12. Saaty, T.L.: The Analytic Hierarchy Process. McGraw-Hill, New York (1988)
13. Buyya, R., et al.: Modeling and simulation of scalable cloud computing environments and the CloudSim toolkit: challenges and opportunities. In: International Conference on High Performance Computing and Simulation, Leipzig, Germany, pp. 1–11, June 2009
14. Buyya, R., et al.: Cloud computing and emerging IT platforms: vision, hype, and reality for delivering computing as the 5th utility. Future Gener. Comput. Syst. 25(6), 599–616 (2009)

15. Wickremasinghe, B., et al.: CloudAnalyst: a cloudsim-based visual modeller for analysing cloud computing environments and applications. In: 24th IEEE International Conference on Advanced Information Networking and Applications, Perth, Australia, pp. 446–452 (2010)
16. Calheiros, R., et al.: CloudSim: a toolkit for modeling and simulation of cloud computing environments and evaluation of resource provisioning algorithms. J. Softw. Pract. Experience **41**(1), 23–50 (2011)

Applications

Real-Time Task Scheduling Algorithm for Cloud Computing Based on Particle Swarm Optimization

Huangning Chen[1] and Wenzhong Guo[2(✉)]

[1] College of Mathematics and Computer Science, Fuzhou University, Fuzhou, China
chan406@126.com
[2] Fujian Provincial Key Lab of the Network Computing
and Intelligent Information Processing, Fuzhou University, Fuzhou, China
fzugwz@163.com

Abstract. As a new computing paradigm, cloud computing is receiving considerable attention in both industry and academia. Task scheduling plays an important role in large-scale distributed systems. However, most previous work only consider cost or makespan as optimized objective for cloud computing. In this paper, we propose a soft real-time task scheduling algorithm based on particle swarm optimization approach for cloud computing. The optimized objectives include not only cost and makespan, but also deadline missing ratio and load balancing degree. In addition, to improve resource utilization and maximize the profit of cloud service provider, a utility function is employed to allocate tasks to machines with high performance. Simulation results show the proposed algorithm can effectively minimize deadline missing ratio, maximize the profit of cloud service provider and achieve better load balancing compared with baseline algorithms.

Keywords: Cloud computing · Task scheduling · Real-time · Virtual machine · Particle swarm optimization

1 Introduction

Cloud computing, as a new computing paradigm [1,2], is receiving considerable attention in both industry and academia. Scalability, virtulization [3], on-demand services and ubiquitous access are main features of cloud computing. Cloud computing provides the following three services: infrastructure as a service (IaaS), platform as a service (PaaS) and software as a service (SaaS) [4]. Task scheduling is one of the crucial technologies in cloud computing. The overall performance of cloud system depends on that the scheduling algorithm is superior or not. How to efficiently and rationally allocate the finite, heterogeneous and geographically distributed resources to meet the end-user's requirements is an urgent issue for cloud service provider.

As a typical problem of the field of high performance computing, task allocation and scheduling has been addressed in a variety of applications, such as

© Springer International Publishing Switzerland 2015
W. Qiang et al. (Eds.): CloudCom-Asia 2015, LNCS 9106, pp. 141–152, 2015.
DOI: 10.1007/978-3-319-28430-9_11

multiprocessor system, social networks and grid computing [5,6]. The researches on task allocation and scheduling in distributed systems can not directly apply in cloud due to the commercial-oriented nature of cloud computing. Recently, more and more attention has been pay to the issue on task scheduling in cloud. In [7], a new task scheduling model is proposed to solve load imbalance problem between virtual machines (VM). However, they only aim at load balancing and minimizing execution time. Ramezani et al. [8] proposed an algorithm to transfer extra tasks from an overloaded VM instead of migrating the entire overloaded VM. But, our method is quite different from their's for achieving load balance. In [9], authors propose a lookahead genetic algorithm (LAGA). The proposed algorithm utilizes the reliability-driven reputation to optimize both the makespan and the reliability of a workflow application. An energy-aware scheduling algorithm named EARH for real-time and independent tasks is proposed in [10]. The EARH employs a rolling-horizon optimization policy. However, both [9,10] fail to explicitly discuss the cost of cloud service provider. In [4,11–13], authors employ particle swarm optimization approach to guide the process of task scheduling. However, their optimized objective is either makespan or cost. Moreover, the system in [4] is hybrid cloud. But hybrid cloud is still in its infancy because formal inter-cloud agreement is hardly to set. Therefore, there is still considerable room for further research on task allocation and scheduling in cloud. In this paper, we propose a soft real-time task scheduling algorithm based on particle swarm optimization (RTA-PSO) approach for cloud computing. In order to improve resource utilization and maximize the profit of cloud service provider within deadline constraint, a utility function is employed to allocate tasks to machines with high performance.

The rest of this paper is organized as follows. In Sect. 2, problem description and formulation is given. Section 3 presents our algorithm. Section 4 provides the simulation results and analysis. Finally, we make conclusions and discuss the future work in Sect. 5.

2   Problem Description and Formulation

2.1   Problem Description

The symbols used in this paper are listed in Table 1. In this paper, a task is an entity which fulfills a given specific function and is the smallest granularity of task scheduling. Resources in cloud generally refer to virtual machine (vm) which is running separately on physical machine. Due to the virtualization technology, utilization of resources has been greatly improved. Given a set $T = \{t_1, t_2, t_3..., t_m\}$ of m independent tasks which have different arrival time and strict deadline constraint. Considered a group $VM = \{vm_1, vm_2, vm_3, ..., vm_n\}$ of n virtual machines running on the cloud system. Given a task $t_i \in T$, a_i and d_i are used to denote the arrival time and deadline time of task t_i, respectively. f_i and $allo(t_i)$ represent the expected finish time of t_i and serial number of vm task t_i allocated to, respectively. Because virtual machines we considered are heterogeneous, thus the execution time and cost of tasks are different. Matrixes

Table 1. Symbols in this paper

Symbol	The meaning of symbols
vm_i	The ith virtual machine (vm)
t_i	The ith task
a_i	The arrival time of t_i
d_i	The deadline time of t_i
et_{ij}	The execution time of t_i on vm_j
$cost_{ij}$	The cost of t_i on vm_j
f_i	The expected finish time of t_i
est_{ij}	The earliest start time of t_i on vm_j
$allo(t_i)$	The number of vm t_i is allocated to
$ET = (et_{ij})_{m \times n}$	The execution time matrix
$Cost = (cost_{ij})_{m \times n}$	The cost matrix
$EST = (est_{ij})_{m \times n}$	The earliest start time matrix of t_i

$ET = (et_{ij})_{m \times n}$ and $Cost = (cost_{ij})_{m \times n}$ refer to the execution time and cost of t_i executed on vm_j, respectively. Besides, tasks in this paper are deadline-sensitive. Thus, we calculate the earliest start time of a task on vm to which the task allocated. We use matrix $EST = (est_{ij})_{m \times n}$ to denote the earliest start time of task t_i allocated to vm_j. Then, our goal is to rationally and efficiently allocate the m tasks to the n vms to minimize task makespan, balance the loads and maximize the profit of cloud service provider under deadline constraint.

2.2 Problem Formulation

The total execution time for m tasks can be defined as

$$Time = \sum_{i=1}^{m} et_{i,allo(t_i)} \tag{1}$$

Similarly, the total cost for m tasks can be express as:

$$Tot\_Cost = \sum_{i=1}^{m} cost_{i,allo(t_i)} + pena \times fail\_num \tag{2}$$

where $pena$ denotes the penalty cost of a fail task and $fail\_num$ refers to total failure tasks. In order to approach the realistic operation cost of providers, we introduce this simple punishment mechanism, i.e. the right part of formula (2).

Due to the limited capacity a vm has, load balance should be considered in a superior task scheduling algorithm. Here, the degree of load balance can be represented as:

$$Bal = \sum_{i=1}^{n} |b_i - \bar{b}| \tag{3}$$

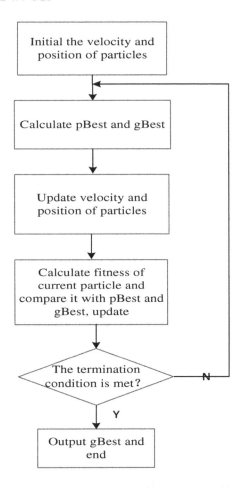

Fig. 1. Standard PSO flowchart.

where b_i, the load of vm_i, denotes the time that vm_i successfully executes all tasks in its task queue. $\bar{b}$ denotes the average load of vms in the system.

Finally, the failure ratio of task scheduling can be expressed as:

$$FR = \frac{fail\_num}{m} \qquad (4)$$

where m is the total number of tasks. According to the above description, the problem can be formulated as follows.

$$Minimize(Time, Tol\_Cost, Bal, FR)$$
$$S.T. \forall t_i \in T, est_{i,allo(t_i)} + et_{i,allo(t_i)} \leq d_i \quad (if\ y_i = 1) \qquad (5)$$

where y_i equal to 1 means that task t_i was executed successfully. Otherwise, task t_i failed.

3 Proposed Algorithm

3.1 Particle Swarm Optimization Flowchart

Figure 1 is a standard Particle swarm optimization (PSO) flowchart. PSO is a relatively recent heuristic optimization approach proposed by Kennedy and Eberhart [14]. Ease of implementation and fast convergence are the advantages of PSO over many other optimization algorithms. A great number of experimental results show that PSO can solve nearly all kinds of optimization problems that can be solved by genetic algorithm and other optimization approaches, thus it is indeed a powerful and vital optimization tool.

3.2 Particle Encoding and Fitness Function Design

Task allocation and scheduling is a NP-Complete problem. As the description in last section, task scheduling in cloud is also a multi-objective optimization problem. In this paper, PSO is tailored to address task scheduling problem in cloud environment, i.e. discrete particle swarm optimization (DPSO). We adopt binary encoding by using an array X and V to represent position and velocity of a particle [15, 16]. Both X and V have $m \times n$ elements.

$$x_{ij} = \begin{cases} 1, & if\ i^{th}\ task\ can\ executed\ on\ j^{th}\ vm \\ 0, & else \end{cases} \tag{6}$$

At each iteration, the position and velocity of particles are updated according to the following formulas:

$$V^{t+1}(i) = wV^t(i) + c_1 r_1 (pBest(i) - X^t(i)) \\ + c_2 r_2 (gBest - X^t(i)) \tag{7}$$

$$X^{t+1}(i) = \begin{cases} 1, & if\ rand() < sigmoid(V^{t+1}(i)) \\ 0, & else \end{cases} \tag{8}$$

where t refers to the current iteration times, $X(i)$ and $V(i)$ is the position and velocity of i-th particle, $pBest(i)$ is the best position of particle i, while $gBest$ is the best position of whole population, w is inertia weight, c_1 and c_2 are acceleration factors, r_1 and r_2 represents two numbers randomly generated in the range of [0,1], $rand()$ is the random function for generating random number in the range of [0,1], and $sigmoid(x) = 1/(1 + e^{-x})$ is data normalized function which maps large number onto smaller one.

An appropriate inertia weight w can strike a better balance between global and local search. Therefore, we manipulate DPSO with a linearly descending method [17]:

$$w = w_{\max} - (w_{\max} - w_{\min}) \times \frac{Cur\_iter}{Max\_iter} \tag{9}$$

where $w_{\max}$ and $w_{\min}$ are maximal and minimal value of w, respectively. Cur_iter and Max_iter refer to the current iteration times and maximal iteration times, respectively.

In Sect. 2, we have described that task allocation in cloud environment is a multi-objective optimization problem. In this paper, we transform the problem into a single-objective optimization problem by means of weighted sum approach. The fitness function is shown as follows.

$$Fitness = \alpha \times Time \times (1 + FR) + \beta \times Bal$$
$$+ \gamma \times Tot\_Cost \tag{10}$$

where α, β and γ are weighted parameters. And $\alpha + \beta + \gamma = 1$.

3.3   DPSO-Based Task Scheduling Algorithm Description

Algorithm 1 shows the pseudo code of DPSO-based Task Scheduling procedure. i refers to the serial number of current particle. j and m denotes the serial number of task and the total number of tasks, respectively. $cur\_particle\_fit$ represents the fitness value of the current particle. The smaller the value is, the better position of current particle is (according to Eq. 10). Generally, the termination condition is the maximum iteration times or a good enough solution. If the termination condition is met, the solution will be obtained and the run is terminated. Algorithm 2 is the pseudo code of Allocate_task operation in Algorithm 1. par_num is serial number of current particle. $task\_num$ is serial number of task which is waiting for allocating. And vm_num is serial number of virtual machine which is waiting for choosing. The algorithm chooses a vm with better comprehensive capacity to perform the task. Then, it updates needed informations and puts the task into task queue of the vm choosed. If there is no vm to undertake the task, then, vm_num is set as -1.

Algorithm 3 is the pseudo code of Choose_vm operation in Algorithm 2. In order to improve resource utilization and maximize the profit of cloud service provider within deadline constraint, a utility function is employed to allocate tasks to vm with high performance. Here, the weighted sum approach is employed as well. We allocate a task to the vm according to the formula below.

$$U(i, j) = wt1 \times UB(i, j) + wt2 \times UT(i, j)$$
$$+ wt3 \times UC(i, j); \tag{11}$$

where wt_1, wt_2 and wt_3 are weighted parameters, $U(i, j)$ is the utility function of task, the smaller the value of $U(i, j)$ is, the better vm_i execute t_j comprehensively. $UB(i, j)$, $UT(i, j)$ and $UC(i, j)$ denote the load degree, time consumption degree and cost degree of vm_i which executing t_j, respectively compared with other vms which take part in t_j. The current load b_i, time consumption et_{ji} and $cost_{ji}$ of vm_i are mapped in the range of 0 to 1 by using a linear data standardization function, $f(x) = (x - Min)/(Max - Min)$.

4   Simulation Results and Analysis

4.1   Experiment Setup

A distributed computing environment is simulated to evaluate the performance of our proposed scheduling algorithm. We compare real-time task scheduling

---

Algorithm 1. DPSO-Based Task Scheduling Algorithm

 Input: $Task, Particles$
 Output: $gBest$
1 **foreach** *particle* P_i **do**
2 | $pBest_i = Generate\_initial\_position(P_i)$

3 **foreach** $pBest$ *of particle* P_i **do**
4 | $gBest = Max(pBest_1, pBest_2, ...)$

5 **repeat**
6 | $j \leftarrow 1$;
7 | **while** $j \leq m$ **do**
8 | | *Select* the task t_j;   `/* has sorted by priority */`
9 | | $Calculate\_est(t_j)$;
10 | | $Allocate\_task(t_j)$;
11 | | $j++$;
12 | **foreach** *particle* P_i **do**
13 | | $Calculate\ cur\_particle\_fit$;
14 | | **if** $cur\_particle\_fit < pBest_i\_fit$ **then**
15 | | | $Update(pBest_i)$;
16 | | **if** $cur\_particle\_fit < gBest\_fit$ **then**
17 | | | $Update(gBest)$;
18 | **if** *the termination condition is met* **then**
19 | | Output $gBest$; Break;
20 | **else**
21 | | **foreach** *particle* P_i **do**
22 | | | $Update(P_i\_velocity)$;
23 | | | $Update(P_i\_position)$;
24 **until** *the termination condition is met*;

---

algorithm based on particle swarm optimization (RTA-PSO) approach with real-time task scheduling algorithm based on genetic algorithm (RTA-GA) which is redesigned from look-ahead genetic algorithm (LAGA) [9]. Min-Min [6,18] algorithm is a traditional heuristic approach on task allocation. To further verify the effectiveness of RTA-PSO, we extend Min-Min algorithm (name it RTA-MM) and compare it with our proposed algorithm. Generally, the parameters of simulation are set as follows: both values of et_{ij} and $cost_{ij}$ are selected from a uniform distribution in [50, 350]. α, β and γ are set as 0.3, 0.2 and 0.5, respectively. $wt1$, $wt2$ and $wt3$ are set as 0.1, 0.4 and 0.5. Max_iter, w_{max} and w_{min} are set as 50, 0.9 and 0.4. c_1 and c_2 are set as 2. $pxover$ and $pmutate$ are set as 0.8 and 0.15, respectively. In this paper, we compare RTA-PSO with RTA-GA and RTA-MM on deadline missing ratio (DMR), task makespan, total cost and load balance degree.

Algorithm 2. Allocate_task operation

Input: $task\_num, par\_num$
Output: vm_num
1 $vm\_num \leftarrow -1$;
2 Task t;
3 $t \leftarrow task[task\_num]$;
4 $vm\_num \leftarrow Choose\_vm(par\_num, task\_num)$;
5 **if** $(vm\_num == -1)$ **then**
6     $cur\_failnum[par\_num] + +$;
7     $particle\_scheme[task\_num][vm\_num] \leftarrow -1$;
8   return;
9 **else**
10   *Update* the *makespan* of current particle;   /* update information */
11   *Update* the *workload* of current particle;
12   *Update* the *cost* of current particle;
13   *Update* information of task t;
14   *Put* t into vm's task list;

Algorithm 3. Choose_vm operation

Input: $task\_num, par\_num$
Output: vm_num
1 int $ret\_num \leftarrow -1$;
2 **foreach** *Virtual Machine* VM_i **do**
3   *Find* Max_Cost$(VM_i, task\_num)$;
4   *Find* Min_Cost$(VM_i, task\_num)$;
5   *Find* Max_Load$(VM_i, task\_num)$;
6   *Find* Min_Load$(VM_i, task\_num)$;
7   *Find* Max_Exec_Time$(VM_i, task\_num)$;
8   *Find* Min_Exec_Time$(VM_i, task\_num)$;
9 **foreach** *Virtual Machine* VM_i **do**
10   *Calculate* U$(i, task\_num)$;   /* according to formula(11). */
11 **foreach** *Virtual Machine* VM_i **do**
12   *Find* minimal U$(i, task\_num)$; $ret\_num = i$;
13 return ret_num;

4.2 Effect of Task Deadlines

First, we carry out a group of experiments to observe the performance impact of task deadlines on RTA-PSO, RTA-GA and RTA-MM. Without loss of generality,

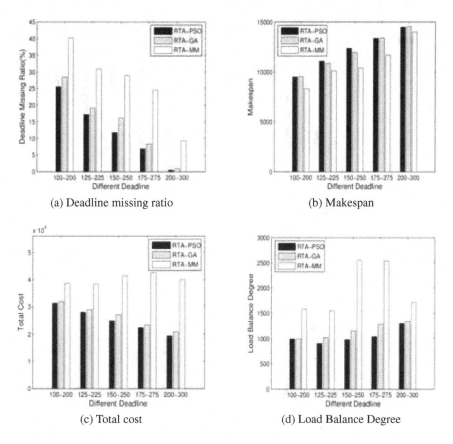

(a) Deadline missing ratio

(b) Makespan

(c) Total cost

(d) Load Balance Degree

Fig. 2. Effect of task deadline.

the number of tasks is set as 200 and the deadline of tasks is distributed uniformly in [100, 200], [125, 225], [150, 250], [175, 275] and [200, 300] for 5 times experiments.

From Fig. 2(a), (b), (c), and (d), we can find that the deadline missing ratio of three algorithms are dropping with the increasing of deadline time. RTA-MM only chooses the vm which has the smallest execution time to perform a task. Thus, RTA-MM has the best performance on task makespan, but performs not well on other indexes. RTA-PSO strikes a good balance between global exploration and local exploitation and well balances the multiple objectives. Therefore, it performs better on each index especially deadline missing ratio. RTA-GA performs better than RTA-MM on every metric except makespan, but there is still a gap between RTA-GA and RTA-PSO. Because, RTA-GA weeds out too many individuals making diversity of the population more and more monotonous and likely to get stuck into local optimal.

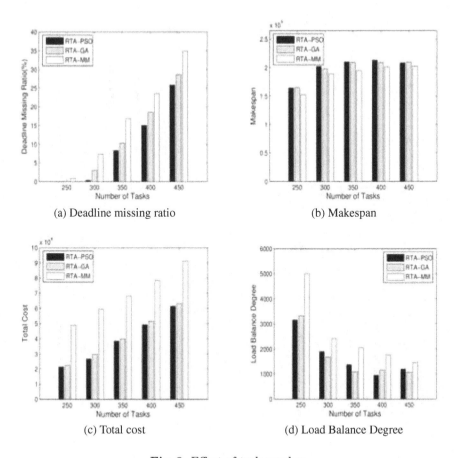

Fig. 3. Effect of task number.

4.3   Effect of Task Number

To observe performance impact of different number of tasks, without loss of generality, sixty vms are tested with 250, 300, 350, 400, 450 tasks, respectively. The deadline of each task randomly distributed over [200, 300].

As Fig. 3(a), (b), (c), and (d) shown, regarding makespan, RTA-MM is affected most with task number increment while RTA-PSO is affected the least and RTA-GA is medium. RTA-PSO performs well on each metric especially deadline missing ratio and total cost. RTA-GA is closed to RTA-PSO except deadline missing ratio. However, RTA-MM performs not well on every index.

5   Conclusion

This paper presents a novel soft real-time task scheduling algorithm based on particle swarm optimization approach (RTA-PSO) for cloud computing. To maximize profit of cloud service provider, minimize deadline missing ratio and makespan, and well balance workload, we construct a DPSO method to solve this multi-objective problem. Meanwhile, a utility function is designed to exploit the capacity of cloud resources more thoroughly. To evaluate the performance of RTA-PSO, we conduct extensive simulations to compare RTA-PSO with RTA-GA and RTA-MM. The simulation experiments show the proposed algorithm is effective. In the future, we will focus on providing fault-tolerant mechanism to improve the reliability of cloud system.

Acknowledgment. Thank anonymous reviewers for their valuable suggestions. This work is partly supported by the National Natural Science Foundation of China under Grant No. 61103175, the Fujian Province Key Laboratory of Network Computing and Intelligent Information Processing Project under Grant No. 2009J1007.

References

1. Buyya, R., Garg, S.K., Calheiros, R.N.: SLA-oriented resource provisioning for cloud computing: challenges, architecture, and solutions. In: International Conference on Cloud and Service Computing, Hong Kong, China, pp. 1–10 (2011)
2. Rodriguez, M.A., Buyya, R.: Deadline based resource provisioningand scheduling algorithm for scientific workflows on clouds. IEEE Trans. Cloud Comput. **2**(2), 222–235 (2014)
3. Siddhisena, B., Warusawithana, L., Mendis, M.: Next generation multi-tenant virtualization cloud computing platform. In: IEEE 13th International Conference on Advanced Communication Technology, Seoul, Korea, pp. 405–410 (2011)
4. Zuo, X.Q., Zhang, G.X., Tan, W.: Self-adaptive learning PSO-based deadline constrained task scheduling for hybrid IaaS cloud. IEEE Trans. Autom. Sci. Eng. **11**(2), 564–573 (2014)
5. Chang, R.S., Lin, C.Y., Lin, C.F.: An adaptive scoring job scheduling algorithm for grid computing. Inf. Sci. **207**(10), 79–89 (2012)
6. Shivle, S., Castain,R., Siegel, H.J., et al.: Static mapping of subtasks in a heterogeneous ad hoc grid environment. In: Proceedings of the 18th International Parallel and Distributed Processing Symposium (2004)
7. Liu, Z., Wang, X.: A PSO-based algorithm for load balancing in virtual machines of cloud computing environment. In: Tan, Y., Shi, Y., Ji, Z. (eds.) ICSI 2012, Part I. LNCS, vol. 7331, pp. 142–147. Springer, Heidelberg (2012)
8. Ramezani, F., Lu, J., Hussain, F.K.: Task-based system load balancing in cloud computing using particle swarm optimization. Int. J. Parallel Program. **42**(5), 739–754 (2014)
9. Wang, X.F., Yeo, C.S., Buyya, R., et al.: Optimizing the makespan and reliability for workflow applications with reputation and a look-ahead genetic algorithm. Future Gener. Comput. Syst. **27**(8), 1124–1134 (2011)
10. Zhu, X.M., Yang, L.T., Chen, H.K.: Real-time tasks oriented energy-awarescheduling in virtualized clouds. IEEE Trans. Cloud Comput. **2**(2), 168–180 (2014)
11. Beegom, A.S.A., Rajasree, M.S.: A particle swarm optimization based pareto optimal task scheduling in cloud computing. In: Tan, Y., Shi, Y., Coello, C.A.C. (eds.) ICSI 2014, Part II. LNCS, vol. 8795, pp. 79–86. Springer, Heidelberg (2014)

12. Ramezani, F., Lu, J., Hussain, F.: Task scheduling optimization in cloud computing applying multi-objective particle swarm optimization. In: Pautasso, C., Zhang, L., Fu, X., Basu, S. (eds.) ICSOC 2013. LNCS, vol. 8274, pp. 237–251. Springer, Heidelberg (2013)

13. Guo, L.Z., Shao, G.J., Zhao, S.G.: Multi-objective task assignment in cloud computing by particle swarm optimization. In: IEEE International Conference on Wireless Communications, Networking and Mobile Computing, Shanghai, China, pp. 1–4 (2012)

14. Kennedy, J., Eberhart, R.C.: Particle swarm optimization. In: Proceedings of IEEE Int'l Conference on Neural Networks, Piscataway, NJ, pp. 1942–1948 (1995)

15. Guo, W.Z., Gao, H.L., Chen, G.L., Yu, L.: Particle swarm optimization for the degree-constrained MST problem in WSN topology control. In: The International Conference on Machine Learning and Cybernetics, Baoding, China, pp. 1793–1798 (2009)

16. Guo, W.Z., Xiong, N.X., Vasilakos, A.V., et al.: Distributed k-connected fault-tolerant topology control algorithms with PSO in future autonomic sensor systems. Int. J. Sens. Netw. **12**(1), 53–62 (2012)

17. Shi, Y., Eberhart, R.C.: A modified particle swarm optimizer. In: Proceedings of the 1998 IEEE International Conference on Evolutionary Computation, Anchorage, Alaska, pp. 69–73 (1998)

18. Tian, Y., Boangoat, J., Ekici, E., et al.: Real-time task mapping and scheduling for collaborative in-network processing in DVS-enabled wireless sensor networks. In: Proceedings of the 20th International Parallel and Distributed Processing Symposium, Island, Greece (2006)

Making GPU Warp Scheduler and Memory Scheduler Synchronization-Aware

Jianliang Ma[1]([✉]), Tianzhou Chen[1], and Minghui Wu[2]

[1] College of Computer Science Zhejiang University,
Hangzhou, Zhejiang, People's Republic of China
{majl,tzchen}@zju.edu.cn
[2] Zhejiang University City College,
Hangzhou, Zhejiang, People's Republic of China
mhwu@zucc.edu.cn

Abstract. Modern GPU applications often need to synchronize thousands of threads for correctness. The warp scheduling algorithm, memory coalescing and memory scheduling algorithm etc. may cause different execution schedules for the warps in the same Cooperative Thread Array (CTA). So when synchronization happens, waiting is required and synchronization cost is introduced. In this paper, we examine the synchronization cost of multiple GPU applications in three metrics. With synchronization information in CTA boundary, the warps still running in the CTA can know their lagging degrees. We promote the warp scheduling priority and memory scheduling priority for these warps and their memory requests to accelerate the execution speed of these warps, making warp scheduler and memory scheduler synchronization-aware. The experiments show that the synchronization-aware warp scheduling algorithm reduces the synchronization metrics to 86.66 %, 92.12 % and 85.63 % compared with the baseline and improves the GPU performance by 5.76 %. For memory intensive benchmarks, the synchronization-aware memory scheduling algorithm improves the system performane by 6.81 %. The combination of these two schedulers can further improve the GPU performance by 6.46 %.

1 Introduction

The development of general purpose Graphics Processing Units (GPGPUs) [15] and high-level parallel programming models such as CUDA [21] and OpenCL [24] have led to the increasing adoption of the GPU for running data parallel workloads. The Single-Instruction, Multiple-Thread (SIMT) nature of GPUs makes them run a group of threads (warp in CUDA terminology or wavefronts in OpenCL) in lockstep. So the threads in the warp have the same execution progress. But this is not sure for Cooperative Thread Array (CTA). When a CTA reaches a synchropoint, that is, at least one warp of the CTA has reached the synchropoint, we call the CTA as *Synchronization waiting CTA* (Sw-CTA). Otherwise, it is called *non-Sw-CTA*. Since the warps in a CTA often run unevenly,

© Springer International Publishing Switzerland 2015
W. Qiang et al. (Eds.): CloudCom-Asia 2015, LNCS 9106, pp. 153–164, 2015.
DOI: 10.1007/978-3-319-28430-9_12

the first arrived warps have to wait for the rest warps in the Sw-CTA, causing *synchronization cost*. Besides, current programming models such as CUDA
and OpenCL only support the thread synchronization in CTA boundry. Since
multiple CTAs can be assigned and run on a Streaming Multiprocessor (SM)
simultaneously, the warps of these CTAs are interleaved to increase resource
utilization and instruction throughput. So the synchronization of the warps in a
CTA is further disturbed.

We observe that the synchronization cost of different synchropoints and different applications varies a lot. Large synchronization cost will cause severe dis-
ruption to applications performance. To our knowledge, although several work
have considered the synchronization [10,12,16], they never attempt to optimize
its cost. In this paper, we try to reduce the synchronization cost by warp scheduling optimization and memory scheduling optimization, using *synchronization
information* (sync-info). The sync-info is appended in each *Synchronization waiting warp* (Sw-warp). The Sw-warps in the SM are prioritized and are *warp
scheduling synchronization-aware* (sync-w). So the time for the synchronization
can be reduced. If a Sw-warp executes a memory instruction, then each of the
memory requests produced by this warp is called *Synchronization waiting request*
(Sw-request). The sync-info is also appended in each Sw-request to promote their
priority. So the memory access time of the Sw-request can also be reduced by
making the *memory scheduling synchronization-aware* (sync-m).

The experimental results show that by using sync-w, the synchronization
cost is reduced to 86.66 %, 92.12 % and 85.63 %, according to the three metrics.
By using sync-m, the synchronization cost is reduced to 71.09 %, 80.24 % and
69.69 % for the memory intensive applications. They both improved the GPU
performance by 5.76 % and 2.81 % in average, respectively. Finally, the combination of these two optimization further improves the GPU performance.

2   Background and Methodology

2.1   Baseline GPU Architecture

Our baseline GPU architecture consists of multiple SMs. Each SM has several
private caches and a software managed shared memory. All SMs connect to a
shared L2 through an on-chip interconnection network. The L2 is banked and
each bank corresponds to a memory channel. Each memory channel has an independent GDDR5 memory scheduler. We use GPGPU-Sim (v.3.2.2) [3], a cycle
accurate GPU Simulator to fulfill a Fermi-like GPU architecture as baseline. The
detailed configuration parameters of our baseline GPU are showed in Table 1.

GPU Thread Synchronization: CUDA applications call the __syncthreads()
function when coordinated communication between the threads of the same
CTA is needed to avoid data hazards. Once the function is called, the threads
have to stall at the barrier until the rest threads in the same CTA arrive.

Warp Scheduling Algorithm: A greedy-then-oldest (GTO) warp scheduler
is employed in the baseline GPU. GTO runs a single warp until it stalls and

Table 1. Baseline GPU architecture configuration

#SM	15 * 2
SM configuration	700 MHz, SIMT width = 32
Resources per SM	Max. 1536 threads (48 warpsSM), 48 KB shared memory, 32768 registers
Caches per SM	16 KB 4-way L1 data cache, 12 KB 24-way texture cache, 8 KB 2-way constant cache, 2 KB 4-way I-cache, 128 B cache block size
Warp scheduling	Greedy-then-oldest (GTO) [23]
Interconnection	1 dimension butterfly (30 SMs + 6 L2 Banks), 700 MHz
L2 Cache	768 KB 8-way 6 banked L2 shared cache, 120-cycle latency
Memory model	6 GDDR5 Memory Controllers (MCs), FR-FCFS scheduling 16 DRAM-banks/MC, 924 MHz memory clock, 340-cycle latency
Hynix GDDR5 timing [9]	$t_{CL} = 12$, $t_{RP} = 12$, $t_{RC} = 40$, $t_{RAS} = 28$, $t_{CCD} = 2$, $t_{RCD} = 12$, $t_{RBD} = 6$

then it picks the oldest warp. The age of a warp is determined as the time it is assigned to the SM.

Memory Scheduling Algorithm: Commonly used First-ready FCFS (FR-FCFS) [22,25] is used as our baseline GPU memory scheduling algorithm. This scheme is targeted at improving DRAM row hit rates, so (1) row-hit requests are prioritized over other requests; (2) older requests are prioritized over younger requests.

2.2 Benchmarks

Since not all GPU applications need thread synchronization, we only choose the synchronization demanded applications in this paper. In total, we employ 16 applications from suite [3], CUDASDK, Rodinia application suite [5], and Mars [8] as listed in Table 2. Each application runs at most 1 billion GPU instructions.

Table 2. Benchmarks

Suite	Applications (abbreviations)
ISPASS09	AES, LPS, NQU, RAY
CUDASDK	dxtc(DXTC), matrixMUL(MM)
Rodinia	Backprop(BP), b+tree(B+), heartwall(HW), hotspot(HS), LUD, pathfinder(PF), srad(SRAD)
Mars	Kmeans(KM), PageViewCount(PVC), SimilarityScore(SS)

3 Motivation

3.1 Synchronization Cost Sources

There are multiple runtime factors make large CTA synchronization cost. Firstly, the branch instructions make warp divergence that executing different directions, leading to asynchronous warp execution. Secondly, warp scheduling algorithms usually interlace warps of all CTAs to increase throughput, so the warps in a CTA will run discontinuously. Thirdly, current memory schedulers do not distinguish the Sw-request and non-Sw-request, so they are processed in an interleaved manner.

3.2 Synchronization Cost Analysis

To represent the synchronization cost accurately, we propose three metrics, namely *Worst Synchronization Waiting Time (WSWT)*, *Average Synchronization Waiting Time (ASWT)* and *Average Synchronization Cost (ASC)*. For a GPU application, we assume that it has SN synchropoints, n_i is the warp count on ith synchropoint, W_{ij} is the jth warp in ith synchropoint. Then W_{i0} is the first warp that reaches ith synchropoint and W_{in_i} is the last warp that reaches ith synchropoint. Then WSWT, ASWT and ASC can be calculated as Eqs. 1, 2 and 3, respectively, where $W_{in_i} - W_{i0}$ is the synchronization cost of synchropoint n_i.

$$WSWT = MAX(W_{in_i} - W_{i0}) \tag{1}$$

$$ASWT = \frac{1}{SN} \sum_{i=0}^{SN} (W_{in_i} - W_{i0}) \tag{2}$$

$$ASC = \frac{1}{SN} \sum_{i=0}^{SN} \sum_{j=0}^{n_i} (W_{in_i} - W_{ij}) \tag{3}$$

Figure 1 shows the ASWT, WSWT and ASC of the benchmarks. We find that different benchmarks have various ASWT, WSWT and ASC. For example, ASWT value varies from 35.46 of NQU to tens of thousands of PF and SS. The differences are also huge for WSWT and ASC. The closer the value of WSWT and ASWT, the more balance the synchronization lies. MM and LUD are of this kind as their kernels are relatively regular. In this paper, we focus on the warp scheduling algorithm and the memory scheduling algorithm to reduce the synchronization cost.

4 Design Methodology

4.1 Synchronization-Aware Warp Scheduling (sync-w)

The idea of sync-w is reducing the synchronization cost by prioritizing the Sw-warps. When a warp reaches a synchropoint, the barrier will record the number

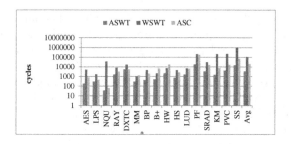

Fig. 1. ASWT, WSWT and ASC of GPU applications

of warp that reached the synchropoint in the sync-info. For a warp that reachs the synchropoint, we say it is a Sw-warp and the CTA it belongs to is a Sw-CTA. Since sync-w only concerns the Sw-warps, it has to depend on a current one. We will see that this is also true for sync-m. To demonstrate the idea of sync-w, we use GTO as baseline warp scheduling algorithm where the warps are sorted according to their initialized time.

We add an ordering rule on GTO for sync-w to prioritize the Sw-warps in scheduling. First, higher sync-info Sw-warps are prioritized over lower Sw-warps and all Sw-warps are over other warps; Secondly, older warps are prioritized over other warps. And for the Sw-warps of same sync-info, older warps are prioritized over other warps. Figure 2 shows scheduling examples of baseline GTO scheduler and sync-w scheduler. Wx_y in the figure stands for the yth warp of CTA x. There are two CTAs (CTA a and CTA b) in the SM, both of them consist of two warps (Warp 0 and Warp 1). At time T1, Wb_0 arrives at a synchropoint and is blocked. In GTO warp scheduler, the warp with oldest age Wa_0 will be issued next. Wb_0 can be resumed faster when Wb_1 arrives at the synchropoint earlier. In sync-w, Wb_1 becomes a Sw-warp. By prioritizing Wb_0, the sync-w scheduler achieves higher ready warps than that of GTO scheduler.

Higher ready warps usually provide higher possibility to achieve higher performance. For example, when all the warps are waiting for long operations, sync-w is more hopeful to overlap the latency with more ready warps to issue. Since all Sw-warps in the same CTA have the same value of sync-info, they will be issued in short time interval, reducing synchronization cost. A side-effect of sync-w is that it may delay other critical warps and disturb following dependent warps. The experiment shows that this may even degrade the performance in very few benchmarks. But in most cases, the sync-w is beneficial.

4.2   Making Memory Scheduling Synchronization-Aware

The idea of sync-m is prioritizing the Sw-requests to accelerate the data fetch speed for Sw-warps. When a memory instruction is executed, we append the sync-info in each generated memory request. The sync-info is only for global memory but not used in the cache system currently. The memory scheduler in the memory controller will order the requests according to their sync-info value. We implement the sync-m based on FR-FCFS. The larger value of sync-info denotes that the Sw-request is more emergent and it should have higher priority. So the sync-m's ordering rules as: (1) row-hit requests are prioritized over other requests; (2) requests with higher sync info value are prioritized over other requests; (3) older requests are prioritized over younger requests.

Figure 3(A) and (B) illustrates scheduling examples of baseline scheduler and sync-m scheduler, respectively. Each block in the figure represents a warp, in which gray block denotes Sw-warp and white block denotes other warp. Figure 3(A) shows that the baseline memory scheduler processes the requests in an interleaved manner in order to maximum memory throughput. However, as warp A_n is a Sw-warp but warp B is not, the long service latency of warp A_n will lead warp A_1 to A_{n-1} to stall as well and it does not apply to warp B. Contrarily speaking, reducing the stall time of warp $A1$ to warp A_n at the expense

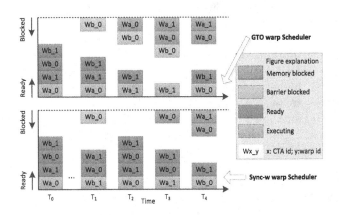

Fig. 2. Sync-aware warp scheduler operation example

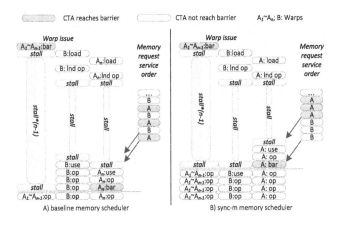

Fig. 3. Sync-aware memory scheduler operation example

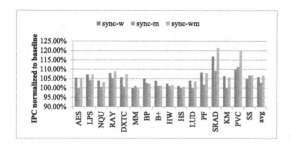

Fig. 4. Performance of sync-w, sync-m and sync-wm

of increasing the stall time of warp$B$ is beneficial. So in sync-m (as shown in Fig. 3(B)), the requests of warpA_n are scheduled and serviced before warp$B$. In addition, in stead of interleaving requests from different warps, sync-m substantially squeezes the requests of warpA_n and further reduces the memory stall of warpA_1 to warpA_n.

A side effect of sync-m is reducing the inter-warp interference. The service interval of a warp is the number of memory requests of other warps scheduled when servicing the requests from this warp [4]. The smaller the interval is, the less inter-warp interference there will be, but the memory bandwidth utilization falls at the same time. Assume that the average value of this interval of FR-FCFS is m. Then WG [4] keeps it 0 by strictly grouping the requests of a warp together. The service interval in sync-m is between 0 and m since all memory requests with the same sync-info value are grouped together.

5   Experimental Results

5.1   Performance Gain

Figure 4 illustrates the sync-w performance. We can see that the performance improvement lies between -0.03% and 16.89%, with 5.76% in average. The most improved applications are SRAD and PVC, with 16.89% and 9.71% improvement, respectively. That is because: (1) SRAD and PVC have most synchronization counts. (2) The synchronization cost of them is also high as shown in Fig. 1. (3) The warps_in_CTA value is 8 in both applications, which in not small. The worst performance occurs in MM and HS, which even decreases compared with the baseline. One reason is that MM and HS have already low synchronization cost in baseline.

Cache system weakens sync-w efficiency as it diminishes the amount of Sw-requests arriving at MC. The Fig. 5 shows that *LPS, BP, SRAD, PVC* and *SS* have average memory queue length larger than four, and are memory intensive benchmarks. Other benchmarks are memory non-intensive benchmarks. Memory insensitive applications do not suffer from bad memoy scheduling algorithm. Any memory scheduling algorithm will degrade into FCFS for memory non-intensive benchmarks.

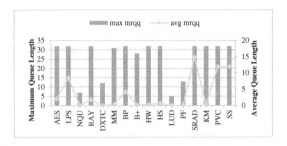

Fig. 5. Maximum and average of memory requests queue length

Sync-wm is the combination of sync-w and sync-m. We think its performance should be higher than both sync-w and sync-m, as they mutually promote each other. For example, MM and HS has low performance in sync-w, but sync-m makes this up, so the performance of sync-wm is better than that of sync-w and even better than that of baseline. The highest and overall performance of sync-wm is higher than sync-w and sync-m in general. The experiment result shows that the performance gain of sync-wm is between 0.02 % and 21.47 %, with 6.46 % in average.

5.2   Impact on Synchronization Cost

Figure 6 is the ASWT, WSWT and ASC with sync-w and sync-m normalized to baseline result in Fig. 1. Since sync-m only improves LPS, BP, SRAD, PVC and SS, we only list their changes in Fig. 6. For sync-w, we can see from the figure that most benchmarks reduce their ASWT, WSWT and ASC compared to baseline. The averages are 86.66 %, 92.12 % and 85.63 % respectively. For WSWT, some benchmarks are larger than baseline because Sw-warps with higher sync-info make low Sw-warps delayed. However, as WSWT only stands for the worst case instead of whole situation as ASWT and ASC does, we will pay more attention on the relationship between ASWT, ASC and IPC.

Generally speaking, the variation of synchronization cost is inversely related to the performance for all benchmarks by combing the result of Fig. 4. The more

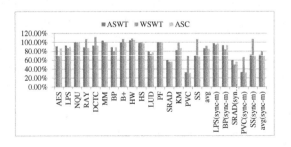

Fig. 6. Impact on synchronization cost, ASWT, WSWT and ASC

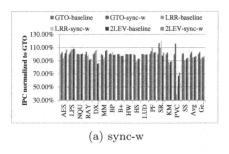
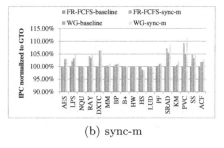

(a) sync-w   (b) sync-m

Fig. 7. Scalability of sync-w and sync-m

synchronization cost reduced, the more performance gained, such as SRAD and PVC. MM and HS increase their ASWT and ASC, so their performances show degradation. One exception is LUD. The synchronization cost of LUD reduces obviously, but its IPC changes little. We find the performance bottleneck of LUD is the 31 kernels that tested warps_per_CTA to be 1, which our sync-w cannot improve. For the rest 15 kernels with warps_per_CTA as 8, their performances improve by about 7.04 %. The result of *LPS(sync-m)*, *BP(sync-m)*, *SRAD(sync-m)*, *PVC(sync-m)* and *SS(sync-m)* shows that the synchronization cost for sync-m is also reduced.

5.3   Scalability

In this section, we show the scalability of our idea with other scheduling algorithms. We test Loose Round-Robin (LRR) and two-level (2LEV) with sync-w and WG [4] with sync-m. LRR travels all the warps on the SM in round-robin to find the first ready warp. To make LRR synchronization-aware, it chooses the ready warp with highest sync-info. The 2LEV scheduler checks a warp window size of 6 in loosen round-robin, and exchange blocked warps in the window with warps out of the window in strict round-robin. It is similar to the scheduler described in Narasiman et al. [17] with a fetch group size of 8. To make 2LEV synchronization-aware, it prioritize the Sw-warps in the warp window.

Figure 7(a) shows the performance change after applying synchronization-aware with GTO, LRR and 2LEV warp scheduling algorithms. All the results are normalized to GTO-baseline. They all improve their performance by modest magnitudes with a similar performance trend. For example, *SRAD* and *PVC* improve most in the three warp scheduling algorithms. So we conclude that sync-w has good scalability. Figure 7(b) shows the performance of synchronization-aware FR-FCFS and WG. The baseline WG usually outperforms FR-FCFS, which is compatible with the result of [4]. We find that the performance variation trend is similar in FR-FCFS and WG. That is, the performance speedup by applying sync-m on FR-FCFS and WG are similar. So the result shows good scalability of sync-m in the two memory scheduling algorithms.

6   Related Works

Jog et al. [12] proposed a coordinated CTA-aware scheduling policy including warp scheduling cache utilization and prefetching, as well as DRAM bank-level parallelism. But it does not concerns the internal threads of CTA as in our work. Hsien-Kai et al. [16] focus on cache capacity-aware thread scheduling that minimize the synchronization cost without cache contention. The synchronization cost is only a constraint, which vary greatly from our work. Minseok et al. [17] introduce CTA scheduling ahead of warp scheduling and propose two CTA scheduling methods to leverage corresponding warp schedulers. Jablin et al. [10] proposed an automatic speculative global scheduling algorithm for GPU by creating constraints to preserve correntness for synchronization. But they have not considered the cost of synchronization.

Lee et al. [18] proposed a similar idea to solve the warp scheduling problem under the synchronization aware scenario. But we adopt a differnt way to identifying the warp criticality. Jianmin et al. [6] proposed a novel warp scheduling algorithm that flexibly uses the round-robin feature of time slice to utilize GPU parallelism. Narasiman et al. [20] proposed a two-level warp scheduler that splits the concurrently executing warps into groups to improve memory latency tolerance. Gebhart and Johnson et al. [7] proposed a two-level warp scheduling technique that focuses on reducing the energy consumption in GPUs. Kayiran et al. [14] modulated the available thread-level parallelism by intelligent CTA scheduling. Jog et al. [13] proposed a prefetching-aware warp scheduling policy to separate in time the scheduling of consecutive warps such that they are not executed back-to-back. Rogers et al. [23] proposed an efficient cache sensitive warp scheduling policy by limiting the number of actively-scheduled warps.

There is plenty of memory scheduling research for traditional CMP architectures. But in GPU/GPGPU field, the memory research is still in infancy. Lakshminarayana et al. [19] discussed the memory scheduling algorithms and executing characteristics of GPU applications and developed a DRAM scheduling policy according to the work in [2]. Chatterjee et al. [4] proposed a warp-group scheduling algorithm and optimized it on multi-channel, memory bandwidth and write-drain aspects. Besides, there also few works focus on CPU-GPU heterogeneous systems. Ausavarungnirun et al. [1] proposed an effective and practical 3-stage memory scheduler (SMS). Jeong et al. [11] proposed a QoS-aware memory controller for the GPU.

7   Conclusion and Future Work

The synchronization cost will influence the GPU application performance severely. In this paper, we promote a synchronization-aware warp scheduling algorithm sync-w and a synchronization-aware memory scheduling algorithm sync-m by prioritizing for Sw-warps and Sw-requests, respectively. The experiment shows that speed of Sw-warps is accelerated and the synchronization cost is reduced. In the future, we will discuss the cache system design method based on synchronization information. We believe that the concern of thread synchronization will grow in future GPU design.

Acknowledgements. This paper is supported by the National Natural Science Foundation of China under Grant No. 61379035, the National Natural Science Foundation of Zhejiang Province under Grant No. LY14F020005, Open Fund of Mobile Network Application Technology Key Laboratory of Zhejiang Province, Innovation Group of New Generation of Mobile Internet Software and Services of Zhejiang Province.

References

1. Ausavarungnirun, R., Chang, K.K.W., Subramanian, L., Loh, G.H., Mutlu, O.: Staged memory scheduling: Achieving high performance and scalability in heterogeneous systems. SIGARCH Comput. Archit. News **40**(3), 416–427 (2012)
2. Lakshminarayana, B.N., Lee, J., Kim, H., Shin, J.: Dram scheduling policy for GPGPU architectures based on a potential function. IEEE Comput. Archit. Lett. **11**(2), 33–36 (2012)
3. Bakhoda, A., Yuan, G., Fung, W., Wong, H., Aamodt, T.: Analyzing CUDA workloads using a detailed GPU simulator. In: 2009 IEEE International Symposium on Performance Analysis of Systems and Software, ISPASS 2009, pp. 163–174, April 2009
4. Chatterjee, N., O'Connor, M., Loh, G.H., Jayasena, N., Balasubramonian, R.: Managing dram latency divergence in irregular GPGPU applications. In: Proceedings of the International Conference for High Performance Computing, Networking, Storage and Analysis, SC 2014, pp. 128–139. IEEE Press, Piscataway (2014)
5. Che, S., Boyer, M., Meng, J., Tarjan, D., Sheaffer, J., Lee, S.H., Skadron, K.: Rodinia: a benchmark suite for heterogeneous computing. In: 2009 IEEE International Symposium on Workload Characterization, IISWC 2009, pp. 44–54, October 2009
6. Chen, J., Tao, X., Yang, Z., Peir, J.K., Li, X., Lu, S.L.: Guided region-based GPU scheduling: utilizing multi-thread parallelism to hide memory latency. In: 2013 IEEE 27th International Symposium on Parallel Distributed Processing (IPDPS), pp. 441–451, May 2013
7. Gebhart, M., Johnson, D.R., Tarjan, D., Keckler, S.W., Dally, W.J., Lindholm, E., Skadron, K.: Energy-efficient mechanisms for managing thread context in throughput processors. SIGARCH Comput. Archit. News **39**(3), 235–246 (2011)
8. He, B., Fang, W., Luo, Q., Govindaraju, N.K., Wang, T.: Mars: a mapreduce framework on graphics processors. In: Proceedings of the 17th International Conference on Parallel Architectures and Compilation Techniques, PACT 2008, pp. 260–269. ACM, New York (2008)
9. hynix: "hynix gddr5 sgram part h5gq1h24afr" (2009). www.hynix.com/datasheet/pdf/graphics/H5GQ1H24AFR(Rev1.0).pdf
10. Jablin, J.A., Jablin, T.B., Mutlu, O., Herlihy, M.: Warp-aware trace scheduling for GPUs. In: Proceedings of the 23rd International Conference on Parallel Architectures and Compilation, PACT 2014, pp. 163–174. ACM, New York (2014)
11. Jeong, M.K., Erez, M., Sudanthi, C., Paver, N.: A QoS-aware memory controller for dynamically balancing GPU and CPU bandwidth use in an MPSoC. In: Proceedings of the 49th Annual Design Automation Conference, DAC 2012, pp. 850–855. ACM, New York (2012)

12. Jog, A., Kayiran, O., Chidambaram Nachiappan, N., Mishra, A.K., Kandemir, M.T., Mutlu, O., Iyer, R., Das, C.R.: Owl: cooperative thread array aware scheduling techniques for improving GPGPU performance. SIGPLAN Not. **48**(4), 395–406 (2013)

13. Jog, A., Kayiran, O., Mishra, A.K., Kandemir, M.T., Mutlu, O., Iyer, R., Das, C.R.: Orchestrated scheduling and prefetching for GPGPUS. In: Proceedings of the 40th Annual International Symposium on Computer Architecture, ISCA 2013, pp. 332–343. ACM, New York (2013)

14. Kayiran, O., Jog, A., Kandemir, M., Das, C.: Neither more nor less: optimizing thread-level parallelism for GPGPUS. In: 2013 22nd International Conference on Parallel Architectures and Compilation Techniques (PACT), pp. 157–166, September 2013

15. Keckler, S., Dally, W., Khailany, B., Garland, M., Glasco, D.: Gpus and the future of parallel computing. IEEE Micro **31**(5), 7–17 (2011)

16. Kuo, H.K., Yen, T.K., Lai, B.C., Jou, J.Y.: Cache capacity aware thread scheduling for irregular memory access on many-core GPGPUs. In: 2013 18th Asia and South Pacific Design Automation Conference (ASP-DAC), pp. 338–343, January 2013

17. Lee, M., Song, S., Moon, J., Kim, J., Seo, W., Cho, Y., Ryu, S.: Improving GPGPU resource utilization through alternative thread block scheduling. In: 2014 IEEE 20th International Symposium on High Performance Computer Architecture (HPCA), pp. 260–271, February 2014

18. Lee, S.Y., Wu, C.J.: Caws: criticality-aware warp scheduling for GPGPU workloads. In: Proceedings of the 23rd International Conference on Parallel Architectures and Compilation, PACT 2014, pp. 175–186. ACM, New York (2014)

19. Lakshminarayana, N.B., Kim, H.: Workshop on Language, Compiler, and Architecture Support for GPGPU (2010)

20. Narasiman, V., Shebanow, M., Lee, C.J., Miftakhutdinov, R., Mutlu, O., Patt, Y.N.: Improving GPU performance via large warps and two-level warp scheduling. In: Proceedings of the 44th Annual IEEE/ACM International Symposium on Microarchitecture, MICRO-44, pp. 308–317. ACM, New York (2011)

21. NVIDIA: "nvidia cuda c programming guide v4.2" (2012). docs.nvidia.com/cuda/

22. Rixner, S., Dally, W.J., Kapasi, U.J., Mattson, P., Owens, J.D.: Memory access scheduling. SIGARCH Comput. Archit. News **28**(2), 128–138 (2000)

23. Rogers, T.G., O'Connor, M., Aamodt, T.M.: Cache-conscious wavefront scheduling. In: Proceedings of the 2012 45th Annual IEEE/ACM International Symposium on Microarchitecture, MICRO-45, pp. 72–83. IEEE Computer Society, Washington (2012)

24. Stone, J.E., Gohara, D., Shi, G.: OpenCL: a parallel programming standard for heterogeneous computing systems. IEEE Des. Test **12**(3), 66–73 (2010)

25. Robinson, T., Zuravleff, W.: Controller for a synchronous dram that maximizes throughput by allowing memory requests and commands to be issued out of order (1997). Google Patents

A Novel Grid Based K-Means Cluster Method for Traffic Zone Division

Yunpeng Zheng, Gang Zhao$^{(\boxtimes)}$, and Jian Liu

School of Information Management,
Beijing Information Science and Technology University, Beijing 100192, China
{zhengpeng911001,liujianspace999}@126.com,
zhaogang@bistu.edu.cn

Abstract. Traffic zone division plays an important role in analyzing traffic flow and the trend of city traffic. A traditional method based on sampling investigation has the shortcomings of high cost, long period and low sampling precision. With the development of traffic control and management methods, some cluster methods for location points are proposed to be used in the division of traffic plot. However, simple clustering analysis often need to detect the boundary of traffic zones, and the boundary of the division result is not clear, furthermore, abnormal data has great influence on results. In order to make the traffic zone division results clear and accurate and reduce the cost of the division, this paper proposes a novel grid based K-Means cluster method for traffic zone division by using Taxi GPS data. The experiment used GPS data of Nanjing city taxies and automatically divided traffic zones in Nanjing City area, which verified the validity of this partition method. The experimental results show that this classification method is effective and gives a good reference value for the analysis of city traffic flow and trends.

Keywords: Traffic zone · Cluster · K-Means · GPS · GIS · Grid division · Traffic trends

1 Introduction

Division of traffic zones is one of the main means to analyze the city traffic. The accurate identification of city traffic zones can help people to know urban road conditions and the distribution of both traffic resources and transportation demands.

GPS has become a powerful ubiquitous sensor in our daily life. Nowadays, it has been embedded into many smart phones, vehicles, and other devices. GPS devices are playing a front-line role to continuously create and record the digital footprints of the carrier. Many daily facets of the carrier can be inferred from these spatio-temporal data. Since the most commonly used GPS devices by average population are smart phones and vehicles, which are extensively deployed in metropolitan areas, they can thus be a very rich data source for understanding city dynamics and revealing hidden social and economic "realities". Previous works along this line include similar place identification (Mobile Landscapes) [1], mobility pattern analysis [2–4].

Public transportation is one of the most popular and important application fields of GPS. In modern cities, many public transportation vehicles, such as taxis, have been

© Springer International Publishing Switzerland 2015
W. Qiang et al. (Eds.): CloudCom-Asia 2015, LNCS 9106, pp. 165–178, 2015.
DOI: 10.1007/978-3-319-28430-9_13

equipped with GPS devices. In the beginning, GPS devices equipped in taxis were mainly used for localization, navigation, scheduling and planning. As more and more taxis are equipped with GPS sensors and wireless communication units, immense amount of taxi status and GPS trajectory data can be collected. Many interesting research issues have been explored based on the large-scale taxi GPS trajectories, such as hot-spots and traffic condition detection [5–7], and taxi mobility intelligence mining [8–11].

In this paper, we intend to combine the concepts of mesh and cluster, proposed the division method based on K-Means transportation grid cell and proved the effective-ness of the method with empirical study.

2   Related Work

2.1   The Concept of Traffic Zone

Transportation zone refers to several zones that merges many traffic sources on a comprehensive understanding of traffic because it is impossible to study traffic sources separately. There are man-made space delimited ranges to facilitate the relevant study instead of actual boundaries of the zones [12]. The traffic zone is a collection of nodes or connections that have a certain degree of traffic association and traffic similarity, and changes with the association and traffic similarity over time, reflecting the city road network traffic temporal variation characteristics [13]. When division of traffic zone was first proposed in the field of transportation planning, its main purpose was to define the location of the traffic start and end points of the city road network and predicted the amount of traffic between each traffic zones according to the forecasting model. The area and boundaries of traffic zone boundary delimitation will directly affect the traffic survey, analysis, workload and the accuracy of prediction.

2.2   Principles in Traffic Zone Division

Division of traffic zone is a good way to analyze city traffic network because different traffic zones have similar traffic characteristics and strong traffic relevance. Division of traffic zones is closely related to the city's population, size, economic characteristics, industrial structures, etc. [14], and reflects the attractiveness of a city to a certain extent. In general, the division of traffic zone should follow the following principles [15]:

1. Homogeneity, make the land use, economy, society and other characteristics in the same zone it consistent as much as possible.
2. Use a rail, river or other natural barriers as the partition boundaries.
3. Try not to break the division of administrative regions in order to take advantage of existing SAR government statistics.
4. Consider the road network structure, the district gravity center may be used as the junction of the road network.
5. Appropriate zones, medium-sized cities ≤ 50 months, large cities no more than l00 to 150.
6. Appropriate population, about 10,000 to 20,000 people.

2.3   Division Method

Considering characteristics and division principles of the traffic zones, domestic and foreign scholars proposed the following cell division method of transportation.

Control-oriented division of traffic zones. Control-oriented division of cell traffic is mainly to control the intersections in the same zones in order to improve the overall effectiveness of the urban road network. Based on the analysis of road network topology characteristics, the zone division algorithm is formed combining the dynamic and static division partitioning strategy [16].

The division of traffic zone based on proportion of region traveling. The main purpose of the traffic division is to describe different types of traffic and travel conditions and the flow of traffic sources, while traveling inside the region cannot reflect the flow of traffic sources, and therefore scientific and reasonable traffic cell radius is needed to control region traveling proportion to an appropriate extent, in order to meet the demands of the traffic survey, analysis, forecast accuracy. For these reasons, we proposed a traffic cell radius calculation method using region traveling proportion as constraints based on the distance of residents travelling [17].

Phone traffic clustering analysis. Phone traffic cluster analysis method is using characteristics of mobile phone traffic time distribution to analyze the distribution and division of urban activities and land use and use divided land use characteristics to further map out traffic zone [18].

Sector segmentation. For common sector segmentation to segment traffic zone, the first is to identify the possible city roads node, and then divided forces fan circles for each node. Each circle is a subdivided region traffic zone [19].

Comparing the study of the various traffic cell division methods and using city taxi GPS data, we proposed a Grid Based K-Means cluster method for traffic zone division.

3   Grid Based K-Means Cluster Method for Traffic Zone Division

3.1   K-Means Clustering Analysis

K-Means clustering algorithm was composed of Steinhaus 1957 [20], Lloyd 1957 [21], McQueen 1967 [22] respectively in different fields of their proposed independently. The main idea of the algorithm is: Given n d dimensional data points $\{X_1, X_2, \cdots, X_k\}$, Purpose is to put the N data points into $k(k \leq n)$ clusters $S = \{S_1, S_2, \cdots, S_k\}$, $A = \{a_1, a_2, \cdots, a_k\}$ is the corresponding cluster center. Making cluster in the data point to the cluster center distance squared sum (SSE) Minimum. Mathematically expressed as

$$\arg\min_s \sum_{i=1}^{k} \sum_{X \in Si} \|X - a_i\|^2 \tag{1}$$

The method of measuring distance according to the problem to be solved is selected, the most common way for Euclidean distance metric distance

$$d_{(X_1,X_2)} = \sqrt{\sum_{i=1}^{n} (X_{i1} - X_{i2})^2} \qquad (2)$$

Wherein i represents the i-th value of the X dimension.

K-Means clustering algorithm work flow: First, select k initial points as the centroid, usually a randomly generated initial centroid. Then, assign each point in the data set into one cluster, in particular, for each point to find the nearest centroid, and assign it to the corresponding to the centroid of the cluster. Then, update the cluster centroid is the average of all points in the cluster, calculate the value of SSE. Finally, when the SSE's value is not changed, the algorithm ends.

K-Means algorithm has the characteristics of easy implementation, simple, efficient, since it has been proposed that the algorithm at home and abroad has been widely used in various fields, such as natural language processing, soil, archeology and so on [23–25].

3.2   Data Preprocessing and Extraction

Generally speaking, taxi GPS data are consist of many sample records, each record includes the current time, vehicle ID, latitude and longitude, speed, passenger status (1/0 said with/without passengers) and orientation information.

For example, Nanjing's taxi GPS data are gathered with the speed of 19 million per day and data sending frequency is about 60 s, the data including taxi GPS location information, time and passenger capacity and other relevant information. The data fields and their meanings are shown in Table 1.

Table 1. The GPS data field and its meaning

Data filed	Meaning
ID	unique ID of the taxi in the dataset
VehicleSimID	Vehicle Number of the taxi
GPSSpeed	current speed of the taxi
GPSTime	sampling timestamp in the format of "YYYY-MM-DD HH:MM:SS"
GPSLongitude	current longitude of the taxi
GPSLatitude	current latitude of the taxi
PassengerState	current status (occupied/vacant) of the taxi

Because of signal errors and equipment failures, GPS data transmission may exist deviation, For example, location information and the vehicle passenger conditions are not correct. Therefore, it is necessary to preprocess the original data.

Passengers Getting On or Off Position Extraction. Passenger states which are arranged in chronological order can be drawn as the following sequence:

$$\overbrace{0,0,0,0,0,0,0,0,0,0,0,0,0,0,}^{\text{vacant}} \uparrow \underbrace{1,1,1,1,1,1,1,1,1,1,1,1,1,1,1,1,1,1,1,1,1,1,}_{\text{occupied}} \downarrow \overbrace{0,0,0,0,0,0,0,0,0,0,0,0,0,0}^{\text{vacant}}$$

When passenger state changes from 0 to 1, it means the passenger get on the taxi. When passenger state changes from 1 to 0, it means the passenger get off the taxi.

Passenger status information is given on the driving route: Tour $= \{t_1, t_2, \cdots, t_n\}$ $(t_n \in \{0, 1\})$, and the corresponding positions: $p_0 = \{p_1, p_2, \cdots, p_n\}$.

Get on positions:

$$P_{SUE} = \{p_n : t_n = 1, t_{n-1} = 0, t_{n+1} = 1\} \tag{3}$$

Get off positions:

$$P_{SDE} = \{p_n : t_n = 1, t_{n-1} = 1, t_{n+1} = 0\} \tag{4}$$

Then the useful taxi locations for traffic zone division can be expressed as:

$$\mathbf{P} = P_{SUE} \cup P_{SDE} \tag{5}$$

In addition, the taxi stopped locations where taxi is vacant or occupied will not be considered. Because those locations can't reflect city transportation trends and can't direct connect the passenger with traffic incidents, it is a wise way to exclude these locations for city traffic zone division.

Sliding Window Data Detection. The rate of speed change is always in the range of variation within a smooth window when taxi is running. Wrong the GPS data will make the measurement speed is too high and it's usually have a bad influence for traffic zones division, so the abnormal data need to be removed with correct ways.

Definition:

$$f(p_i, p_j) = \begin{cases} -1, V_{(p_i, p_j)} > \partial, (i \neq j) \\ 1, V_{(p_i, p_j)} \leq \partial, (i \neq j) \end{cases} \tag{6}$$

Equation (7) represents the evaluation function of travel paragraph on road running state.

Among them:

$$V_{(p_i, p_j)} = \left| \frac{L_{(p_i, p_j)}}{T_{p_i} - T_{p_j}} \right| \tag{7}$$

$L_{(p_i, p_j)}$ represents Manhattan distance between p_i and p_j. T_{p_i} represents vehicle GPS timestamp at pi. ∂ is a threshold which represents the taxi average driving speed in a city. Through the analysis of city taxi operational data, we take $\partial = 120$ km/h. If $V_{(p_i, p_j)} > \partial$, it represents the route is unusual, if $V_{(p_i, p_j)} \leq \partial$, it represents the route is normally.

Threshold is set to 120 km/h according to urban traffic regulation. Records with velocity greater than the threshold indicate that there is an abnormal movement, therefore they should be removed. We define W_{p_i} as the abnormal velocity indicator for the pi record, which is defined as the sum of the neighboring $f(p_i, p_j)$

$$W_{p_i} = \sum_{j=i-r}^{i+r} f(p_i, p_j) \tag{8}$$

r is the width of the sliding window. If $W_{p_i} < 0$, it means data record with position pi is abnormal and should be removed. We heuristically set r to 3.

Outside the Region Data Filtering. In order to ensure the accuracy of traffic zones division for target city, position data that does not belong to the scope of target area need to be removed. For example, Nanjing is located in the lower reaches of the Yangtze River and it is in the southwestern of Jiangsu Province. Geographical coordinates of latitude $31°14''$ E to $32°37''$, longitude $118°22''$ to $119°14''$. The data records that aren't in Nanjing's geographical scope will not be considered for traffic zone division.

3.3   The Mesh and Traffic Zone Candidate Set

By using traditional K-Means to compute traffic zone with original data points, it always need to handle big amount of points and the boundaries are usually not clear. A separate algorithm for edge detection needs to be used to solve this issue. In order to make the traffic zone division is accurate and reduce the influence of bad GPS records, we put forward the method of grid division for target area and generate traffic zone candidate set as input data of K-Means algorithm.

Given the target area $D = (x_{min}, x_{max}, y_{min}, y_{max})$, x_{min} represents the minimum longitude, x_{max} represents the maximum longitude, y_{min} represents the lowest latitude, y_{max} represents the maximum latitude. Target area D is divided with the grid can be expressed as:

$$D = \{A_{ij} : \forall r_x, r_y, \ 0 \leq i \leq (x_{min} - x_{max})/r_x, \ 0 \leq j \leq (y_{min} - y_{max})/r_j\} \tag{9}$$

r_x represents the minimum span of longitude, latitude r_y represents the minimum span.

In order to reduce the influence of outlier data for division result, we generate traffic small candidate sets in the following ways.

Given the original taxi GPS position data set $P = \{p_{(x,y)} : p \in D\}$, $p_{(x,y)}$ represents a location, region in D, (x, y) represents their latitude and longitude. Then the candidate set for traffic zone:

$$D_0 = \{A_{ij} : \sum_{p \in A_{ij}} p \geq \lambda\} (\lambda \text{ is the threshold}) \tag{10}$$

3.4   Grid Based K-Means Cluster Method for Traffic Zone Division

Grid Based K-Means cluster method for traffic zone division process with the flowing steps, and entire division method is expressed as Table 2.

Table 2. Algorithm for Grid Based K-Means Cluster Method for Traffic Zone Division

---

Input: The target region $D = (x_{min}, x_{max}, y_{min}, y_{max})$, The GPS data set $P =$

$\{p_{(x,y)}: p \in D\}$ Traffic zone is divided into k, the accuracy of E

Output: K traffic zone of the target region $S = \{S_1, S_2, \cdots, S_k\}$

Step1. Get the target area grid collection D with equation (9)

Step2. Generation of candidate traffic grid unit whit equation (10)

Step3. Initialize the centroid generated from D_0

$$M_0 = \left\{ m_1^{(0)}, m_2^{(0)}, \cdots, m_k^{(0)} \right\} \tag{11}$$

Step4. Assigning data D to a cluster and get k clusters $S_t = \left\{ S_1^{(0)}, S_2^{(0)}, \cdots, S_k^{(0)} \right\}$:

$$S_i^t = \left\{ g: \left\| g - m_i^{(t)} \right\|^2 \leq \left\| g - m_j^{(t)} \right\|^2, \forall j, 1 \leq j \leq k, g \in D_0 \right\} \tag{12}$$

Step5. Update the centroid, get $M_{t+1} = \left\{ m_1^{(t+1)}, m_2^{(t+1)}, \cdots, m_{k1}^{(t+1)} \right\}$, wherein:

$$m_i^{t+1} = \frac{\sum_{g \in S_i^{(t)}} g}{\left| S_i^{(t)} \right|} \tag{13}$$

Step6. Calculate the sum of square error SSE

$$SSE^{(t)} = \sum_{i=1}^{k} \sum_{g \in S_i} \| g - m_i \|^2 \tag{14}$$

When $SSE^{(t+1)} - SSE^{(t)} > E$, Execution Step4,5,6, otherwise continue Step7

Step7. The $\{S_1, S_2, \cdots, S_k\}$ output to the map, and return to S.

---

1. Do data preprocessing for taxi GPS data records;
2. According to the target city's longitude and latitude range, we carry on the grid division of the target city and generate traffic zone candidate set.
3. Using the K-Means clustering algorithm for spatial clustering analysis of the candidate set and get k traffic zone;
4. Output the result to the GIS map and traffic zones with clear boundaries will be shown in that map.

3.5   Performance Comparison

In order to verify the best performance in time used and cluster boundaries of Grid Based K-Means, we conducted some tests. The test randomly generated 100 two-dimensional data points for ten experiments; each experiment ran ten times and computed the average time used by using traditional K-Means and Grid Based K-Means. The experiment platform was Windows 7 32-bit operating system, 2 G of memory and parameters: K = 10, E = 0.01.

Figure 1 compares the performance of traditional K-Means and Grid Based K-Means with same data set (100 data points). Figure 2 compares the performance of traditional K-Means and Grid Based K-Means with different data set. Obviously, the Grid Based K-Means method achieves the best performance in time used. It is well known that K-Means cluster method need to compute the distance between each data point whit cluster center for each loop. In this experiment, traditional K-Means cluster method must to compute the distance for 100 data points, but Grid Based K-Means method only handles the basic grid cell which has fewer amounts than 100 data points. In this way, Grid Based K-Means can reduce the time cost.

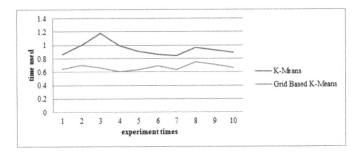

Fig. 1. Comparison between K-Means and Grid Based K-Means in time used with same data

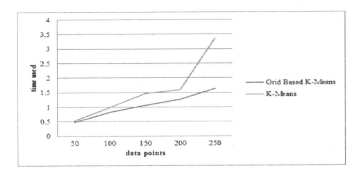

Fig. 2. Comparison between K-Means and Grid Based K-Means in time used with different data

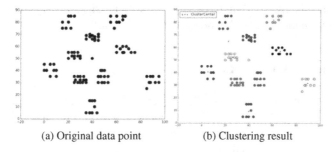

(a) Original data point   (b) Clustering result

Fig. 3. Traditional K-Means cluster method.

Figure 3 shows the traditional K-Means cluster method for 100 data points, it is clear that data points are divided to 10 clusters, but the boundaries aren't clearly. Typically, separate boundary detection algorithms need to be used to calculate cluster boundaries and those algorithms always cost more time for large-scale data set.

Figure 4 describes the clustering process of Grid Based K-Means method, and the clustering result show us clear boundaries. Original data points are divided into the candidate set which has irregular shape. When Grid Based K-Means method getting

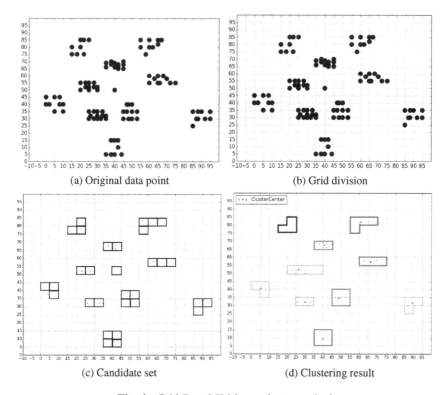

(a) Original data point   (b) Grid division

(c) Candidate set   (d) Clustering result

Fig. 4. Grid Based K-Means cluster method

the clustering results by handling basic grid cell in candidate set, the cluster boundaries are generated without separate boundary detection algorithms.

4   Simulation Experiment

In order to verify the effectiveness of Grid Based K-Means clustering method for traffic zone division, we built GIS-based experiments environment and selected Baidu Maps API solutions as GIS platform. By programming with Python, we analyzed the taxi GPS data which was stored in the database and divided traffic zones and outputted them on the map.

Experimental data was selected from a company's taxi GPS data of Nanjing in September 1, 2010 we divided traffic zones for Nanjing. Based on latitude and longitude range of Nanjing, we obtained Nanjing GIS maps, as shown in Fig. 5. Longitude range is E longitude 118.55° to 119.05°, Latitude range is N latitude 31.7° to 32.2°.

Fig. 5. The map of Nanjing

Figure 6 is a geographical area meshing of Nanjing. Figure 6 divided Nanjing into 64 × 54 grid cells; each cell crosses the longitude 009259252° and latitude 0078125°.

Preprocess the raw GPS data, and erase the wrong data. Table 2 shows the amount of data records before and after processing taxi GPS data.

When $\lambda = 20$, select the candidate set of traffic zones according to Eq. (3). As shown in Fig. 7, the black grid cells are the candidate set.

By using Grid Based K-Means clustering algorithm, we find out k traffic zones from candidate sets and output them onto GIS map (E = 0.0001, k = 8). Table 3 shows the analysis of Nanjing traffic zones by using Grid Based K-Means method, "Traffic

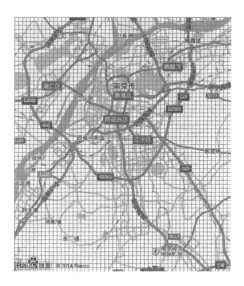

Fig. 6. A geographical area meshing of Nanjing.

Table 3. The amount of data before and after processing taxi GPS data

Preprocessing	Before processing	After processing
1. Passengers Getting On or Off Position Extraction	18668073	432283
2. Sliding Window Data Detection	432283	430354
3. Outside the Region Data Filtering	430354	425006

zone" is the traffic cell identity, "Transportation heat" is the average on and off times in the divided traffic zones, "Main political zone" is the main administrative districts in the divided traffic zones, "Cross river" is the river distribution of traffic zones, the "regional hub" is the transportation hub of traffic zones Table 4.

Figure 8 is the division result of Nanjing by using Grid Based K-Means traffic zone division method.

Figure 8 shows 8 traffic zones and the traffic zones boundaries are clear enough. By detecting zone, some hot places are collected and are shown in Table 3, Through Table 3, each zone have some important traffic hub and is divided by rivers or other nature barriers. It means that the result of Grid Based K-Means match with Nanjing geographical features and fit the standards of traffic zone division. The result of simulation experiment indicates that Grid Based K-Means traffic zone division method is feasible by using taxi GPS data. What's more the result reflect the transportation trends for Nanjing at special period. It's also instructive for government taking measures to reduce traffic congestion.

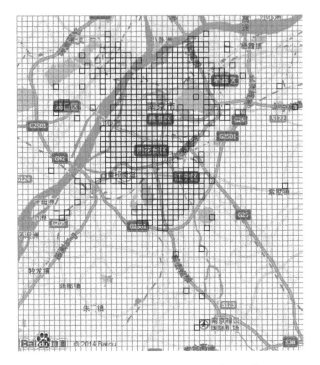

Fig. 7. Traffic small candidate set in Nanjing

Table 4. Analysis of Nanjing traffic zone division result

Traffic zone	Transportation heat	Political zones	Cross river	Regional hub
A	483.15	Jianye District, Yuhuatai District	No	Nanjing Olympic Sports Center
B	2991.30	Gulou District	No	The government of Jiangsu Province
C	3378.41	Jiangning District (North)	Qinhuai River	NanJing South Railway Station
D	4777.64	Yuhuatai District Qinhuai District	No	Tourism Center
E	9205.73	Qixia District	No	Nanjing East Station
F	7257.88	Pukou District	No	Nanjing North Station
G	10421.81	Jiangning District (South)	Qinhuai River	Nanjing Lukou International Airport
H	5009.78	Xuanwu District	No	NanJing Railway Station

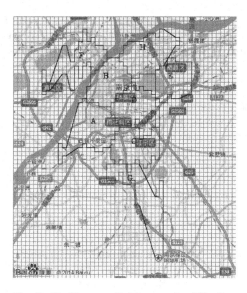

Fig. 8. Traffic zone division result of Nanjing

5 Conclusions

In order to solve problems of the existing methods, like inaccurate results, no clear demarcation of the border, and high dividing cost, etc., this paper presents cluster analysis method based on grid K-Means, which includes following procedures, selecting city taxi GPS data as the division basis, preprocessing the taxi GPS data, doing clustering analysis to divide the traffic zone of the target area, and finally outputting the result to the visual display of GIS platform. As shown in test data, with grid based K-Means traffic zone division method, boundaries are clear and dovetail into the principle of traffic zone division. The results are more accurate taking into account the pre-processing of GPS data. This method has a good reference value for the analysis of urban traffic flow and traffic trends.

There are also some disadvantages of traffic zone division method based on grid K-Means. Clustering partitioning algorithm considers only the spatial position of the cluster, and did not join economic attributes of the grid area, such as residential, commercial, etc. It is also our further improvement direction.

References

1. Ratti, C., Pulseli, R.M., Williams, S., Frenchman, D.: Mobile landscapes: using location data from cell phones for urban analysis. Environ. Plann. B Plann. Des. **33**(5), 727–748 (2006)
2. González, M.C., Hidalgo, C.A., Barabási, A.L.: Understanding individual human mobility patterns. Nature **453**, 779–782 (2008)
3. Ashbrook, D., Starner, T.: Using GPS to learn significant locations and predict movement across multiple users. Pervasive and Ubiquitous Comput. **7**(5), 275–286 (2003)

4. Farrahi, K., Gatica-Perez, D.: Learning and predicting multimodal daily life patterns from cell phones. In: Proceedings of the 11th International Conference on Multimodal Interfaces and the 6th Workshop on Machine Learning for Multimodal Interface, pp. 227–280 (2009)

5. Reades, J., Calabrese, F., Sevtsuk, A., Ratti, C.: Cellular census: explorations in urban data collection. IEEE Pervasive Comput. 6(3), 30–38 (2007)

6. Chang, H., Tai, Y., Hsu, J.Y.: Context-aware taxi demand hotspots prediction. Int. J. Bus. Intell. Data Min. 5(1), 3–18 (2010)

7. Phithakkitnukoon, S., Veloso, M., Bento, C., Biderman, A., Ratti, C.: Taxi-Aware map: identifying and predicting vacant taxis in the city. In: de Ruyter, B., Wichert, R., Keyson, D. V., Markopoulos, P., Streitz, N., Divitini, M., Georgantas, N., Mana Gomez, A. (eds.) AmI 2010. LNCS, vol. 6439, pp. 86–95. Springer, Heidelberg (2010)

8. Wong, K.I., Wong, K.C., Bell, M.G.H., Yang, H.: Modeling the bilateral micro-searching behavior for urban taxi services using the absorbing market chain approach. J. Adv. Transp. 39, 81–104 (2005)

9. Yang, H., Fung, C.S., Wong, K.I., Wong, S.C.: Nonlinear pricing of taxi services. Transp. Res. Part A Policy Pract. 44(5), 337–348 (2010)

10. Liu, L., Andris, C., Ratti, C.: Uncovering cabdrivers' behavior patterns from their digital traces. Environment and Urban Systems, Corrected Proof, Available online 16 August (2010, in Press)

11. Liu, L., Andris, C., Biderman, A., Ratti, C.: Uncovering taxi driver's mobility intelligence through his trace. IEEE Pervasive Comput. 160, 1–17 (2009)

12. China Institute of Highway Traffic Engineering Manual. Traffic Engineering Handbook. China Communications Press, Beijing (2001)

13. Li, X.D., Yang, X.G., Chen, H.J.: Study on traffic zone division based on spatial clustering analysis. Comput. Eng. Appl. 45(5), 19–22 (2009)

14. Yang, B., Lu, H.: Improvement of the method about the partition of traffic zone. Traffic Transp. 23(7), 23–26 (2007)

15. Xu, J.Q.: Traffic Engineering Subjects. China Communications Press, Beijing (2003)

16. Zhong, Z., Huang, W., Ma, W.: Design of traffic zone division algorithm for coordinated control with the implementation. In: The Fourth Session of the China Intelligent Transport Annual Meeting Proceedings (2008)

17. Ma, C.Q., Wang, R., Wang, Y.P., et al.: Calculating method of traffic zone radius in city based on inner trip proportion. J. Traffic and Transp. Eng. 7(1), 68–72 (2007)

18. Zhou, W., Dong, X., Xu, W., et al.: Mobile phone information based transportation cell mapping method preliminary study. In: The Fourth Session of the China Intelligent Transport Annual Meeting Proceedings (2008)

19. Yang, Z.Z., Wang, L., Cheng, G.: Method of traffic analysis zone partition for traffic impact evaluation. J. Highway Transp. Res. 6 24(6), 102–105,117 (2007)

20. Steinhaus, H.: Sur la division des corps matériels en parties. Bull. Acad. Polon. Sci. (in French) 4(12), 801–804 (1957)

21. Lloyd, S.P.: Least squares quantization in PCM. IEEE Trans. Inf. Theor. 28(2), 129–137 (1982)

22. MacQueen, J.B: Some methods for classification and analysis of multivariate observations. In: Proceedings of 5th Berkeley Symposium on Mathematical Statistics and Probability, vol. 1, pp. 281–297. University of California Press (1967)

23. Wilpon, J.G., Rabiner, L.R.: A modified K-Means clustering algorithm for use in isolated word recognition. IEEE Trans. Acoust. Speech Signal Process. 33(3), 587–594 (1985)

24. McBratney, A.B., De Gruijter, J.J.: A continuum approach to soil classification by modified fuzzy K-Means with extra grades. J. Soil Sci. 43(1), 159–175 (1992)

25. Simek, J.F.: A K-Means approach to the analysis of spatial structure in upper paleolithic habitation sites, pp. 400–401. B.A.R., Oxford (1984)

Color Image Fusion Researching
Based on S-PCNN and Laplacian Pyramid

Xin Jin, Rencan Nie, Dongming Zhou[(⊠)], and Jiefu Yu

Information College, Yunnan University, Kunming 650091, China
`18487219630@163.com, zhoudm@ynu.edu.cn`

Abstract. In this paper we propose an effective color image fusion algorithm based on the simplified pulse coupled neural networks (S-PCNN) and the Laplacian Pyramid algorithm. In the HSV color space, after regional clustering feature of H components by S-PCNN, and then achieved the fusion of the H component from each source image using Oscillation Frequency Graph. At the same time, decomposing S, H component by Laplacian Pyramid algorithm, and then using different fusion rules to fusion S, H component. Finally, inversing HSV transform to get RGB color image. The experiment indicates that the new color image fusion algorithm is more efficient both in the subjective aspect and the objective aspect than other commonly color image fusion algorithm.

Keywords: Simplified pulse coupled neural networks · Laplacian pyramid · Feature extraction · Color image fusion

1 Introduction

Image fusion is a branch of information fusion, and it is becoming a popular research field [1,2]. But now the researching in color image fusion field is comparatively few, however, the identifiable degree of human vision to color information is far higher than gray image [3]. With the improvement of sensor technology, color image fusion will be received more and more attentions.

Color image is the combination of different brightness and different colors. Color image is comprised of several components. Therefore, color image fusion is the fusion of each color space component. There are some common algorithm, such as weighted method, HIS [4] and PCA [3,5] transform method, it is easy to implement but the performance is not good. In addition, there are some image fusion methods based on multi-resolution analysis. Their first step are image transform, and then recombine the coefficient of the transformed image, at last the fused image can be obtained by inverse transformation. According to the different ways of decomposition, these can be divided into pyramid transform; wavelet transform and multi-scale geometric transform, these kinds of algorithms are normally used in pixel-level image fusion. Pixel level image fusion is the bottom layers in image fusion grade, and each pixel is given by the corresponding pixel of other source image [5].

© Springer International Publishing Switzerland 2015
W. Qiang et al. (Eds.): CloudCom-Asia 2015, LNCS 9106, pp. 179–188, 2015.
DOI: 10.1007/978-3-319-28430-9_14

PCNN is a new kind of artificial neural network model with a deep background in biology, known as the third generation of artificial neural networks. It has been widely used in image processing [5,6], pattern recognition [7], multiple constrained QoS route [8], and show outstanding properties in these fields. Its neurons capture feature will cause the brightness of the surrounding neurons similar to capture the ignition. It can automatically couple and transmit information. Compared with other artificial neural network it has an incomparable advantage over other traditional artificial neural network. Laplacian Pyramid algorithm is a type of image processing method with the characteristic of multi-scale, multi-resolution, multiple-level decomposition. It can decompose the important image features (such as edge, texture, etc.) into different decomposed layers according to different scales, compared with simple image fusion algorithm, it can obtain better fusion effect, and it has been widely used in image processing field [9].

In this paper, a new color image fusion algorithm is proposed based on simplified pulse coupled neural network (S-PCNN) and Laplacian Pyramid algorithm. It firstly converts RGB color image to HSV color space. And then, H component is input into S-PCNN model to fusion new H component by comparing the ignition times of S-PCNN neurons, at the same time S and V components are fused by Laplacian Pyramid algorithm to get new S and V components. Finally, the three new components of HSV color image space are inversed transform to RGB color image space to obtain a new color image. The experimental results indicate that new algorithm is more effective to save the color information of the source color image than other common color image fusion algorithms, and it contains more edges, texture and detail.

2   HSV Color Space

RGB color image contains almost all basic colors that can be perceived by human vision, however, the correlation among the components are very strong, it makes RGB color image very difficult to deal with that the color of the image will be changed if a component changes. In hue saturation value model, H, S and V denote the hue, saturation, and value, respectively. This color system is closer than the RGB color system to human experience and perception of the color. Hue is also called the color attribute, expressed in angle, which ranges from 0 to 360; saturation represents the purity of the color, value represents the brightness of the image, which is a measure of gray, ranging from binary 0 to 1.

HSV image can be obtained by the RGB transform. The values of R, G and B correspond to unique H, S and V values, as the values of H, S and V components depend on the values of R, G and B in RGB color space (R stands for red, G stands for green, B stands for blue). The conversion formulas of RGB color space to HSV color space as (1) to (3), where R, G and B are normalized values.

$$H = \begin{cases} 0, & S = 0 \\ 60 \times \frac{G-B}{S \times V}, & \max(R, G, B) = R \bigcap G \geq B \\ 60 \times \frac{2+(B-R)}{S \times V}, & \max(R, G, B) G \\ 60 \times \frac{4+(R-B)}{S \times V}, & \max(R, G, B) = B \\ 60 \times \frac{6+(G-B)}{S \times V}, & \max(R, G, B) = R \bigcap G < B \end{cases} \quad (1)$$

$$S = \frac{\max(R, G, B) - \min(R, G, B)}{\max(R, G, B)}, \quad (2)$$

$$V = \max(R, G, B). \quad (3)$$

In HSV space to RGB space conversion, let $i = H/60$, $f = H/60 - i$, where i is the divisor of H divided by 60; f is the remainder of H divided by 60. And let $P = V(1 - S)$, $Q = V(1 - Sf)$, $T = V[1 - S(1 - f)]$, formula of HSV color space to RGB color space see (4) [10].

$$\begin{cases} R = V, G = T, B = P, & i = 0 \\ R = Q, G = V, B = P, & i = 2 \\ R = P, G = V, B = T, & i = 2 \\ R = P, G = Q, B = V, & i = 3 \\ R = T, G = P, B = V, & i = 4 \\ R = V, G = P, B = Q, & i = 5 \end{cases} \quad (4)$$

3   S-PCNN Model

PCNN model has three fundamental parts: the receptive field, the modulation field and the pulse generator. S-PCNN (Simplified PCNN) model [11] is composed as the same as original PCNN model, but the input of F channel is only related to image gray value and has no relationship with external coupling and exponential decay characteristics. We regarded each pixel of original images as a neuron of the neutral network. The receptive field receives input signals from two channels, one is the linking input L_{ij}, and the other is feeding input F_{ij}. In this paper, we use S-PCNN. The variables of each neuron N_{ij} above satisfy the following (5) to (9):

$$F_{ij}(n) = S_{ij}, \quad (5)$$

$$L_{ij}(n) = e^{-\alpha^L} L_{ij}(n-1) + V_{ij}^L \sum_{kl} W_{ijkl} Y_{ijkl}(n-1), \quad (6)$$

$$U_{ij}(n) = F_{ij}(n)[1 + \beta L_{ij}(n)], \quad (7)$$

$$\theta_{ij}(n) = e^{-\alpha^\theta} \theta_{ij}(n-1) + V_{ij}^\theta Y_{ij}(n-1), \quad (8)$$

$$Y_{ij}(n) = \begin{cases} 1, & U_{ij}(n) > \theta_{ij}(n) \\ 0, & otherwise \end{cases}, \quad (9)$$

where the subscripts i, j denote the (i, j)th pixel in the corresponding sub-image, $L(i, j)$, $U(i, j)$, $Y(i, j)$, (i, j) are the number of input connections, internal activity items, pulse output and dynamic threshold of each neuron, respectively. α^L is time decay constant of linking input and α^θ was the time decay constant. V^θ and V^L are the amplification coefficient; n represents the iteration time. W_{ij} is the synaptic weights of the neuron j connected with the other neurons in the linking branch; the number of neurons in the network is equal to the number of pixels in the image to be fused. Each pixel is connected to a unique neuron and each neuron is connected with the surrounding neurons. Need to point out that, a neuron fires means an S-PCNNs neuron generates a pulse. If the neuron satisfies the condition $U_{ij}(n) > \theta_{ij}(n)$, a pulse would be produced by neurons. We call an ignition. The total number of ignitions represents image information of the corresponding point after N iterations, the ignition mapping of all neurons firing times are used as output of PCNN. More details for PCNN will be found in [11].

4   Laplacian Pyramid Algorithm in Image

Laplacian pyramid image transform is obtained by Gauss in pyramid. Therefore, Laplace pyramid decomposition has two steps: first to get Gauss pyramid, and then get Laplacian pyramid [9,12].

4.1   Gaussian Pyramid Decomposition

Let the original image as G_0, the bottom of the Gaussian pyramid denoted by G_0. Firstly, Gaussian low-pass filter is used to the original image (That is the convolution of Gaussian operator and the original image), and then downsampling to the filtered image, so we can get the first layer of the Gaussian pyramid. When we repeat the above process of Gaussian low-pass filter, and down sampling that we will get next Gaussian pyramid layer. So repeating the operation to obtain several layers to constitute a Gaussian pyramid. G_1 denotes the first layer of Gaussian pyramid, see (10).

$$G_l(i, j) = \sum_{m=-2}^{2} \sum_{n=-2}^{2} \omega(m, n) G_{l-1}(2i + m, 2j + n),$$

$$(1 \leq l \leq N, 0 \leq i < R_l, 0 \leq j < C_l) \tag{10}$$

where N denotes the number of Gaussian pyramid layer, and R_i and C_i respectively denote the rows and columns of the lth layer Gaussian pyramid, $\omega(m, n)$ is 5×5 matrix window function with low pass resistance performance, see (11).

$$\omega = \frac{1}{256} \begin{bmatrix} 1 & 4 & 6 & 4 & 1 \\ 4 & 16 & 24 & 16 & 4 \\ 6 & 24 & 36 & 24 & 6 \\ 4 & 16 & 24 & 16 & 4 \\ 1 & 4 & 6 & 4 & 1 \end{bmatrix}, \tag{11}$$

where G_0, G_1, ..., G_N make up Gaussian pyramid, G_0 is the same as original image, and G_N is the top layer.

4.2 Laplacian Pyramid Decomposition

The image size we get by Gaussian pyramid algorithm is 1/4 of above layer. Then dilation used in Gauss pyramid by using interpolation method, making the size of G_l in the lth layer is the same as the size of $G_{(l-1)}$ in the $(l-1)$th after dilation, see (12) to (14).

$$G_l^*(i,j) = 4 \sum_{m=-2}^{2} \sum_{n=-2}^{2} \omega(m,n) G_l(\tfrac{i+m}{2}, \tfrac{j+n}{2}),$$
$$(0 < l \leq N, 0 \leq i < R_l, 0 \leq j < C_l) \tag{12}$$

where,

$$G_l^*(\frac{i+m}{2}, \frac{j+n}{2}) = \begin{cases} G_l(\frac{i+m}{2}, \frac{j+n}{2}), \text{when } \frac{i+m}{2}, \frac{j+n}{2} \text{ are integer} \\ 0, \hspace{3.5cm} \text{otherwise} \end{cases}. \tag{13}$$

Let,

$$\begin{cases} LP_l = G_l - G_{l+1}^*, & 0 \leq l < N \\ LP_N = G_N, & l = N \end{cases}. \tag{14}$$

N denotes the top of Laplacian Pyramid layer, LP_l denotes the lth layer of Laplacian Pyramid. $LP_0, LP_1, \ldots, LP_N$ make up Laplacian Pyramid. The layer of Laplacian pyramid is the difference between this layer of Gaussian pyramid and upper layer of the interpolation enlarged image, the process amounts to band pass filter.

4.3 Laplacian Pyramid Image Reconstruction

Equation (15) can be obtained by (14):

$$\begin{cases} G_N = LP_N, & l = N \\ G_l = LP_l + G_{l+1}^*, & 0 \leq l < N \end{cases}. \tag{15}$$

From (15) can be known that Laplace pyramid layers are gradually enlarged by interpolation, making the size of the layer is the same as the lower layer, so that we can reconstruct the original image by adding the layer and the lower layer.

5 Color Image Fusion Based on S-PCNN and Laplacian Pyramid

In this paper, S-PCNN we used in color fusion algorithm to extract of image features is a neural network model with biological characteristics and it can extract the information of the images texture, edge, regional distribution and has a good effect in the gray areas of image fusion. It was known as the ignition characteristics of S-PCNN, it recorded the ignition or not corresponding neurons to the each pixel in the iteration of S-PCNN, this can generate a binary image.

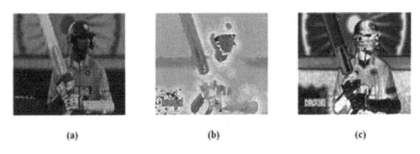

(a)   (b)   (c)

Fig. 1. (a) Source image, (b) OFG of H component, (c) OFG of S component

These binary images effectively express the feature information in the image (such as texture, edge and regional distribution, etc.). After the globally statistics of the pulse of the neurons, an oscillation frequency graph (OFG) could be obtained, it can be described by (16), where N denotes the iteration times, IO^N denotes an OFG, Y_{ij} denotes the pulse output of the neuron (i, j).

$$IO^N(i,j) = \sum_{K=1}^{N} Y_{ij}(K). \tag{16}$$

Figure 1(a) is source image, Fig. 1(b) and (c) show OFG of H and S components in HSV color space after 200 times iteration in S-PCNN model.

Laplacian pyramid image decomposition decompose source image into different spatial frequency bands. The features and details on different bands are different, so the image fusion rules will affect the quality of image fusion considerably. In this paper we use different rules on different bands so that it can fusion the feature and detail of various source images into a new image. Energy can reflect the quality of the image when Laplacian pyramid layer range from 0 to N, all by the regional energy way select coefficient. First, calculate the area energy, so we use regional energy to select coefficient. First, calculate the regional energy, see (17).

$$ARE(i,j) = \sum_{-p}^{p}\sum_{-q}^{q} \omega(p,q)|LA_N(i+p,j+q)|, \tag{17}$$

$$BRE(i,j) = \sum_{-p}^{p}\sum_{-q}^{q} \omega(p,q)|LB_N(i+p,j+q)|, \tag{18}$$

where, $p = 1$, $q = 1$, $\omega = \frac{1}{16}\begin{bmatrix} 1 & 2 & 1 \\ 2 & 4 & 2 \\ 1 & 2 & 1 \end{bmatrix}$, the other layers image fusion sees (19).

$$LF_l(i,j) = \begin{cases} LA_l(i,j), & ARE(i,j) \geq BRE(i,j) \\ LB_l(i,j), & ARE(i,j) < BRE(i,j) \end{cases}, \quad 0 \leq l < N. \tag{19}$$

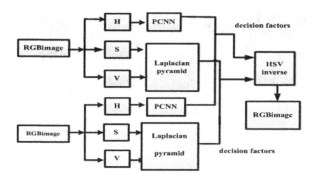

Fig. 2. Algorithm flow chart

After getting all levels of the pyramid image fusion LF_1, LF_2, ..., LF_N, we can restructure the fused image by (16) to obtain the final fused image. S-PCNN is sensitive to detail, edge and other information of the image. And Laplacian pyramid high-layer contains the detail and texture character. So in this paper we use S-PCNN model to choose Laplacian pyramid high-layer coefficient. Because of the correlation among the R, G and B components are very strong, its fusion effect is not satisfied. So in this paper, the image is transformed into HSV space from RGB color space to get the H, S and V components. And then H component is sent into S-PCNN model to OFG of the H component, we will get the fused H component by comparing the ignition times of S-PCNN neurons between the source image, in the meantime S and V components are fused by Laplacian pyramid algorithm. The color image fusion algorithm processes are shown in Fig. 2.

6   Results and Analysis

To verify the validity of the algorithm presented in the paper, we take two groups of experimental color images with different focus position to test. The first group of color image is cups (see Fig. 3). Image A focuses on the left and image B focuses on the right, and there are many words as details. The parameters in the S-PCNN model are: $W = [0.5, 1, 0.5; 1, 0, 1; 0.5, 1, 0.5]$, $V^L = 1$, $\beta = 0.9$, $\alpha^\theta = 0.2$, $\alpha^L = 1$, $V^\theta = 20$ and the iteration $N = 800$. The fusion images using different methods are showed in Fig. 4. According to Fig. 4, from the subject aspect, the contrast of the fusion image generate from the method in this paper is better than the others. The algorithm in this paper does well in extracting the characteristics of the source images, and the fused image is closer to the natural color. It contains more edges, texture and detail. We can conclude that the method in this paper is an effective method for getting extraordinary contrast fusion image. In order to verify the effectiveness of this method in object aspect, Table 1 shows the evaluation of the fused image quality with space frequency (SF), standard deviation (STD), entropy (EN), average gradient (AV), mean value (M).

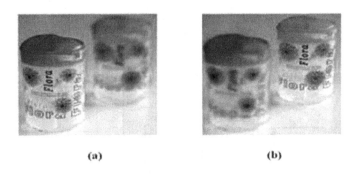

Fig. 3. (a) Source image A, (b) Source image B

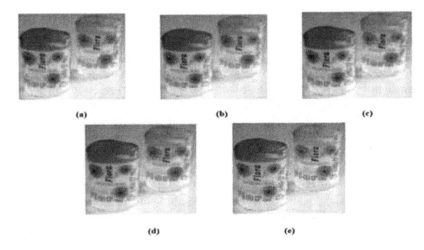

Fig. 4. Fusion images from different methods. (a) Weighted method, (b) PCA method, (c) Wavelet transforms method, (d) PCNN method, (e) This paper

Table 1. Fusion quality using different method

Method	SF	STD	EN	AV	M
Weighted	12.57	65.01	7.45	4.20	116.97
PCA	12.76	65.17	7.44	4.23	166.67
WT	13.19	65.39	7.49	4.31	163.33
PCNN	17.62	66.27	7.45	5.27	168.50
This paper	20.10	60.95	7.44	5.85	169.33

As we can see in Table 1, it is indicated that the fusion image generated by this method contains much more information. The SF of this method is much larger than other methods. The AV and M are also much better than other methods; EN is as large as other methods, only STD is less than other methods.

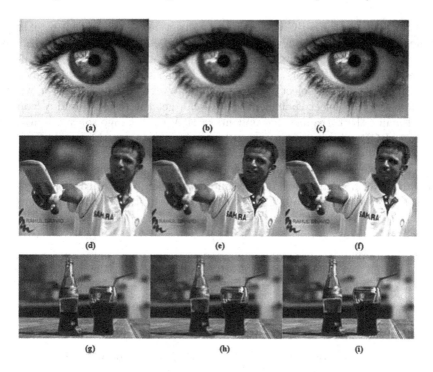

Fig. 5. Second group image experiment. (a) Source image A, (b) Source image B, (c) Fusion image, (d) Source image A, (e) Source image B, (f) Fusion image, (g) Source image A, (h) Source image B, (i) Fusion image

More experimental results are shown in Fig. 5, source image A focuses on the left and image B focuses on the right, and there are a lot of textures in the source images. It can be seen in Fig. 5 that the edge of the fusion image is clear, and it retains most of the textures in the source images, besides the details are also well preserved. This method can extract the main features from the source images; it shows that the method in this paper also achieves effective results in these groups of color images.

It can be concluded that the subjective evaluation and objective evaluation were the same conclusion. Overall, the method presented in this paper was better than the traditional method obviously. Be compared with the other methods, this method reflected better performance and visual effect in term of definition and detail.

7   Conclusion

We propose an effective color image fusion algorithm based on S-PCNN model and Laplacian pyramid algorithm. It is also combined with HSV color space which is suitable for human vision. Because of the characteristic of color image region clustering in the S-PCNN model, we can get new H component by extracting the feature of H component in the image, such as the texture, edge, regional distribution. Laplacian pyramid algorithm can decompose the important image features into different layer according to different scales, in the high-layer we use S-PCNN model to select the coefficient, however in the low layer we use regional energy to select coefficient. So we can get the fusion S and V components. Finally, the three new components of HSV color image space are inversed transform RGB color image space to obtain the fusion color image. The experimental results show that the color image fusion algorithm proposed in this paper can fuse different focus position of the color image, and the new image contains more information about color, texture and detail. Compared with the traditional algorithms, this method embodies better fusion performance in information, SF, AV, and M on the objective aspect.

Acknowledgements. Our work is supported by the National Natural Science Foundation of China (No. 61365001, No. 61463052), Natural Science Foundation of Yunnan Province (No. 2012FD003).

References

1. Zhu, L., Sun, F., Xia, F.L., et al.: Review on image fusion research. Transduces Microsyst. Technol. **33**(2), 14–18 (2014). (in Chinese)
2. Vladimir, P., Vladimir, D.: Focused pooling for image fusion evaluation. Inf. Fusion **22**, 119–126 (2015)
3. Xia, Y., Qu, Sh.R.: Color image fusion framework based on improved (2D) 2PCA. Acta Optica Sin. **34**(10), 10100011–10100018 (2014). (in Chinese)
4. Sabalan, D., Hassan, G.: MRI and PET image fusion by combining IHS and retina-inspired models. Inf. Fusion **11**, 114–123 (2010)
5. Duan, X.H., Cao, J.J., Liu, J.: Application research of modified PCNN model in multispectral and panchromatic images fusion. Mod. Electron. Tech. **37**(3), 55–60 (2014). (in Chinese)
6. Monica Subashini, M., Sahoo, S.K.: Pulse coupled neural networks and its applications. Expert Syst. Appl. **41**, 3965–3974 (2014)
7. Jin, X., Nie, R.C., Zhou, D.M.: An improved iris recognition algorithm based on PCNN. Comput. Sci. **41**(11A), 110–115 (2014). (in Chinese)
8. Nie, R.C., Zhou, D.M., Zhao, D.F., et al.: CPCNN and its application to multiple constrained QoS route. J. Commun. **31**(1), 65–72 (2010). (in Chinese)
9. Ma, X.X., Peng, L., Xu, H.: PCA based Laplacian pyramid in image fusion. Comput. Eng. Appl. **48**(8), 211–213 (2012). (in Chinese)
10. Pekel, J.F., Vancutsem, C., Bastin, L., et al.: A near real-time water surface detection method based on HSV transformation of MODIS multi-spectral time series data. Remote Sens. Environ. **140**, 704–716 (2014)
11. Zhao, C.H., Shao, G.F., Ma, L.J., et al.: Image fusion algorithm based on redundant-lifting NSWMDA and adaptive PCNN. Optik **125**, 6247–6255 (2014)
12. Wen, D.H., Jiang, Y.S., Zhang, Y.Z., et al.: Modified block-matching 3-D filter in Laplacian pyramid domain for speckle reduction. Opt. Commun. **322**, 150–154 (2014)

Rationalizing the Parameters of K-Nearest Neighbor Classification Algorithm

Jian Liu[✉], Gang Zhao, and Yunpeng Zheng

Beijing Information Science and Technology University, Beijing 100192, China
{liujianspace999,15210846693}@126.com, zhaogang@bistu.edu.cn

Abstract. With arrival of big-data era, data mining algorithm becomes more and more important. K nearest neighbor algorithm is a representative algorithm for data classification; it is a simple classification method which is widely used in many fields. But some unreasonable parameters of KNN limit its scope of application, such as sample feature values must be numeric types; Some unreasonable parameters limit its classification efficiency, such as the number of training samples is too much, too high feature dimension; Some unreasonable parameters limit the effect of classification, such as the selection of K value is not reasonable, such as distance calculating method is not reasonable, Class voting method is not reasonable. This paper proposed some methods to rationalize the unreasonable parameters above, such as feature value quantification, Dimension reduction, weighted distance and weighted voting function. This paper uses experimental results based on benchmark data to show the effect.

Keywords: KNN · Data mining · Rationalize parameters · Quantify feature value · Dimension reduction · Weighted euclidean distance · Weighted voting function

1 Introduction

K nearest neighbor algorithm is a well-known machine learning classification algorithm [1]; its main idea is classify the new sample through judging class of the K train samples closest in distance to new sample. Despite its simplicity, KNN classification algorithm has been successful in a large number of classification problems, such as text classification [2], plagiarism investigation [3], prediction analysis [4], pattern recognition, image processing [5]. In conclusion, the KNN algorithm still occupies a very important position in data mining.

This paper mainly researches how to rationalize some important parameters of the classical KNN algorithm. This paper discusses the parameters in samples pretreatment and algorithm process two aspects.

Rationalizing the parameters of samples pretreatment mainly has three aspects: compress samples size, quantify feature, reduce feature dimension. Rationalizing the parameters of algorithm process contains selection of K value, selection of distance

© Springer International Publishing Switzerland 2015
W. Qiang et al. (Eds.): CloudCom-Asia 2015, LNCS 9106, pp. 189–202, 2015.
DOI: 10.1007/978-3-319-28430-9_15

metric measure and class voting method. And experimental results based on UCI and KDD benchmark data sets will prove these measures can improve classification effect and efficiency of KNN algorithm. This paper uses the leave one out cross validation [6] as the validation method. Classification accuracy is the most important factor to evaluate effect of KNN algorithm.

2   KNN Algorithm

K-nearest neighbors algorithm (or KNN for short), is a relatively simple classification algorithm in machine supervised learning, a famous pattern recognition statistical method of data mining [7–11].

In KNN classification, all neighbors to the new sample have been correctly classified. If most of the k nearest neighbors to the new sample is belong to certain class, the new sample will be classified to this class, and has the characteristics of the samples in this class. The method to make the classification decision based on only one or several neighbor samples.

The main process of the classical KNN classification algorithm is as follows:

Step1:   Calculate the distance between the test sample and every training sample, the distance calculating method is the Euclidean distance.
Step2:   Choose the k minimum distance samples as the k nearest neighbors.
Step3:   Majority votes of its neighbors will determine which class it belongs to.

Although KNN algorithm has excellent performance in many fields, many unreasonable parameters cause its limitations. This paper mainly discusses how to rationalize parameters in sample parameters and algorithm parameters two aspects.

3   Rationalize the Sample Parameters

3.1   Sample Compression

KNN algorithm is a lazy algorithm. Too big sample size will greatly increase the complexity of the algorithm, reducing the efficiency of classification. So reducing the number of sample points has a considerable significance. In sampling process, we try to follow the following principles:

(1)   Proportionate sampling [12]: Try not to change sample quantitative proportion of each class. This paper uses interlaced sampling to choose samples. For example, supports there are 3 classes of sample data (class A: 600 samples; class B: 300 samples; class C: 150 samples). We choose the odd numbered samples (also may be the even numbered samples) of each class, so the result is class A: 300 samples, class B: 150 samples; C: 75 samples. We do not change the proportion of each class.

(2) Class Sample number threshold: The number of each class sample must be greater than a threshold. In KNN algorithm, it is unfair to a class with too small sample size. This threshold is at least greater than k/n (k is the number of nearest neighbors, n is the number of classes). If a class sample quantity is less than k/n, this kind of sample will lose its significance in k nearest neighbor voting; because it is impossible to judge a new sample is belong to this class. For example, support there are 3 classes of sample data (class A: 400 samples; class B: 100 samples; class C: 30 samples), and we set the threshold = 25. According to interlaced sampling, we choose the odd numbered samples of each class; half number of class C is 15 that do not reach the threshold 25, so we do sampling the class C, the sampling result is class A: 200 samples, class B: 50 samples; C: 30 samples. The number of each class is greater than threshold.

This paper use KDD1999 intrusion detection dataset as experiment data, we choose kddcup.data_10_percent as original data document in the experiment [13]. This data set has 5 classes, a total of 494020 lines (normal: 97278; probe: 4107; Dos: 391458; u2r: 52; r2l: 1126); it is a very large sample dataset. Because the number of u2r is too low, we only select Dos, normal, probe, r2l four classes in the experiment. Based on the principles above, ensuring every class quantity is greater than the threshold = 40, we extract every class sample by Interlaced sampling.

After several times of sampling, six final training samples dataset content as shown in Table 1.

Table 1. Six training sample dataset extracted from the KDD1999

samples lines	dos	normal	probe	r2l
7743	6118	1520	65	40
3924	3059	760	65	40
2015	1530	380	65	40
1060	765	190	65	40
583	383	95	65	40
345	192	48	65	40

In the experiment, this paper uses one out cross validation to test KNN algorithm classification accuracy. To the six training sample dataset, the result as shown in Fig. 1:

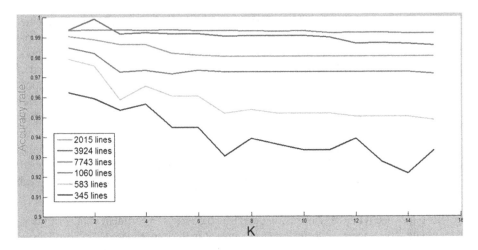

Fig. 1. The classification accuracy rate in six training sample dataset

The average time cost by six training samples dataset in the algorithm process is as shown in Table 2:

Table 2. Average time of six training samples dataset cost in the classification process

Sample dataset	The average time(s)
7743-lines	58
3924-lines	13
2015-lines	3
1060-lines	0.472
583-line	0.413
345-lines	0.208

From Fig. 1 we can see, with the decrease in the number of samples, the sample dataset performance becomes worse, 7743-lines sample dataset do the best perform-ance in the aspect of classification accuracy rate. But there is only an extremely small gap between the 7743-lines and the 3924-lines. In Table 2, the time 7743-lines cost is far more than the 3924-lines. Obviously, 7743-lines are not the cost-effective choice. If the demand of accuracy rate is not high, for example, if 97 % is an acceptable accuracy rate, we can choose 2015-lines as the training sample dataset.

Above all, we can conclude that sample compression can greatly improve the clas-sification efficiency of the KNN algorithm. In the condition of ensuring high correct classification rate, compressing samples is very necessary.

3.2   Quantify Non-Numeric Feature

In KNN algorithm, distance calculation requires feature value must be numeric type. Obviously not all the samples feature value are numeric type, so we must quantify the sample feature value.

To the dataset that we can judge every class qualitatively (for example, if a flower dataset has two classes: beautiful and ugly, obviously class beautiful is better than class ugly), we use the following method: Give a numeric value to every class. We use $(C_1, C_2...C_k)$ as the numeric value vector to each class. $(V_1, V_2... V_l)$ is the numerical value vector to a non numerical feature. V_i can be calculated by the following formula:

$$
\begin{aligned}
V_1 &= \frac{1}{N_1} \sum_{i=1}^{k} n_{1i} C_i \\
V_2 &= \frac{1}{N_2} \sum_{i=1}^{k} n_{2i} C_i \\
&\quad ... \\
V_l &= \frac{1}{N_l} \sum_{i=1}^{k} n_{li} C_i
\end{aligned}
\tag{1}
$$

In the formula, N_k represents the number of the value k in this feature, n_{ki} represents the number of the value k which belongs to class i.

In order to verify the rationality of this quantitative method, this paper selects the UCI benchmark dataset car to validation. The car dataset contains a total of 1728 samples, 6 features (all feature value is discrete, and most of them are not numeric type).

There are 4 kinds of classification results: unacc, acc, good, vgood. According to our understanding of these four classes, we can easily judge: unacc < acc < good < vgood. So we establish a value vector C = {1, 2, 3, 4} to four classes. According to the formula above, calculate result to all feature values are as shown in Table 3:

Table 3. The result of quantifying feature value in car dataset

Buying		Maint		Doors		Persons		Lug_boot		Safety	
nonu	num	nonu	num	nonu	num	nonu	num	nonu	num	nonu	num
vh	1.167	vh	1.167	2	1.326	2	1	small	1.255	low	1
h	1.25	h	1.333	3	1.417	4	1.625	med	1.448	med	1.448
m	1.553	m	1.553	4	1.458	4+	1.620	Big	1.542	high	1.797
l	1.690	l	1.606	5+	1.458						

In Table 3, the first line contains all feature names in car database. The row "nonu" contains original non-numerical feature value in each feature. The row "num" contains the quantized value to each feature. This paper uses one out cross validation to test KNN algorithm classification accuracy of the quantized feature value car dataset, the result as shown in Fig. 2:

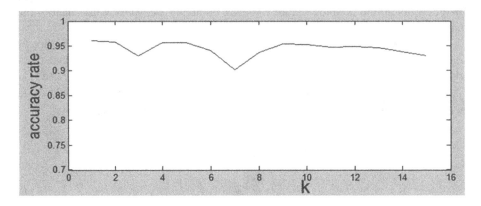

Fig. 2. KNN classification effect on quantized feature value car dataset

The K value from 1 to 15, all the correct rate are above 90 %, the highest correct rate when K = 1 is 96.06 %, the lowest correct rate when K = 7 is 90.22 %. In conclusion, the dataset with numeric feature value calculated by this quantity method makes a very good performance in the KNN algorithm.

3.3   Dimension Reduction of Sample Feature

Too high sample feature dimensions will greatly reduce the efficiency and the classification accuracy rate of the KNN algorithm [14]. And some sample dataset have a large number of features, but not every feature is necessary to classification, they have little impact on classification, we call these feature useless feature. If we can eliminate these useless features, the KNN classification algorithm effect will be greatly improved. So features selection is very necessary. In order to select the appropriate feature, this paper is mainly divided into the following two steps:

(1)   Eliminate features with extremely low variance: Variance reflects the concentration degree of a set of data distribution, if a feature variance is too small, it shows the feature value does not have too big wave. In KNN classification algorithm, if a feature variance is extremely low, the distance between two samples is close to zero, it is Meaningless to choose the K nearest neighbor. So, we can set a variance threshold, and then calculate variance of every normalized feature values. If the variance of a feature does not reach this threshold, we will eliminate this feature.

(2)   Chi square test to eliminate features [18]: The basic thought of Chi square test is calculating the deviation value between actual value and the theoretical value, determine whether the theoretical value is appropriate. In the classification algorithm, if a feature's deviation value between theoretical value and observation value

is relatively large in each of class, the feature has a relatively large impact on the classification. For a sample set, we first compute theoretical value of each feature to each class. Assuming the sample database is made of n samples, m class, l features. The theoretical value of feature j can be calculated by the following formula:

$$E_{ij} = p_i \sum_{k=1}^{n} x_k \tag{2}$$

In the formula, E_{ij} represents the theoretical value from feature j to the class i, P_i represents class i sample's proportion in the total sample, x_k represents the feature j value in the line k.

The observed value of feature j can be calculated by the following formula:

$$O_{ij} = \sum_{k=1}^{n^{(i)}} x_k^{(i)} \tag{3}$$

In the formula, O_{ij} represents the observation value from feature j to the class i, $n^{(i)}$ represents the number of sample in class i, $x_k^{(i)}$ represents feature j value of sample k in class i.

The deviation value of feature j can be calculated by the following formula:

$$S_j = \sum_{i=1}^{m} \frac{(O_{ij} - E_{ij})^2}{E_{ij}} \tag{4}$$

Through the formula above, we can get the final deviation vector $\{S_1, ..., S_l\}$ to each feature. We will eliminate one or more features with the smallest the deviation value S_j.

In order to verify the effect of above dimension reduction methods, we still use kdd1999 data which has been reduced to 3924 lines as the sample set for validation. The number of features of the original data sets is 41.

First step, we set a very low threshold = 0.000001, the purpose is to eliminate the feature with very low variance, in this step we eliminate 4 features, then there are 37 features left.

Second step we use chi square test eliminate q features with lower deviation. We set q = 7, 12, 22, then the features dimension is reduced to 30 dimensions, 25 dimensions and 15 dimensions. Classification results of the KNN algorithm as shown in Fig. 3:

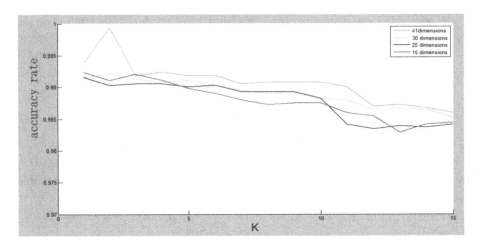

Fig. 3. KDD1999 (3924-lines) KNN classification results after dimension reduction

The average time of different feature dimension cost in the KNN classification process as shown in Table 4:

Table 4. Average time of six training samples dataset cost in the classification process

Feature dimension	The average time(s)
41 features	13
30 features	10
25 features	8
15 features	5

Based on the above experimental results we can see, the classification accuracy rate of 41 dimensions, 30 dimensions, 25 dimensions and 15 dimensions in using the KNN algorithm have not much difference. Considering the average time algorithm consuming, low dimension data have obvious advantages on high dimension data. From the results above, we can conclude that the method for selecting the feature is quite reasonable in KNN algorithm, and feature dimension reduction is meaningful when there is not an extremely strict demand to accuracy rate.

4   Rationalize the Algorithm Parameters

4.1   Rationalize K Value

Selection of k value has a large impact on the correct rate of classification in KNN algorithm, so the method we choose k value should be conservative, k value should not be too large or too small. If k value is too small, the classification results are susceptible to noise influence; if k is too large, neighbors and may contain too many points of other classes. There is no uniform standard method to select k value currently, we usually using cross validation to test k from x to other large value, x is usually greater than 1; x can take appropriate great value when sample size is large. The judge index is accuracy (sometimes considered the cost of algorithm), and finally select the k value which perform best.

4.2   Rationalize Distance Calculating Method

The metric of classic KNN algorithm for distance calculate is the Euclidean distance, Euclidean distance formula is as follows:

$$d(i,j) = \sqrt{\sum_{l=1}^{n} |X_{il} - X_{jl}|^2} \tag{5}$$

In the formula, $i = (X_{i1}, X_{i2} \dots X_{in})$, $j = (X_{j1}, X_{j2} \dots X_{jn})$ represents two data sample point. In Euclidean distance, every feature is equally important in calculating the distance. But the fact is not the case, some features can have a significant effect on classification, and some features have little contribution to the classification. So it is very necessary to determine the weights of each feature.

To the feature weight calculation, there are many researches at present. Methods are roughly divided into the objective determination weight method and subjective weight determination method. Subjective weight determination method mainly has the expert scoring method, Delphi method, AHP method; the objective weight determination method mainly has the entropy method, grey correlation method, artificial neural network weighting method, factor analysis, regression analysis. This paper selects maximum ratio between between-class and inner-class deviation method [15] as the feature weight calculation method, and test its effect on KNN algorithm in the experiment.

Deviation is generally used to indicate the degree of concentration of a set of data, reflect the gap from real value to average value. After the classification, if the samples from the same class are more concentrated and the samples from different class are more discrete, we can judge this classification as a more qualified classification. The ratio of between-class deviation and inner-class deviations is higher, this classification is more reasonable. So this paper use maximum ratio of between-class and inner-class deviation method to determine the weight of each features value. Specific steps are as follows:

If the sample data has a total of K features, assuming their weight vector $W = (W_1, W_2, \dots, W_k)^T$, the sample data has m classes, the number of samples of each class is $n_1, n_2, \dots n_m$, according to the normalized value in sample data, sample values of each type are as follows:

$$
\begin{aligned}
C_1 &: W^T X_1^1, W^T X_2^1, \ldots W^T X_{n_1}^1 \\
C_2 &: W^T X_1^2, W^T X_2^2, \ldots W^T X_{n_2}^2 \\
&\quad\cdots \\
C_m &: W^T X_1^m, W^T X_2^m, \ldots W^T X_{n_m}^m
\end{aligned} \tag{6}
$$

In the formula, C_1, C_2,..., C_m represents m classes, X_j^i represents feature value of sample j in class i.

The mean of class i is as follows:

$$
W^T \bar{X}^i = \frac{1}{n_i} \sum_{j=1}^{n_i} W^T X_j^i \tag{7}
$$

Inner-class sum of deviations squares is as follows:

$$
ISSD = \sum_{i=1}^{m} \sum_{j=1}^{n_i} (W^T X_j^i - W^T \bar{X}^i)^2 =
$$

$$
W^T [\sum_{i=1}^{m} \sum_{j=1}^{n_i} (X_j^i - \bar{X}^i)(X_j^i - \bar{X}^i)^T] W = W^T A W \tag{8}
$$

$$
\text{s.t. } A = \sum_{i=1}^{m} \sum_{j=1}^{n_i} (X_j^i - \bar{X}^i)(X_j^i - \bar{X}^i)^T
$$

The total mean of all the m class is

$$
W^T \bar{X} = \frac{1}{n} \sum_{i=1}^{m} \sum_{j=1}^{n_i} W^T X_j^i \tag{9}
$$

Between-class sum of deviations squares is as follows

$$
CSSD = \sum_{i=1}^{m} n_i (W^T \bar{X}^i - W^T \bar{X})^2 =
$$

$$
W^T [\sum_{i=1}^{m} n_i (\bar{X}^i - \bar{X})(\bar{X}^i - \bar{X})^T] W = W^T B W \tag{10}
$$

$$
\text{s.t. } B = \sum_{i=1}^{m} n_i (\bar{X}^i - \bar{X})(\bar{X}^i - \bar{X})^T
$$

According to the sample data, the weight vector should make

$$
\theta(W) = \frac{CSSD}{ISSD} = \frac{W^T B W}{W^T A W} \tag{11}
$$

maximum. According to the related mathematical derivation, maximum $\theta(W)$ value is largest eigenvalue of formula $|B-\lambda A|$, the corresponding eigenvector is W. W is the feature weight vector. B and A can be calculated by sample dataset.

Combine resulted weight vector with Euclidean distance, get the weighted Euclidean distance formula [16]

$$d(i,j) = \sqrt{\sum_{l=1}^{n} W_1 |X_{il} - X_{jl}|^2} \tag{12}$$

We use it as the distance calculating method in KNN algorithm. Taking dataset datingTestSet as the test dataset, Classification results of the KNN algorithm as shown in Fig. 4:

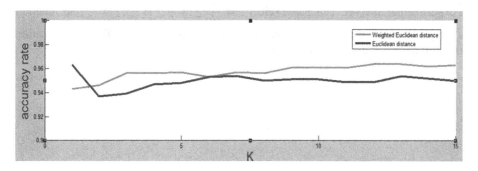

Fig. 4. Classification results of KNN algorithm of weighted Euclidean distance and Euclidean distance

From Fig. 4, we can see weighted Euclidean distance (weight vector calculated by maximum ratio of between-class and inner-class deviation method) perform just a little better than the non-weighted Euclidean distance in KNN classification algorithm.

Weighted Euclidean distance (Weight vector calculated by maximum ratio of between-class and inner-class deviation method) performs barely satisfactory in KNN algorithm. With the increase number of feature, the ability to distinguish of Euclidean distance becomes worse. There are mainly two kinds of solutions: 1. Dimension reduction, which has been mentioned in the previous chapter. 2. With other distance calculating method, such as the Manhattan distance, cosine angle distance [17] etc. This is our next step research to rationalize distance calculating method.

4.3 Rationalize Method for Judging the Class

The voting method of classical KNN algorithm is select K nearest neighbors and each neighbor can vote one. All of the neighbors are equal. In reality, however, K neighbors have the different distance to the target sample point. According to KNN algorithm principal, the nearest neighbor should have the highest status, and has highest weight in voting. Therefore, the weight voting would be more proper. The weight is negative correlation with distance between the neighbor and target sample point. In that way, the weighting function is monotone decreasing function that regards distance as the independent variable. Each class value should be calculated as following formula:

$$V_j = \sum_{i=1}^{n_j} f(D_{ji}) \tag{13}$$

V_j represents total weighted voting value obtained by class j, n_j represents the number of class j nearest neighbor, D_{ji} represents the distance between neighbor i in class j and target point, $f(D)$ is the weighted voting function that regards distance as the independent variable.

Select the weighted voting function f should be based on the test results of sample dataset; this paper selects the UCI benchmark database User Knowledge Modeling Data as experimental data set. This paper select $f = \frac{1}{\sqrt{D}}, f = \frac{1}{D}, f = \frac{1}{D^2}, f = \frac{1}{e^D}$ four function as weighted voting function, f = 1 means non weighted voting, the final classification results of the KNN algorithm as shown in Fig. 5:

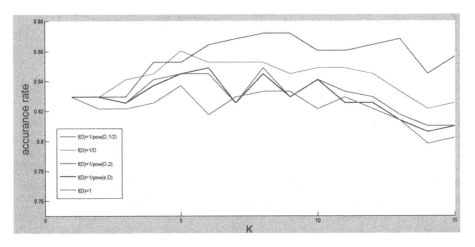

Fig. 5. Classification results using different weighted voting function in KNN algorithm

The effect can be seen in figure, base on User Knowledge Modeling this data set, KNN algorithm with the classic non weighted voting method performs worst, KNN algorithm that uses $f = \frac{1}{D^2}$ as vote weighting function performs best.

The result also proves that a reasonable weighted voting function can obviously improve the classification accuracy of KNN algorithm.

5   Conclusions

Undoubtedly, KNN algorithm was regarded as one of the most classical machine learning classification algorithm. During big data era, KNN algorithm plays a more important role in data mining. In this paper, we discussed some important parameters. We also partly changed those parameters based on classical KNN algorithm and utilized some benchmark data to test, making those parameters more rational. Therefore, the effect and efficient of KNN algorithm could be improved. Meanwhile, some parameters were not discussed thoroughly, for example, the Similarity distance other than Euclidean distance. The testing for such distance was the direction of future research.

Acknowledgement. *This work was supported by the National Natural Science Foundation of China (Grant No. 61272513) and Beijing Municipal Science and Technology Project (Grant No. D151100004215003).*

References

1. Larose, D.T.: k-nearest neighbor algorithm. In: Discovering Knowledge in Data: An Introduction to Data Mining, pp. 90–106 (2005)
2. Soucy, P., Mineau, G.W.: A simple KNN algorithm for text categorization. In: Proceedings of the IEEE International Conference on IEEE Data Mining, ICDM 2001, pp. 647–648 (2001)
3. Bandara, U., Wijayarathna, G.: A machine learning based tool for source code plagiarism detection. Int. J. Mach. Learn. Comput. **1**(4), 337–343 (2011)
4. Chung, C.H., Parker, J.S., Karaca, G., et al.: Molecular classification of head and neck squamous cell carcinomas using patterns of gene expression. Cancer Cell **5**(5), 489–500 (2004)
5. Costa, J.A., Hero, A.O.: Manifold learning using Euclidean k-nearest neighbor graphs (image processing examples). In: Proceedings of the IEEE International Conference on Acoustics, Speech, and Signal Processing (ICASSP 2004), vol. 3, pp. iii-988–iii-991. IEEE (2004)
6. Kohavi, R.: A study of cross-validation and bootstrap for accuracy estimation and model selection. IJCAI **14**(2), 1137–1145 (1995)
7. Parvin, H., Alizadeh, H., Minati, B.: A modification on k-nearest neighbor classifier. Glob. J. Comput. Sci. Technol. **10**(14), 37–41 (2010)
8. Parvin, H., Alizadeh, H., Minaei-Bidgoli, B.: MKNN: modified k-nearest neighbor. In: Proceedings of the World Congress on Engineering and Computer Science, pp. 831–834 (2008)
9. Li, L., Weinberg, C.R., Darden, T.A., et al.: Gene selection for sample classification based on gene expression data: study of sensitivity to choice of parameters of the GA/KNN method. Bioinformatics **17**(12), 1131–1142 (2001)
10. Yahia, M.E., Ibrahim, B.A.: K-nearest neighbor and C4.5 algorithms as data mining methods: advantages and difficulties. In: ACS/IEEE International Conference on Computer Systems and Applications, p. 103. IEEE (2003)
11. Mylonas, P., Wallace, M., Kollias, S.D.: Using k-nearest neighbor and feature selection as an improvement to hierarchical clustering. In: Vouros, G.A., Panayiotopoulos, T. (eds.) SETN 2004. LNCS (LNAI), vol. 3025, pp. 191–200. Springer, Heidelberg (2004)
12. Midzuno, H.: On the sampling system with probability proportionate to sum of sizes. Ann. Inst. Stat. Math. **3**(1), 99–107 (1951)
13. Sabhnani, M., Serpen, G.: Application of machine learning algorithms to KDD intrusion detection dataset within misuse detection context. In: MLMTA 2003, pp. 209–215 (2003)

14. Li, T., Zhang, C., Ogihara, M.: A comparative study of feature selection and multiclass classification methods for tissue classification based on gene expression. Bioinformatics **20**(15), 2429–2437 (2004)
15. Chen, W.: A method to determine weight according samples. China's High. Educ. Eval. **4**, 018 (2003)
16. Hechenbichler, K., Schliep, K.: Weighted k-nearest-neighbor techniques and ordinal classification (2004)
17. Qian, G., Sural, S., Gu, Y., et al.: Similarity between Euclidean and cosine angle distance for nearest neighbor queries. In: Proceedings of the 2004 ACM Symposium on Applied Computing, pp. 1232–1237. ACM (2004)
18. Liu, H., Setiono, R.: Chi2: Feature selection and discretization of numeric attributes. In: 2012 IEEE 24th International Conference on Tools with Artificial Intelligence, pp. 388–388. IEEE Computer Society (1995)

CUDAGA: A Portable Parallel Programming Model for GPU Cluster

Yong Chen[1,2], Hai Jin[1], Dechao Xu[2], Ran Zheng[1(✉)], Haocheng Liu[1], and Jingxiang Zeng[1]

[1] Services Computing Technology and System Lab,
School of Computer Science and Technology,
Huazhong University of Science and Technology,
Wuhan 430074, China
{hjin,zhraner}@hust.edu.cn
[2] China Electric Power Research Institute, Beijing 100192, China

Abstract. GPU cluster is important for high performance computing with its high performance/cost ratio. However, it is still very hard for application developers to write parallel codes on GPU. MPI is mostly used for parallel programming, and data locality and communication must be specified explicitly by developers. Moreover, data transmission between CPU and GPU must also be processed with CUDA codes. CUDAGA, a new parallel programming model for GPU cluster with CUDA, is presented to provide portable interfaces for commu-nication on GPUs. GA (*Global Arrays*), a portable shared-memory programming model for distributed memory computers, is the base to facilitate parallel pro-gramming and maintain transparent global arrays on GPUs. Experiments show that CUDAGA can decrease parallel programming difficulties, but ensures better performance for some specific applications.

Keywords: GPU cluster · Parallel computing · Programming model · Global arrays

1 Introduction

Graphics Processing Unit (GPU) [1] has been used widely for general-purpose computation. The applications accelerated by GPU can achieve high performances and GPU cluster [2] has been explored successfully for specific applications, such as power flow calculation, flow simulation and molecular modeling. In most GPU cluster, the computing kernel is normally programmed with CUDA (*Compute Unified Device Architecture*) [3] and the communications are programmed with MPI (*Message Passing Interface*). Therefore, developers need to be good at programming with CUDA and MPI. Moreover, proper GPU device should be chosen and data locality and communication should be explicitly specified. It is too hard to manage both computing kernels on GPU and the communications executed with MPI.

© Springer International Publishing Switzerland 2015
W. Qiang et al. (Eds.): CloudCom-Asia 2015, LNCS 9106, pp. 203–216, 2015.
DOI: 10.1007/978-3-319-28430-9_16

Shared memory model are adapted on GPU cluster by many researchers to reduce programming complexity, but some of them get poor performances because of extra costs on data consistency, and some are only suitable for graphics processing. The device choosing problem is ignored when there is more than one GPU in one node. Moreover, basic libraries for public functions are missing, so that it is hard for application developers to write high-efficiency programs quickly. It will be better to provide standard libraries with some common functions, so that developers can pay more attentions on designing better application algorithms but not the details on how to realize those basic functions on GPUs.

CUDAGA, a portable parallel programming model for GPU cluster, is proposed to reduce developers' work in this paper. GA-based communication mechanisms and CUDA memory management are combined, and many types of interfaces are provided for simple programming. Compared with MPI-based method, CUDAGA can release application developers from low-level burdens on data transfer, and the performances can become higher with the combination of GA [4] and CUDA.

This paper is organized as follows. Section 2 provides related work about parallel programming on GPU cluster. Section 3 proposes the new CUDAGA parallel programming model of CUDAGA. Section 4 introduces the details and key issues of CUDAGA. Experimental results are presented in Sects. 5 and 6 concludes the paper.

2   Related Work

GPU-to-GPU communication interface is necessary to simplify parallel development in cluster. There are two major paradigms: message passing and shared memory models [5]. MPI is the dominant interface for message passing model and OpenMP is designed for shared memory model.

MPI is widely used on CPU cluster and currently applied on GPU cluster. CaravelaMPI [6] provides unified and transparent interface to manage both communication and GPU execution, which focuses on traditional runtime environment and asynchronous message passing interface to overlap communication with computation [7]. cudaMPI [8], à GPU-to-GPU communication model, combines MPI communication interfaces with CUDA memory copy functions. However, the model is based on MPI, so that it is still difficult for parallel programming.

Since shared-memory model is widely considered easier to develop than message passing, many researchers have sought to realize a shared-memory model for GPU cluster. A *distributed shared-memory* (DSM) model [9] is implemented through the virtualization of distributed texture memories. However, its performance is very bad because of its high latency overhead of data-consistency maintenance.

Combining the advantages of distributed memory model and shared memory, GA toolkit [10] provides an interface to allow data distribution, whose programming syntax is similar with the programming on a single processor. Zippy [11] adapts GA programming model on GPU cluster and combines it with traditional graphics processing program [12]. It reduces the difficulty for parallel programming on GPU cluster and provides a debug tool for debugging parallel program. However, it is only suitable for graphics processing. Therefore, it is necessary to design a parallel programming model for general-purpose computation.

As we all know, both GA and CUDA are all good at disposing array data structure. For the above reasons, CUDAGA is proposed, which combines GA and CUDA to decrease the difficulty for parallel programming on GPU cluster.

3 Parallel Programming Model on GPU Cluster with CUDA

CUDAGA is similar to the process of GA, and the implementation of CUDAGA is on the upper layer of CUDA and GA. This section presents CUDAGA as a parallel programming model on GPU cluster.

3.1 Layered Architecture of CUDAGA

Distributed Arrays Layer is user-oriented and implemented based on *Library Layer*, in which CUDA runtime API and APIs in GA are included, shown in Fig. 1.

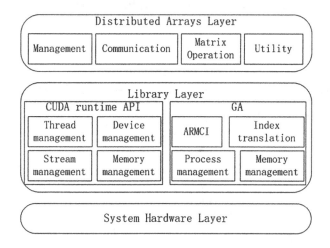

Fig. 1. Layered architecture of CUDAGA

There are four modules on Distributed Arrays Layer: management, communication, matrix operation and utility module. The functions of the four modules are implemented through combining CUDA runtime API and the APIs in GA. CUDA runtime API includes the functions of thread management, device management, stream management and memory management. ARMCI (*Aggregate Remote Memory Copy Interface*), memory management, index translation and process management are included in GA. Of course, a lot of optimization can be operated during the combination to improve the performance of CUDAGA. The bottom layer is System Hardware Layer, including the switches to connect NVIDIA GPUs among all cluster nodes.

Application developers can invoke related interfaces of CUDAGA to do various operations on GPU. Data must be synchronized in global array from GPU to CPU and the environment must be cleaned correspondingly when the programs on GPU are finished. After starting the operations on GPU, the changes of data on CPU and GPU will not bother with each other, because the storages of global array on CPU and GPU are independent. CUDAGA can help choosing proper GPU [13] for each process when initializing environment. Data can be put to or got from global array through the invoking of communication functions.

3.2   Two-Level Communication Model

Two-level parallelism is designed for the communication among different GPUs: data transfer between different CPU memories and data transfer between CPU and GPU. The bandwidth of memory accessing is non-uniform in two-level parallelism. The bandwidth inside GPU memory is the largest, and the bandwidth between CPU and GPU is larger than that between different CPUs. The bandwidth between CPU and GPU is decided by PCIE and the bandwidth between different CPUs is decided by switch. It is better to improve the performance by designing good algorithm to decrease data transfer in low-bandwidth environment.

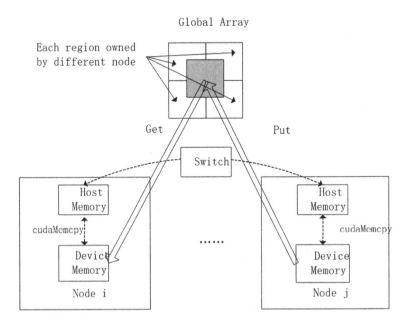

Fig. 2. Communication model of CUDAGA

CUDAGA hides the details of data communication, so that developers can manage data transfer with the functions of *get* and *put* in Fig. 2. The cost of data transfer between CPU and GPU is reduced to improve the performance. Moreover, some interfaces, such as management, matrix operations and utility functions, are provided in CUDAGA to help developers to execute some common operations.

As we all know, developers must dispose communication interfaces for CPU-to-CPU communication and CUDA memory management for CPU-to-GPU/GPU-to-CPU when developing parallel applications. CUDAGA provides direct GPU-to-GPU communication interface, in which some transfer details are hidden, such as memory copy from GPU to CPU in remote node, data copy among different nodes and memory copy from CPU to CPU in one node. At the same time, matrix functions can be called directly to speed up linear algebra computation.

The main target of CUDAGA is to release developers from the work of implementing effective communication, so that they can pay more attentions on the realization of applications themselves. With these function packages, developers can write fewer codes to implement the same operation to make the code more concise. At the same time, some optimization is done in the package to improve the performance.

4   Key Issues and Technologies in CUDAGA

Some key issues must be solved during the development of a parallel application on GPU cluster: how to select a suitable GPU device, how to simplify the programming in GPU cluster, how to debug or monitor on GPU cluster.

4.1   Execution Management

The initialization and finalization of environment are implemented in management module of CUDAGA. Device allocation of each process is done during the initialization, there are two advantages: it can avoid duplicated allocation by different processes, and all devices can be allocated for utilization. But it can not deal with the situation that the number of GPU units is less than the number of running process.

Global array is created by creation function on CPU and GPU. A corresponding temp global array exists on CPU to participate the maintenance of data consistency with global array on GPU. The following operations are executed transparently in global arrays of GPU and corresponding temp global arrays of CPU, so that developers only need to write the codes of synchronizing data from CPU to GPU or from GPU to CPU. The operations on CPU and GPU will not be bothered each other because of synchronization mechanism.

4.2   GPUs Communication

There are three types of data transfers in GPU cluster: memory copy of GPU-to-CPU and CPU-to-GPU, message communication on different nodes.

Traditionally, MPI function is used for message communication between CPUs, and *cudaMemcpy* is used for memory copy between CPU and GPU. It will place even greater burdens of application developers. Data transfers are implemented with a GA-similar communication mode in CUDAGA, and the communication between different GPUs can easily managed by *Get* and *Put* functions. As shown in Fig. 3, data enclosed in black line is shared by node i and j, which can be managed by communication functions. The middle area enclosed by dotted lines is the actual implementation method transparent to developers. Temp GA is created with the creation of CPU GA, and data between CPU GA and GPU GA is synchronized through temp GA. The data in temp GA is always the same with GPU GA, so that data synchronization from GPU to CPU can be done only by one step. All the operations on GPU GA will not affect the data in CPU GA, and data only needs to be synchronized between CPU and GPU at the start and the end of the execution.

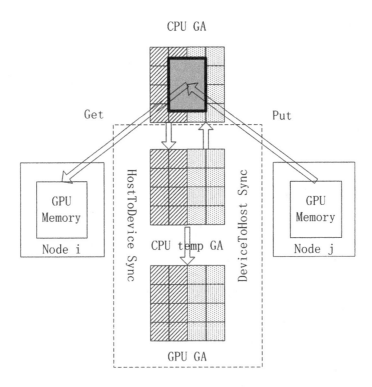

Fig. 3. Data transfer model with CUDAGA on GPU cluster

The implementation details are shown in Fig. 4. Suppose node i gets data from the two nodes and node j puts data to the two nodes. If data is local and continuous, memory copy from GPU array to d_buf is called and its time cost is noted as $dtod$. This is very fast because the bandwidth inside GPU is much bigger than PCI. Of course, data range must be continuous in GPU array to obtain higher performance. Otherwise, data will be transferred from temp GA to h_buf through get, and then from h_buf to d_buf ($htod$) through the operation of memory copy.

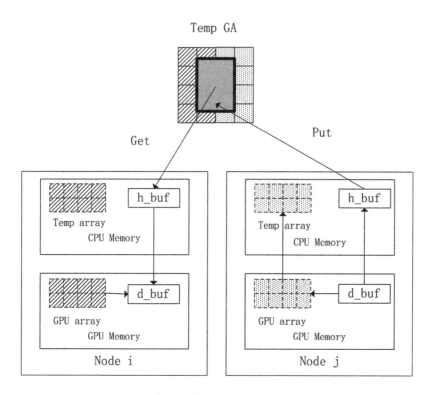

Fig. 4. Inside implementation of data movement

The processing of put function is a little different from get. For put function, if data is local and continuous, two operations are needed: memory copy from d_buf to GPU array (called $dtod$) and synchronization between CPU temp array and GPU array (called $dtoh$) for data consistency. If data is local but not continuous or remote, data must be copied from d_buf to h_buf (called $dtoh$) and put from h_buf to temp global array. Then the processes whose data is from d_buf ($htod$) will be checked out for data synchronization between temp array and GPU array.

To compare with GA+CUDA mode, define T_{get} the time cost of get and T_{put} the time cost of put on GPU cluster. In GA+CUDA mode, the two time costs can be computed with Formulas (1) and (2) respectively.

$$T_{get-GA} = get + htod \tag{1}$$

$$T_{put-GA} = dtoh + put \tag{2}$$

In CUDAGA mode, the two time costs can be computed with Formulas (3) and (4) respectively.

$$T_{get-CGA} = \begin{cases} dtod & \text{(for local and continuous data)} \\ get + htod & \text{(for other data)} \end{cases} \tag{3}$$

$$T_{put-CGA} = \begin{cases} dtod + dtoh & \text{(for local and continuous data)} \\ dtoh + put + htod & \text{(for other data)} \end{cases} \tag{4}$$

As we know, $dtod$ is much faster than $htod$ and $dtoh$. Local get and local put are also very fast while the cost of remote operation is affected by switch bandwidth. $T_{get-CGA}$ is not worse than T_{get-GA} and will improve the performance when data is local and continuous. $T_{put-CGA}$ is approximately equal to T_{put-GA} when data is local and continuous, while $T_{put-CGA}$ needs additional $htod$ overhead. The overhead can be produced by the processes whose data is changed in temp global array. We believe the performance can be improved more than the overhead, because get operation is used more than get operation in most applications.

4.3   Matrix Operation

There are many linear algebra functions in GA, which are included in matrix operation module to avoid duplicate development.

As we know, the bandwidth of PCI Express is 8 GB/s, which is a performance bottleneck when computing kernel is not complicated. These functions are operated on global arrays of GPU, which can achieve good performance without data transfer between CPU and GPU memory. It is very useful for some simple linear calculations. Many temp memory buffers are used during the implementations of these functions, and the overhead of memory allocation or release is small and negligible at each time. However, the sum of overhead is very big when a lot of linear algebra functions are called. It can be reduced by calling unified memory management functions provided in system management module.

Moreover, there are three communication modes between CPU and GPU: ASYNC, PAGEABLE and ZEROCOPY mode. PAGEABLE mode is the most common used mode without any effects on CPU memory. ASYNC mode is used to overlap communication and computation, and ZEROCOPY mode is suitable for integrated graphics cards. The different matric operation functions with the three modules are provided for choosing according to their environments.

4.4   Assistant Support

Some auxiliary functions for monitoring GPU cluster information are included in assistant support module. As we know, program debugging has been difficult in both parallel programming and CUDA programming.

Assistant support can help developers debug in programming. When developing parallel applications, developers can debug their programs or observe the status of GPU cluster with these functions, such as the total number of GPU, GPU id, corresponding process id to GPU.

5   Performance Analysis and Evaluation

The experiments are run on a cluster with five nodes connected by 10 MB/s Ethernet. Four nodes are composed of Intel i7 920 and two NVIDIA GTX 295, and the rest one is equipped with Intel Q9550 and one GTX 260. All nodes run Red Hat Linux with CUDA 2.3, mpich-1.2.7 and GA 4.3. Programming complexity and performance of CUDAGA are analyzed and evaluated, compared with CUDA+MPI. Cannon algorithm and Jacobi iteration are two applications for testing.

5.1   Programming Complexity Analysis

Since CUDAGA helps developers to manage the details of communication on GPU cluster, it is much simpler to program with CUDAGA than CUDA+MPI. Figure 5 shows an example implemented by CUDA+MPI.

We can see that data scattering and gathering are complicated to be implemented with many operations. Data must be scattered from master process to all processes and gathered from all processes back to master process, transferred explicitly between CPU and GPU. It is very complicated for developers to manage the details of MPI and CUDA.

In contrast, developers only need to get data from global array to GPU memory and put data from GPU memory to global array directly if CUDAGA is used. The corresponding example of CUDAGA is shown in Fig. 6. Developers only need to manage device pointers during the communication, because CUDAGA can help to manage temp memory buffer. Data just need to be synchronized on global array between CPU and GPU when program is started or finished on GPU. The range of data is computed in global array before data is accessed.

Comparing the lengths of codes in Figs. 5 and 6, we can see that much less code is necessary by CUDAGA than CUDA+MPI to implement the application. The comparisons of code lengths with CUDA+MPI and CUDAGA for Cannon algorithm and Jacobi iteration are shown in Table 1. About 60 % codes can be saved by CUDAGA, which means developers can spend more time on designing better algorithms but pay less attention on the realization on GPU cluster.

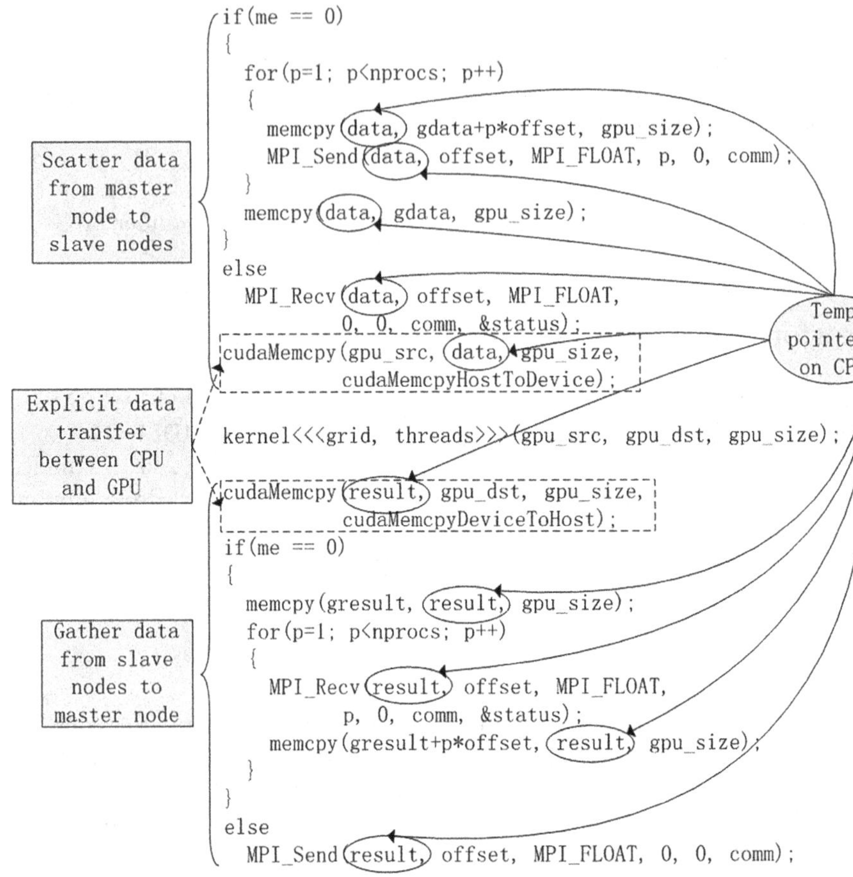

if(me == 0)
{
for(p=1; p<nprocs; p++)
{
memcpy(data, gdata+p*offset, gpu_size);
MPI_Send(data, offset, MPI_FLOAT, p, 0, comm);
}
memcpy(data, gdata, gpu_size);
}
else
MPI_Recv(data, offset, MPI_FLOAT,
0, 0, comm, &status);
cudaMemcpy(gpu_src, data, gpu_size,
cudaMemcpyHostToDevice);
kernel<<<grid, threads>>>(gpu_src, gpu_dst, gpu_size);
cudaMemcpy(result, gpu_dst, gpu_size,
cudaMemcpyDeviceToHost);
if(me == 0)
{
memcpy(gresult, result, gpu_size);
for(p=1; p<nprocs; p++)
{
MPI_Recv(result, offset, MPI_FLOAT,
p, 0, comm, &status);
memcpy(gresult+p*offset, result, gpu_size);
}
}
else
MPI_Send(result, offset, MPI_FLOAT, 0, 0, comm);
Scatter data from master node to slave nodes
Explicit data transfer between CPU and GPU
Gather data from slave nodes to master node
Temp pointers on CPU

Fig. 5. Usage example of CUDA+MPI

cudaGA_Sync(g_a, HostToDevice);
lo[0] = me * (height + 2);
lo[1] = 0;
hi[0] = (me + 1) * (height + 2);
hi[1] = (width + 2) - 1;
ld = width + 2;
cudaNGA_Get(g_a, lo, hi, gpu_src+offset, &ld);
kernel<<<grid, threads>>>(gpu_src, gpu_dst, gpu_size);
cudaNGA_Put(g_b, lo, hi, gpu_dst+offset, &ld);
cudaGA_Sync(g_b, DeviceToHost);
Data synchronization between CPU and GPU
Compute the range of accessing data on global array
Only device pointers are needed
Scatter and gather data directly to GPU by one function call

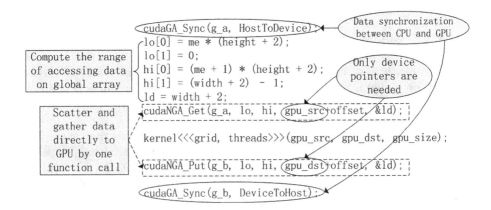

Fig. 6. Usage example of CUDAGA

Table 1. The coding comparisons between CUDA+MPI and CUDAGA (lines)

	Cannon algorithm		Jacobi iteration	
	CUDA+MPI	CUDAGA	CUDA+MPI	CUDAGA
Total communication	172	63	131	58
Operation initialization	26	7	26	7
Data initialization	63	17	29	18
Data collection	44	7	21	4
Kernel computing	39	32	55	29

5.2 Performance Experiments

Since network bandwidth is only 10 MB/s, which is several orders of magnitude
slower than PCI-Express bandwidth and CUDA kernel execution, the cost will
increase with the increase of the number of GPUs. Here we mainly focus on
the comparison of CUDA+MPI and CUDAGA. We believe the performance will
improve as the hardware upgrades.

The tests run on 4 GPUs with five different matrix sizes. As Fig. 7 shows,
the performance of Cannon algorithm implemented by CUDAGA is better than
that by CUDA+MPI. The method implemented by CUDA+MPI involves data
dependency because of data cyclic movement, but data can always get from
global array by CUDAGA to avoid the overhead of data waiting. Moreover,
data does not need to be copied from host to device, which is the bottleneck of
data transfer. Since the bandwidth between CPU and GPU is much lower than
that between GPU and GPU, a lot of time is saved for data transfer be-tween
host and device. Another important factor is that *get* operation is executed more
times than *put* operation. Only *put* operation is called once at the end of Cannon
algorithm. That means more data transfers between CPU and GPU will be saved
in more *get* operations.

The cost of Jacobi iteration implemented by CUDAGA is a little better than
that by CUDA+MPI, shown in Fig. 8.

We can note the feature of matrix: the number of columns is much larger than
that of rows. The overhead of memory copy from host to device is avoided when
maintaining data consistency in CUDAGA, so that it can get much better performance. However, the portion of data communication needs to be big enough to
hide the overhead of data consistency. The communication size of Jacobi iteration
only refers to the boundary of each process, which means that only two columns
of each process participate data communication while all columns of each process
participate maintaining data consistency. So the result implemented by CUD-
AGA is a little better than that by CUDA+MPI. Good performance can be got
when the columns/row ratio is big enough. As the ratio grows, the performance
implemented by CUDAGA will be much better.

In a word, CUDAGA model is suitable for parallel programming of these
applications: (1) array is the basic data structure; (2) most data is involved in

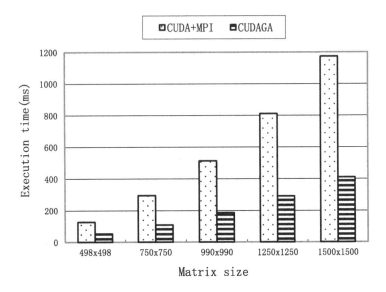

Fig. 7. Execution times of Cannon algorithm with different matrix sizes

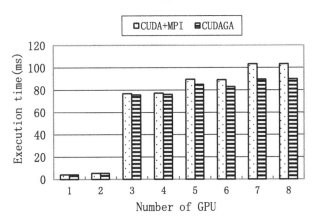

Fig. 8. Execution times of Jacobi iteration for 100*100000 matrix with different GPU numbers

communication; (3) the times getting data from global array is not less than putting. Parallel programming with CUDAGA can not only ensure better performance, but also facilitate better programming effort.

6   Conclusions

As GPGPU is more and more widely used, lots of applications will be implemented on GPU cluster. CUDAGA, a portable parallel programming model for GPU cluster, is presented to combine CUDA with GA library together to ease the program of data transfer. It also helps developers to choose different GPU for each process to ensure resource utilization, and provides auxiliary functions to obtain some information about GPU cluster. Moreover, linear algebra functions can be used to speed up some linear calculations.

We combine CUDA programming interface with blocked communication interface of GA. In the future, CUDAGA will provide asynchronous communication interface to hide overhead of communication and computation. Another important topic is to coordinate multiple GPUs and CPU to work together.

Acknowledgment. This work is supported by the National 973 Key Basic Research Plan of China (No. 2013CB2282036), the Major Subject of the State Grid Corporation of China (No. SGCC-MPLG001(001-031)-2012), the National 863 Basic Research Program of China (No. 2011AA05A118), the National Natural Science Foundation of China (No. 61133008), the National Science and Technology Pillar Program (No. 2012BAH14F02) and the independent innovation project of Huazhong University of Science and Technology.

References

1. Buck, I., Foley, T., Horn, D., Sugerman, J.: Brook for GPUs: stream computing on graphics hardware. ACM Trans. Graph. **23**(3), 777–786 (2004)
2. Volodymyr, V., Jeremy, J. E., Guochun, S.: GPU clusters for high-performance computing. In: Proceedings of IEEE Cluster PPAC Workshop, pp. 1–8. IEEE Computer Society (2009)
3. Hawick, K.A., Leist, A., Playne, D.P.: Regular lattice and small-world spin model simulations using CUDA and GPUs. Int. J. Parallel Prog. **39**(2), 183–201 (2011)
4. Nieplocha, J., Harrison, R.J., Littlefield, R.J.: Global arrays: a non-uniform memory access programming model for high-performance computers. J. Supercomput. **10**(2), 169–189 (1996)
5. Nieplocha, J., Carpenter, B.: ARMCI: a portable remote memory copy library for distributed array libraries and compiler run-time systems. In: Rolim, J., et al. (eds.) IPPS-WS 1999 and SPDP-WS 1999. LNCS, vol. 1586, pp. 533–546. Springer, Heidelberg (1999)
6. Micikevicius, P.: 3D finite difference computation on GPUs using CUDA. In: Proceedings of 2nd Workshop on General Purpose Processing on Graphics Processing Units, pp. 79–84. ACM, New York (2009)
7. William, G.N.D., Lusk, E., Skjellum, A.: A high-performance, portable implementation of the MPI message passing interface standard. Parallel Comput. **22**(6), 789–828 (1996)
8. Orion, S.L.: Message passing for GPGPU clusters: cudaMPI. In: Proceedings of IEEE International Conference on Cluster Computing and Workshops, pp. 1–8. IEEE (2009)
9. Moerschell, A., Owens, J.D.: Distributed texture memory in a multi-GPU environment. In: Graphics Hardware, pp. 31–38 (2006)
10. Nieplocha, J., Harrison, R.J., Littlefield, R.J.: The global array programming model for high performance scientific computing. SIAM News **28**(7), 12–14 (1995)
11. Fan, Z., Qiu, F., Kaufman, A.: Zippy: a framework for computation and visualization on a GPU clusters. Comput. Graph. Forum **27**(2), 341–350 (2008)

12. Strengert, M., Müller, C., Dachsbacher, C., Ertl, T.: CUDASA: compute unified device and systems architecture. In: Proceedings of Eurographics Symposium on Parallel Graphics and Visualization (EGPGV 2008), pp. 49–56. Eurographics Association (2008)
13. Nickolls, J., Buck, I., Garland, M., Skadron, K.: Scalable parallel programming with CUDA. Queue **6**(2), 40–53 (2008)

Node Capability Modeling for Reduce Phase's Scheduling in MapReduce Environment

Chuang Zuo, Qun Liao[✉], Tao Gu, Tao Li, and Yulu Yang

College of Computer and Control Engineering, Nankai University, Tianjin, China
{zuochaung,liaoqun,gutao}mail.nankai.edu.cn,
{litao,yangyl}@nankai.edu.cn

Abstract. MapReduce is a programming model widely used in big data processing. Reduce tasks scheduling in MapReduce is a key issue which affect the performance significantly. Unfortunately, because of the complication of reduce tasks scheduling, there are no acknowledged solution in this issue. Main ideas in optimizing reduce tasks scheduling emphasizes features of computation or data locality. Although few researches tried to explore solutions with theoretical modeling, their models are oversimplified. Aiming to optimizing reduce tasks scheduling, we propose a method of modeling node's computation and communication capability uniformly based on analyzing the procedure of reduce phase theoretically. In the analysis, cost of reduce tasks in intermediate data fetching and processing are integrated. With the proposed model, the optimal load balance of reduce phase is concluded and proved. Evaluations under different environments show that load balance of reduce phase is improved significantly with the scheduling method instructed by the optimal principle.

Keywords: Mapreduce · Load balance · Reduce tasks scheduling · Cloud computing

1   Introduction

Nowadays big data applications are drawing more and more attentions. MapReduce is a popular programming model aiming to simplify the big data processing on large-scale clusters [1]. Recently, MapReduce and its open-source implementation Hadoop [2] are deployed and utilized in many areas by companies and organizations [3–8]. In MapReduce model, an application is divided into large number of map and reduce tasks running on massive connected commodity computers. Methods of scheduling these large number of tasks ad-hoc parallel processing of big data plays an important role in MapReduce clusters which effect the jobs makespan and re-source utilization significantly. Load balance is commonly a main object of tasks scheduling in MapReduce environments which attract many researchers' attention. In typical MapReduce, the map phase is divided into lots of tasks to achieve dynamic load balance. Many researches optimizing the performance of map tasks by improving the scheduling mechanisms [9–12] or data storage strategies [13–15].

© Springer International Publishing Switzerland 2015
W. Qiang et al. (Eds.): CloudCom-Asia 2015, LNCS 9106, pp. 217–231, 2015.
DOI: 10.1007/978-3-319-28430-9_17

However, these mechanisms or strategies can't be applied to reduce tasks scheduling mechanically. The number of reduce tasks which determines the distribution of the MapReduce job's results is constrained by users' requirements. Moreover, each reduce task is responsible for the data with certain range of keys, which means intermediate data with the same key is always allocated to the same task. Thus reduce tasks may have to process quite different size of data. Consequently, load balance of reduce phase is significant and difficult.

Research on optimizing of reduce tasks scheduling get few attentions. Several heuristic solutions are proposed to improve data locality of reduce tasks [16–20]. Although these solutions improve the performance of reduce tasks, they are based on intuitive techniques rather than theoretical analysis. On the theoretical approach, Tan et al. [20] analyzes the reduce task assignment in sequential MapReduce jobs, but it ignores the cost of reduce processing and scheduling inside a MapReduce job. The work of Berlińska and Drozdowski [22] analyzes the map and reduce computations in MapReduce, however, it assumes the data transfer time as a constant, which does not correspond to the fact.

In current MapReduce's scheduling strategy, reduce tasks are allocated to idle nodes randomly, which degrades the performance severely. Without considering the matching of tasks with nodes' capabilities, a reduce task with large amount of intermediate data would be allocated to a slow node. The random scheduling may also affect the data locality of reduce tasks and cause unnecessary intermediate data transmission.

To optimize reduce tasks scheduling, it is necessary to prompt reduce tasks assigned to nodes with matching capability. Thus modeling node capability based on theoretical analysis is one of key points of optimizing reduce phase's load balance. A node's capability is composed with both computation and communication capability. The computation capability can be represented with node's CPU and memory information easily. But the communication capability is hard to estimate, because any node's communication capability is related with specific allocation of intermediate data among reduce tasks.

In this paper, the process of data transmission and processing in reduce phase is analyzed theoretically. Based on the analysis, a node capability model is established. In this model, the computation and communication capability of a node is considered in a combined way. The optimal scheduling for reduce tasks is concluded and proved. The optimal principle suggests that matching reduce tasks with nodes' capabilities can improve the performance of reduce tasks. Simulation results of optimal scheduling show that the load balance of reduce phase can be improved significantly compared with random scheduling methods.

Our contributions include the following:

- A theoretical model to analyze nodes' capabilities is proposed, which combine nodes' computation and communication capability.
- A node capability-aware reduce task scheduling strategy based on the proposed model is presented. And it is proved effective to improve MapReduce's performance in different scenarios.

2 Preliminary and Related Work

2.1 MapReduce

In MapReduce model, a computation process consists of the map and reduce phases. Both map and reduce phase can contain multiple independent sub-tasks which are called map tasks and reduce tasks respectively. Each map task or reduce task is executed by a machine, which is called node in following discussions. As a MapReduce job starts, map tasks are allocated among clusters, these tasks process the input data represented as key/value pairs. Results of map tasks are called intermediate data, and these data are stored as key/value pairs among relevant nodes in clusters. Then the reduce tasks start to execute, they are responsible for fetching the intermediate data from different nodes and generating final results. The works of Dean and Ghemawat [1] and White [23] provide more details about the computation principles of map and reduce phase.

An overview of a MapReduce job processing is presented in Fig. 1. A typical process of a MapReduce job consists of seven steps. In the first two steps, a centralized scheduler allocates map and reduce tasks to relevant nodes. The nodes assigned map/reduce tasks are called mappers/reducers respectively.

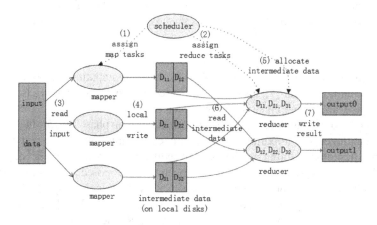

Fig. 1. Execution overview of MapReduce

Step (3) and (4) constitute a whole data flow of a map task. The mapper reads input data and processes them. As a result, intermediate data are generated and stored locally. According to a certain partitioning function these intermediate data are divided according to particular keys. When a map task finishes, the scheduler activates reduce tasks by notifying locations information of intermediate data to reducers in step (5). As shown by step (6), reducers notified read corresponding intermediate data from mappers' local disks. When a reducer gets all assigned intermediate data, it sorts and groups them according to the keys. The sorted pairs of each key are processed with the reduce function provided by user program. The results of reduce tasks are written to output files as shown by step (7).

More details about the procedure of reducers fetching allocated intermediate pairs from mappers are discussed below. The allocations of intermediate data among reducers are determined by the output of map tasks and partitioning function. After execution of a map task, a mapper stores its intermediate data as key/value pairs in local disks. These intermediate data are divided into pieces according to their keys. A reducer only fetches the parts of intermediate data corresponding to certain keys which are assigned to it. Assigning keys to reducers is determined by user. For instance, a piece of intermediate data stored on node i, Dix are stored as key/value pairs with the xth key(s). According to naïve assignment rules, each reducer is assigned a different key. The distribution and transmission of pieces of intermediate data is shown in Fig. 1.

In a MapReduce job, the number of reduce tasks and intermediate keys assignment rules are provided by users. If the user has special requirements on the format of results, for example, particular intermediate keys must be gathered as one result, the reduce tasks can't be set partitioned arbitrarily. This kind of reduce tasks is called fixed reduce tasks in this paper.

It is worthwhile to note that the distribution of intermediate data storage is decided by the input data and map tasks which are defined by users. Modifying the intermediate data storage distribution is impractical. However, the partition of intermediate data can be fixed in reality. This kind of situation happens when user has special requirements on the distribution of results or the amount of keys is less than the number of reduce tasks.

2.2 Related Work

Task scheduling is a crucial issue in MapReduce cluster computing. In current MapReduce programming model [1], a strategy called locality-aware scheduling is employed for map phase, and a random strategy for reduce phase.

Previous studies mainly focus on the optimizing of map phase [9–15]. In [9, 10], the authors recognized the importance of the latency for multiple concurrent jobs, and proposed different scheduling mechanisms to reduce the response time of MapReduce jobs, and maintaining data locality. In [11], the authors proposed NKS (short for Next-K-Node Scheduling) method to achieve data locality in a single MapReduce job, and mainly aims to homogeneous computing environment. In [13–15], data placement schemes are proposed to improve data locality in MapReduce environment.

However, these studies can't apply to our work mechanically, whether scheduling mechanisms or data placement schemes, because the number of reduce tasks is constrained by users and intermediate data can't be allocated randomly. There are mainly two approaches to optimize reduce phase. The first approach is to devise heuristic scheduling strategies [16–20]. In [16], the authors extend Hadoop to offer reliable MapReduce services on opportunistic distributed environment. Their scheduling strategy works in a hybrid resource architecture. In [18], LARTS was proposed to improving MapReduce performance. The basic idea of LARTS is collocating reduce tasks with the maximum required data computed after recognizing input data network locations and sizes. But it only considers the maximum required data which can't represent the whole intermediate data distribution.

The second approach is to analyze the MapReduce job theoretically [20–22]. Theoretical analyses help us make better strategies in reduce task scheduling. In [20], the authors formulate a stochastic optimization framework to improve the data locality for reduce tasks. But it ignores the cost of reduce processing, and is mainly for sequential MapReduce jobs. Berlińska and Drozdowski [22] analyzes the map and reduce computations in MapReduce, but it assumes the data transfer time as a constant in reduce phase, which does not correspond to the fact. In our work, we attach great importance to analyze the data transfer time in reduce phase. This provides richer information for us to model a node's integrated capability.

3   Theoretical Model

Different with other work in optimizing Reduce tasks scheduling which focus on heuristics mainly chasing for high data locality or short computation time. In this work the cost of reduce tasks in computation and communication are considered comprehensively. Each node in MapReduce cluster follows a capability model proposed in this work based on the procedure of a Reduce tasks assignment and execution on any one node. As the capability of node can be known from the proposed model, workloads of each Reduce task also is known as common assumptions, thus assigning reduce tasks to nodes with matched capability is an efficient and reasonable solution of reduce tasks scheduling. This model is based on abstraction of time cost in data fetching and computing of reduce task, and provides unifying computation and communication resource of a node. The modeling method reflects a trade-off in computation and communication in reduce tasks scheduling.

3.1   Assumptions of Nodes in MapReduce

In this paper, each node can be defined as three kinds of role, scheduler, mapper and reducer. With processing unit(s), the scheduler has the ability to schedule tasks. Mapper and reducer can execute assigned tasks with its processing units. Local Storage is essential for mappers to store the intermediate data. Network interface makes any two nodes can communicate with each other.

Two types of communication model between nodes, sequential and concurrent, have been studied so far [24]. In the sequential model, each node can only communicate with one other node at a time. While in the concurrent model, a node can communicate with all other nodes simultaneously [25, 26]. The concurrent communication can be implemented with a processor who uses a processing unit to maintain an output buffer for each communication link [24]. In this paper, it is assumed that processing unit(s) of nodes are able to support concurrent communication.

3.2   Symbols and Terminologies

The mappers are identified by an 'M' followed by a tag number between 1 to m, and reducers are identified by an 'R' followed by a tag number between 1 to r. As nodes can execute both map and reduce tasks in common, a node can act as both mapper and reducer. In the following discussion, mapper/reducer and node are used alternatively according to the context. To facilitate describing the theoretical model of reduce tasks, some symbols and terminologies are defined in Table 1. In this paper, it is assumed that

B_i is same for all reducers. It is reasonable to assume that W_{ij} is zero, if M_i and R_j are on a same node.

Table 1. Terms of a MapReduce job.

Symbol	Description
m	Number of mappers
r	Number of reducers
M_i	A mapper whose tag number is i
R_i	A reducer whose tag number is i
d	Intermediate data unit
D_i	Size of all intermediate data stored in mapper M_i
D_{ij}	Size of intermediate data stored in M_i and processed by R_j
F_{ij}	Proportion of D_{ij} divided by D_i
B_i	Initial time of R_i
P_j	Time of R_j to process a unit of data and write results as output files
W_{ij}	Time needed by transmitting a unit of data between M_i and R_j
T_j	Time needed by R_j to finish allocated reduce tasks in a job
S_j	Moment when R_j starts
Ct_j	Average time for R_j to fetch a unit of data from mappers
PC_j	Time needed by R_j to fetch and process a unit of intermediate data
t_j	Time needed by R_j to fetch intermediate data from all mappers
h_j	Time of R_j to process intermediate data and write result as output files
C_j	R_j's capability
SR	Sorted list of reducers
SR_i	The ith entry of SR
ST	Sorted list of reducer tasks
ST_i	The ith entry of ST
E_k	Size of intermediate data corresponding with reduce task k
E_{ik}	Size of intermediate data processed by reduce task k and stored on M_i
RT_k	Reduce task whose keys are k

3.3   Node Capability Model

As discussed in Sect. 2, reducer R_j is assigned intermediate data E_j, R_j needs to fetch data D_{ij} from every mapper M_i. Because the communication model adopted in this paper, the time needed by R_j to fetch intermediate data can be calculated as (1). It means that data transmission from multiple mappers to R_j happen simultaneously. When the slowest transmission finishes, R_j gets all of the intermediate data it needs.

$$t_j = \max_{1 \leq i \leq m} \left(D_{ij} \times W_{ij} \right) \tag{1}$$

$$Ct_j = \max_{1 \leq i \leq m} \left(D_{ij} \times W_{ij} \right) \Big/ \sum_{i=1}^{m} D_{ij} \tag{2}$$

The average time for R_j to fetch a unit of data from all mappers can be calculated as (2). Under the assumption that the proportion of D_{ij} to D_i remains constant in different mappers, the proportion of D_{ij} to D_i can be represented by a symbol, k. By replacing the D_{ij} with the product of k and D_i, the average time needed by R_j to fetch a unit of intermediate data can be derived. Equation (3) describes an average cost of data transmission of Reducer, R_j fetching a unit of intermediate data.

$$Ct_j = \max_{1 \leq i \leq m} \left(D_i \times W_{ij} \right) \Big/ \sum_{i=1}^{m} D_i \tag{3}$$

So the total time needed by R_j to fetch and process a unit of intermediate data is derived in (4).

$$PC_j = P_j + Ct_j \tag{4}$$

R_j's capability considering both computation power and communication capability is defined in (5).

$$C_j = d \big/ PC_j \tag{5}$$

When the distribution of intermediate data among mappers is known, it is possible to derive a reducer's capability model by analyzing time needed by a reducer to fetch and process a unit of intermediate data. A reducer's computation power and communication capability is considered in a combined way. The capability model adapts different distributions of intermediate data, as long as the proportion of D_{ij} to D_i remains constant in different mappers. This assumption is closely approximate in large clusters executing a MapReduce job with a huge number of map tasks.

3.4   Cost of a Reduce Task

In the execution of a MapReduce job, the scheduler allocates tasks to mappers and reducers. The user program of applications and related input data locations need to be submitted to scheduler. Then scheduler allocates tasks with chosen scheduling algorithm. By default, when the scheduler detects an idle reducer, it actives this reducer by forwarding the program and intermediate data locations. In this paper, it is assumed that every participating reducer gets all needed program and information in the first execution, which takes time B_i. Because B_i is constant among different reducers, it has no impact on the order of reducer's cost. So B_i is ignored in reducer's cost calculation.

As a Reducer, R_i has known the particular key(s) assigned to it, it begins to fetch corresponding data from mappers. It takes time t_j. As long as corresponding intermediate data is received, R_j processes these data and write final results into the output files. It takes time h_j.

$$h_j = \sum_{i=1}^{m} D_{ij} \times P_j \tag{6}$$

According to the process of a reducer, the time when R_j finish processing can be concluded as (7).

$$T_j = S_j + t_j + h_j = S_j + \max_{1 \leq i \leq m} \left(D_{ij} \times W_{ij} \right) + \sum_{i=1}^{m} D_{ij} \times P_j \tag{7}$$

The failures of task execution and backup tasks are not presented directly in the model above. As the backup tasks are used to cope with slow or failed tasks, they can be covered by increasing the size of D_i to an appropriate degree.

4   Optimal Performance of Reduce Tasks

As the combined capability model of reducer presented above, a reducer's capability is defined as combine of its communication capability and computation power. Based on the capability model, the relation between size of intermediate data assigned and nodes' capabilities can be established. The optimal performance of reduce tasks are discussed further. The optimal scheduling for fixed reduce tasks is concluded and proved. To simplify the discussion, S_j in (7) is assumed to be a very small constant. So S_j's impact on the solution of optimal scheduling is ignored in discussion below.

In the reduce phase, scheduler actives reducers to read intermediate data and execute tasks. Each reducer may be allocated more than one reduce tasks in general [1, 23]. It is assumed that the intermediate pairs generated by a mapper are partitioned with partitioning function into r pieces, so each reducer gets only one piece of them.

As a MapReduce job finishes when the slowest reduce task finishes, the objective of optimal schedule is to minimize the time of the slowest reducer in a MapReduce job. The objective function is defined in (8).

$$\min \left(\max_{1 \leq j \leq r} \left(T_j \right) \right) \tag{8}$$

As the combined capability of every reducer can be calculated, all reducers can be sorted by their capabilities in descending order. The sorted list of reducers is represented by SR, the ith entry of SR is represent by SR_i. SR_i's capability satisfies (9). It is worthwhile to note that SR_i and R_i is not the same reducer in general.

$$C_{SR_1} \geq C_{SR_2} \geq \cdots \geq C_{SR_r} \tag{9}$$

All reduce tasks' workloads can be measured by their size of data corresponding. After Reduce task RT_k is assigned to a reducer, the size of data the particular reducer needed to fetch from M_i can be expressed as (10). F_{ik} means the proportion of RT_k's intermediate data to all intermediate data stored on M_i.

$$E_{ik} = D_i \times F_{ik} \tag{10}$$

It is assumed that all intermediate data distributed uniformly among mappers, which means that the proportions of particular keys among different mappers are constant. The intermediate data corresponding with reduce task RT_k, can be calculated as (12).

$$F_{1k} = F_{2k} = \cdots = F_{mk} \tag{11}$$

$$E_k = \sum_{i=1}^{m} E_{ik} = \sum_{i=1}^{m} (D_i \times F_{ik}) \tag{12}$$

All reduce tasks' workloads can be measured by their size of data corresponding. All reduce tasks can be sorted by their workloads in descending order. The sorted list of reduce tasks is represented by ST, the ith entry of ST is represent by ST_i. ST_i's workload satisfies (13).

$$E_{ST_1} \geq E_{ST_2} \geq \cdots \geq E_{ST_r} \tag{13}$$

Theorem 1. Aiming to achieving the objective function defined in (8), the ith reduce task in ST should be assigned to the ith Reducer in SR, number i ranges from 1 to r.

Proof of Theorem 1. It is assumed that T is the time in reduce phase adopting the optimal scheduling defined in Theorem 1. T can be calculated by (14).

$$T = \max_{1 \leq i \leq r} \left(E_{ST_i} / C_{SR_i} \right) \tag{14}$$

Two random reducer SR_s and SR_t are defined who satisfy (15).

$$s < t \tag{15}$$

Exchange their assigned reduce tasks ST_s and ST_t. The new scheduling make the time of reduce phase becomes to T'. Because SR and ST are both in descending order, it is obvious that (16) and (17) are definite.

$$C_{SR_r} \geq C_{SR_t} \tag{16}$$

$$E_{ST_r} \geq E_{ST_t} \tag{17}$$

SR_s and SR_t are chosen randomly, and T is the minimized time of Reduce phase in a MapReduce job. So (18) and (19) can be derived.

$$Max\left\{ E_{ST_r}\Big/C_{SR_t}, E_{ST_t}\Big/C_{SR_r} \right\} \geq Max\left\{ E_{ST_r}\Big/C_{SR_r}, E_{ST_t}\Big/C_{SR_t} \right\} \tag{18}$$

$$T' \geq T \tag{19}$$

Theorem 1 is proved by the derivation from (15) to (19).

5 Performance Evaluation

5.1 Simulation Setup

In this section, A group of simulations are designed to evaluate the optimal scheduling's speedup and load balance. Random scheduling which is adopted by native implement of MapReduce is simulated as contrast. Because proposed scheduling in this paper focus on reduce phase, so the speedup and load balance are both evaluated within reduce phase. And the load balance is represented by the size of intermediate data assigned to each node with different capability. Simulations are implemented upon the support of Matlab. Some basic parameters used in the simulations are set as follows.

The ratio of mappers to reducers is set to be in the discrete uniform distribution. The ratio ranges from 10 to 100. The interval between two adjoining ratios is set to be 10. The number of tasks assigned to a mapper is set to be a random integer between 50 and 150 inclusive. Time of fetching a unit of data from a particular mapper to reducer is set to be a random integer between 5 and 20 inclusive. Time needed by a reducer to process a unit of intermediate data is set to be a random integer between 1 and 50 inclusive. The random numbers used in simulations are generated with the randi function or rand function in Matlab. They are uniformly distributed pseudorandom values in a given range.

Four scenarios of reduce task scheduling in a MapReduce job are simulated. In the first scenario, random scheduling and optimal scheduling are simulated. In the second one, reduce phase time in environment with different deviation of tasks assignment and nodes' capability in optimal scheduling are conducted. Deviation of tasks assignment and nodes' capability is defined to measure the unfitness of nodes' capability and the workloads assigned. It is calculated as (20).

$$\sum_{i=1}^{r} \left| C_{SR_i}\Big/\sum_{i=1}^{r} C_{SR_i} - E_{ST_i}\Big/\sum_{i=1}^{r} E_{ST_i} \right| \tag{20}$$

It can be derived that larger difference between capability of node and corresponding task's data size means larger deviation of tasks assignment and nodes' capability. Modifying proportion of reduce tasks' intermediate data size can get different value of deviation of tasks assignment and nodes' capability. Three situations with different size of tasks is simulated. The size of tasks generated randomly, they share the same nodes' capabilities.

Load balance is evaluated in the latter two scenarios. In the third and the forth scenarios, nodes' computation capability and communication capability are set to be constant respectively. The constant computation capability is set to be 20, and the communication capability is set to be 10. The results of them are shown in Figs. 4 and 5. They describes that how the communication capability and computation capability of nodes impact the assignment of reduce tasks. Those indicate the load balance of reduce phase.

5.2   Performance Evaluation

Figure 2 shows the speedup between optimal scheduling and random scheduling. Higher ratio m to r means each reducer needs to process more intermediate data on average. It is observed that the optimal scheduling only take one third of time in reduce phase adopting random scheduling when the ratio m to r is 100. The relation between performance improvement and the ratio m to r is also depicted in Fig. 2. It shows the higher the ratio is, the more improvement preforms.

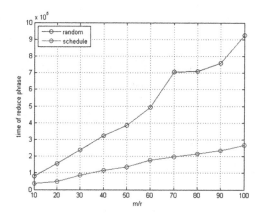

Fig. 2. Comparison of Reduce phase time between optimal scheduling and random scheduling (Color figure online).

The unfitness of nodes' capability causes longer time of reduce phase, although the optimal scheduling offer the best tasks assignment. As shown in Fig. 3, the more deviation of tasks assignment and nodes' capability is, the longer time it takes to finish reduce phase processing. Three situations with different size of tasks which generated randomly provide analogous trend lines. It shows that within the same deviation, bigger size of tasks would cause a longer reduce phase time. It can be concluded that partitioning proportion of reduce tasks' intermediate data size by fitting the proportion of reducer's capability can shorten the time of reduce phase in a MapReduce job.

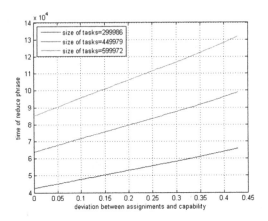

Fig. 3. Relation between reduce phase time, deviation of assignment and nodes' capability (Color figure online).

As shown in Fig. 4, when computation capability is constant, there is a positive correlation communication capability of node and the size of intermediate data assigned to a particular reducer. Higher communication cost in fetching a unit of data means a node has smaller communication capability. The higher communication cost a reducer has, the smaller size of intermediate data is assigned. Results under three different ratios of mappers to reducers appear analogously.

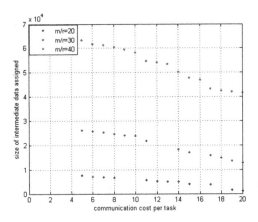

Fig. 4. Relation between intermediate data assignment and nodes' communication capability (Color figure online).

Analogously, when communication capability is constant, the computation capability of a node dominates the size of intermediate data assigned to it. Another three groups of result are depicted in Fig. 5. It suggests that the matching reduce tasks with nodes' capabilities can improve the performance of reduce tasks.

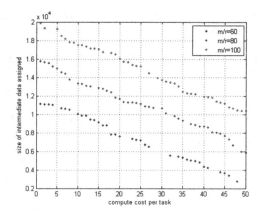

Fig. 5. Relation between intermediate data assignment and nodes' computation capability (Color figure online).

6   Conclusion

Scheduling of reduce tasks in a MapReduce job is an important issue which affects the performance of MapReduce model significantly. In this paper, the procedures of intermediate data fetching and processing are analyzed. A capability model unrelated to particular intermediate data distributions is proposed. In the proposed model, a node's computation power and communication capability are combined. Base on the capability model of nodes, the optimal scheduling of reduce tasks in a MapReduce job is concluded and proved. Simulations also demonstrate the improvement by comparing with random scheduling of reduce tasks.

To simplify the problem, only the situation that a reducer gets one reduce task at most is considered in this paper. Considering situation each reducer can be assigned multiple reduce tasks, the optimal solution becomes difficult to be achieved. This situation should be studied further to make the research suit more situations in reality. Moreover, based on the research in this paper, some heuristic scheduling algorithms can be proposed to optimize scheduling in reduce phase. The distribution of intermediate data among mappers impacts reduce tasks assignment. Research on the methods how to optimize the intermediate data distributions to fit reduce tasks scheduling is another interesting issue.

References

1. Dean, J., Ghemawat, S.: MapReduce: simplified data processing on large clusters. J. Commun. ACM. **51**, 107–113 (2008)
2. Hadoop. http://hadoop.apache.org
3. Applications powered by Hadoop: https://wiki.apache.org/hadoop/PoweredBy

4. Yahoo! Launches World's Largest Hadoop Production Application. https://developer.yahoo.com/blogs/hadoop/yahoo-launches-world-largest-hadoop-production-application-398.html

5. McKenna, A., Hanna, M., Banks, E., Sivachenko, A., Cibulskis, K., Kernytsky, A., Garimella, K., Altshuler, D., Gabriel, S., Daly, M., DePristo, M.A.: The genome analysis toolkit: a Mapreduce framework for analyzing next-generation DNA sequencing data. Genome Res. **20**, 1297–1303 (2010)

6. Kalyanaraman, A., Cannon, W.R., Latt, B., Baxter, D.J.: MapReduce implementation of a hybrid spectral library-database search method for large-scale peptide identification. Bioinformatics **27**, 3072–3073 (2011)

7. Stuart, J.A., Owerns, J.D.: Multi-GPU MapReduce on GPU clusters. In: 2011 IEEE International on Parallel and Distributed Processing Symposium (IPDPS), pp. 1068–1079. IEEE (2011)

8. Srirama, S.N., Jakovits, P., Vainikko, E.: Adapting scientific computing problems to clouds using MapReduce. Future Gener. Comput. Syst. **28**(1), 184–192 (2012)

9. Nguyen, P., Simon, T., Halem, M., Chapman, D., Le, Q.: A hybrid scheduling algorithm for data intensive workloads in a MapReduce environment. In: Proceedings of the 5th International Conference on Utility and Cloud Computing, Chicago, IL, USA, 5–8 November 2012

10. Zaharia, M., Borthakur, D., Sen Sarma, J., Elmeleegy, K., Shenker, S., Stoica, I.: Delay scheduling: a simple technique for achieving locality and fairness in cluster scheduling. In: Proceedings of the 5th European Conference on Computer Systems, Paris, France, 13–16 April 2010

11. Zhang, X., Zhong, Z., Feng, S., Tu, B., Fan, J.: Improving data locality of Mapreduce by scheduling in homogeneous computing environments. In: Proceedings of the 9th International Symposium on Parallel and Distributed Processing with Applications, Busan, Korea, 26–28 May 2011

12. Tang, Z., Zhou, J., Li, K., et al.: A MapReduce task scheduling algorithm for deadline constraints. Cluster Comput. **16**(4), 651–662 (2013)

13. Xie, J., Yin, S., Ruan, X., Ding, Z., Tian, Y., Majors, J., Manzanares, A., Qin, X.: Improving Mapreduce performance through data placement in heterogeneous hadoop clusters. In: Proceedings of IEEE International Symposium on Parallel and Distributed Processing, Workshops and PhD Forum, 19–23 April 2010

14. Abad, C.L., Lu, Y., Campbell, R.H.: DARE: adaptive data replication for efficient cluster scheduling. In: Proceedings of IEEE International Conference on Cluster Computing, Austin, TX, USA, 26–30 September 2011

15. Palanisamy, B., Singh, A., Liu, L., et al.: Purlieus: locality-aware resource allocation for MapReduce in a cloud. In: Proceedings of 2011 International Conference for High Performance Computing, Networking, Storage and Analysis, p. 58. ACM (2011)

16. Lin, H., Ma, X., Archuleta, J., Feng, W., Gardner, M., Zhang, Z.: Moon: Mapreduce on opportunistic environments. In: Proceedings of the 19th ACM International Symposium on High Performance Distributed Computing, Chicago, Illinois, USA, 21–25 June 2010

17. Zaharia, M., Borthakur, D., Sarma, J.S., Elmeleegy, K., Shenker, S., Stoica, I.: Job scheduling for multi-user Mapreduce clusters. Technical report, UCB/EECS-2009-55 (2009)

18. Hammoud, M, Sakr, M.F.: Locality-aware reduce task scheduling for MapReduce. In: 2011 IEEE Third International Conference on Cloud Computing Technology and Science (CloudCom), pp. 570–576. IEEE (2011)

19. Verma, A., Cherkasova, L., Campbell, R.H.: ARIA: automatic resource inference and allocation for Mapreduce environments. In: Proceedings of the 8th ACM International Conference on Autonomic Computing, Karlsruhe, Germany, 14–18 June 2011
20. Tan, J., Meng, S., Meng, X., Zhang, L.: Improving ReduceTask data locality for sequential MapReduce jobs. In: Proceedings of the IEEE INFOCOM, Turin, Italy, 14–19 April 2013
21. Yuan, Y, Wang, D, Liu, J.: Joint Scheduling of MapReduce jobs with servers: performance bounds and experiments
22. Berlińska, J., Drozdowski, M.: Scheduling divisible MapReduce computations. J. Parallel Distrib. Comput. **71**, 450–459 (2011)
23. White, T.: Hadoop: The Definitive Guide. O'Reilly Media, Cambridge (2012)
24. Moges, M., Yu, D., Robertazzi, T.G.: Grid scheduling divisible loads from two sources. Comput. Math. Appl. **58**, 1081–1092 (2009)
25. Piriyakumar, A., Murthy, C.S.R.: Distributed computation for a hypercube network of sensor-driven processors with communication delays including setup time. IEEE Trans. Syst. Man Cybern. Part A: Syst. Hum. **28**, 245–251 (1998)
26. Hung, J., Robertazzi, T.: Scalable scheduling for clusters and grids using cut through switching. Int. J. Comput. Appl. **26**, 147–156 (2004)

Analyzing and Predicting Failure in Hadoop Clusters Using Distributed Hidden Markov Model

Bikash Agrawal$^{(\boxtimes)}$, Tomasz Wiktorski, and Chunming Rong

Department of Computer and Electrical Engineering, University of Stavanger,
Stavanger, Norway
{bikash.agrawal,tomasz.wiktorski,chunming.rong}@uis.no

Abstract. In this paper, we propose a novel approach to analyze and predict failures in Hadoop cluster. We enumerate several key challenges that hinder failure prediction in such systems: heterogeneity of the system, hidden complexity, time limitation and scalability. At first, clustering approach is applied to group similar error sequences, which makes training of the model effectual subsequently Hidden Markov Models (HMMs) is used to predict failure, using the MapReduce programming framework. The effectiveness of the failure prediction algorithm is measured by precision, recall and accuracy metrics. Our algorithm can predict failure with an accuracy of 91 % with 2 days in advance using 87 % of data as training sets. Although the model presented in this paper focuses on Hadoop clusters, the model can be generalized in other cloud computing frameworks as well.

Keywords: Hadoop · Failure prediction · Hidden Markov Model · Failure analysis · Machine learning

1 Introduction

The cluster system is quite commonly used for high performance in cloud computing. As cloud computing clusters grow in size, failure detection and prediction become increasingly challenging [18]. The root causes of failure in such a large system can be due to the software, hardware, operations, power failure and infrastructure that support software distribution [24]. The cluster system dealing with a massive amount of data needs to be monitored and maintained efficiently and economically. There have been many relevant studies on predicting hardware failures in general cloud systems, but few on predicting failures in cloud computing frameworks such as Hadoop [32]. Hadoop is an open-source framework for distributed storage and data-intensive processing, first developed by Yahoo. Hadoop provides an extremely reliable, fault-tolerant, consistent, efficient and cost-effective way of storing a large amount of data. Failure in storing and reading data from the large cluster is difficult to detect by human eyes. All the events and activities are logged into their respective application log files. Logs provide information about performance issues, application functions,

© Springer International Publishing Switzerland 2015
W. Qiang et al. (Eds.): CloudCom-Asia 2015, LNCS 9106, pp. 232–246, 2016.
DOI: 10.1007/978-3-319-28430-9_18

intrusion, attack attempts, failures, etc. Most of the applications maintain their own logs. Similarly, HDFS system consists of DataNode and NameNode logs. The logs produced by NameNode, secondary NameNode, and DataNode have their individual format and content.

The prime objective of the Hadoop cluster is to maximize the job processing performance using data-intensive computing. Hadoop cluster normally consists of several nodes and can execute many tasks concurrently. The job performance is determined by the job execution time. The execution time of a job is an important metric for analyzing the performance of job processing in the Hadoop cluster [4]. As Hadoop is a fault-tolerant system if the nodes fail, then the node is removed from the cluster in the middle of the execution and the failed tasks are re-executed on other active nodes. However, this assumption is not realistic because the master node can crash. Many researchers reported that the master node crash is a single point of failure and needs to be handled [19,31]. Even the failure of the data node results in higher job execution time, as the job needs to re-execute in another node. Failure nodes are removed from the cluster so that the performance of the cluster improves. Prediction methods operate mostly on continuously available measures, such as memory utilization, logging or workload, to identify error pattern. Our analysis in this paper is mostly only on a time of occurrence of different types of error events that ultimately cause failure. This will also helps in root cause analysis [34] for automatic triggering of preventive actions against failures.

We used Hidden Markov Models (HMMs) [3] to learn the characteristics of log messages and use them to predict failures. HMMs have been successfully used in speech, handwriting, gesture recognition, and as well also in some machine failure prediction. HMM is well suited to our approach as we have observations of the error messages, but no knowledge about the failure of the system, which is "hidden". Our model is based on a stochastic process with a failure probability of the previous state. As faults are unknown and cannot be measured, they produce error messages on their detection (i.e. present in log files).

Our prediction model is divided into four main parts; First, identifying error sequences and differentiating types of error from the log files. Second, using the clustering algorithms [11] like K-means [16]. Third, training the model. Given the labeled training data, HMM method is used to evaluate maximum likelihood sequence that is used to update the parameters of our model. Last, predicting failure of the system based on the observation of an error sequences. The main idea of our approach is to predict failures by analyzing the pattern of error events that imitates failure. Experimental results for this method can predict failure with 91 % accuracy for 2 days in advance (prediction time). It also shows that our approach can compute on the massive amount of datasets. Ultimately, our approach can be used to improve the performance and reliability of the Hadoop cluster.

1.1 Related Work

A significant number of studies have been done on the performance evaluation and failure diagnosis of systems using log analysis. However, most of the prediction methods focus only on the system logs, but not on the application logs. Many more studies have been done on predicting hardware failure in the cluster. For example, studies in [21,24,27] provide a proactive method of predicting failure in the cluster, based on system logs. These methods provide failure in hardware level but fail to provide failure of a node in the Hadoop clusters.

Konwinski et al. [17] used X-trace [12] to monitor and improve Hadoop jobs. X-trace allows path-based tracing on Hadoop nodes. Additionally, using X-trace in conjunction with other machine-learning algorithm, they were able to detect failure with high accuracy. Similarly, SALSA [28] is another tool in which system logs are analyzed to detect failure using distributed comparison algorithm. Also, it also shows information on how a single node failure can affect the overall performance of the Hadoop cluster. All of these papers present failure detection algorithm in the Hadoop cluster but lacks prediction algorithm.

Fulp et al. [13] demonstrated that failure prediction in the hard disk using SVMs (Support Vector Machines) with an accuracy of 73 % with two days in advance. On the other hand, Liang et al. [18] uses RAS event logs to predict failure in IBM BlueGene/L. They compare their results with Support Vector Machines (SVMs), a traditional Nearest Neighbor method, a customized Nearest Neighbor method and a rule-based classifier, and found that all were outperformed by the customized Nearest Neighbor method. However, these all provide failure prediction algorithm in the different areas, but still lacks the good accuracy of the model.

Hidden Markov Models have been used in pattern recognition tasks such as handwriting detection [22], gene sequence analysis [5,9], gesture recognition [33], language processing [15,23], hard drive failure [29] or machine failure [27]. HMM is a widely used model due to it's flexibility, simplicity, and adaptivity. Indeed, as mentioned earlier, Hadoop log data have challenging characteristics, which thus require expert knowledge to transform data into an appropriate form.

1.2   Our Contribution

We proposed a novel algorithm for failure prediction algorithm using MapReduce programming framework, thus achieving better scalability and better failure prediction probability. The proposed model is based on distributed HMM through MapReduce framework in a cloud-computing environment. Through this paper, we also present our idea to increase the performance of the Hadoop Cluster by predicting failure. The accuracy of our model is evaluated using performance metrics (precision, recall, F-measure).

1.3   Paper Structure

Section 2 gives an overview of the background. Section 3 introduces the design and approach of our analysis. Section 4 evaluates our algorithm and presents the results. Section 5 concludes the paper.

2   Background

2.1   Hadoop

Hadoop [32] is an open-source framework for distributed storage and data-intensive processing, first developed by Yahoo. It has two core projects: Hadoop Distributed File System (HDFS) and MapReduce [7] programming model. HDFS is a distributed file system that splits and stores data on nodes throughout a cluster, with a number of replicas. It provides an extremely reliable, fault-tolerant, consistent, efficient and cost-effective way to store a large amount of data. The MapReduce model consists of two key functions: Mapper and Reducer. The Mapper processes input data splits in parallel through different map tasks and sends sorted, shuffled outputs to the Reducers that in turn groups and processes them using reduce tasks for each group.

2.2   Hidden Markov Models

HMM [2] is based on Markov Models in which one does not know anything about observation sequences. The numbers of states, the transition probabilities, and from which state an observation is generated are all unknown. It consists of unobserved states. And each state is not directly visible, but output and dependent on the state is visible. HMM typically used to solve three types of problem: detection or diagnostic problem, decoding problem and learning problem. Using forward-backward algorithm [14] solves diagnostic problem. Using Viterbi algorithm [30] solves decoding problems. And using Baum-Welch algorithm [8] solves learning problem.

3   Approach

In this section, we describe how the useful information from different logs are extracted and the use of HMMs to predict failure from those log files.

The proposed method deals with all the log files associated with Hadoop cluster (HDFS): DataNode and NameNode logs. The log files are collected from the different nodes associated with the cluster. The logs generated from 11-node clusters are stored in HDFS system using Apache Flume collector [1]. The log files contain all unwanted and wanted information that makes it difficult for the human to read. For this reason, pre-processing of logs is needed before storing to HDFS system. In the pre-processing steps, all the log messages are extracted, and unwanted and noisy messages are removed. The stored data is further analyzed using HMM model. Failure prediction algorithm is used to detect a failure and ignore defective node before running any task.

HDFS system consists of NameNode and DataNode. NameNode is the master node on which job tracker runs. It consists of the metadata (information about data blocks stored in DataNodes - the location, size of the file, etc.). It maintains and manages the data blocks, which are present on the DataNodes.

The DataNode is a place where actual data is stored. The DataNode runs three main types of daemon: Read Block, Write Block, and Write-Replicated Block. NameNode and DataNode maintain their own logging format. Each node records events/activities related to reading, writing and replication of HDFS data blocks. Each log is stored on the local file-system of the executing daemon. Our analysis is based on some important insights about the information available through Hadoop logs. Block ID in DataNode log provides a unique identifier, which represents the HDFS blocks that consist of raw bytes of the data file.

Before using log messages to build the model, we structured and appended all the log files into systematized forms. Four steps are involved in our approach; pre-processing, clustering, training, and predicting as shown in Fig. 1. In first steps, all useful information, such as timestamp, error status, error type, node ID and user ID, are extracted and new log template is created. Since different logs reside on the local disk in different nodes, it is necessary to collect and attach all the log information into a one-log template. In second steps, we use the clustering algorithm to differentiate various types of errors. With the clustering technique, real error types that propagate to failure are recognized. And the third and fourth steps, is the training and prediction algorithm using HMM model, which is discussed in detail below.

We adopted Hidden Markov models (HMMs) for this approach. HMM applies machine-learning techniques to identify whether a given sequence of the message is failure-prone or not. HMM models parameters can be adjusted to provide better prediction. The sequences of an error event are fed into the model as an input. Each error event consists of a timestamp, error ID and error type, which determine the types of error. Failure and non-failure information are extracted from error sequences to create a transition matrix. HMM is characterized by the following modules: hidden states $X = \{x_1, x_2, x_3\}$, observations state $Y = \{y_1, y_2, y_3\}$, transition probabilities $A = a_{ij} = \{P[q_{t+1} = x_j | q_t = x_j]\}$ and emission probabilities $B = b_{ij}$. By the definition of HMM (λ) is:

$$\lambda = (\pi, A, B) \tag{1}$$

where, A is the transition matrix whose elements give the probability of transitioning from one state to another, B is the emission matrix giving $b_j(Y_t)$ the probability of observing Y_t. π is initial state transition probability matrix.

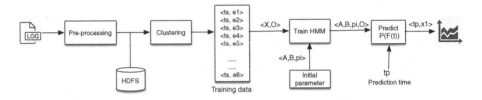

Fig. 1. Workflow of failure prediction.

The observation symbols $O_1 = \{e_1, e_2, e_3, e_4, e_5, e_6\}$ are referred to error events of the system, and failures are represented as hidden state of HMM as shown in Fig. 2. Error patterns are used as training set for HMM model if the model transits to a failure state each time a failure occurs in the training data. Two steps are necessary for obtaining training sequences for the model.

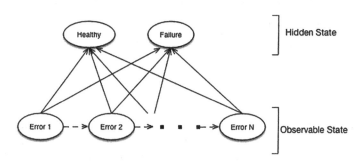

Fig. 2. Mapping failure and errors to Hidden Markov Model.

The first step involves the transformation of error types into a special error symbol in order to classify different types of error. Error event timestamps and error categories form an *error sequence* (event-driven sequence). HMM applies machine-learning techniques in order to identify characteristic properties of an error sequence. We trained the model using past error sequence. The model adjusts its parameters based on those records. The trained model is then applied to the new error sequences to predict failure. Such an approach in machine learning is known as *"supervised learning"*. Let *"e"* represent different types of error in the log files. The series of messages that appear in *"e"* form a time-series representation of events that occurred. In this paper, all categories of *"e"* are identify using *k-means clustering technique* [16]. Six different types of error are distinguished from the given log files and the set e would be $(e_1, e_2, e_3, e_4, e_5, e_6)$. This error set is known as *error sequence* or observation for our model.

In the next step, the model is defined using error sequences. Error sequence consists of failure and non-failure information that has occurred within a sliding window of length Δt as shown in Fig. 3. F is the failure in the system and e_1, e_2, e_3, e_4 represent the error events in the log files. Failure t_p is predicted based on Δt error sequence. HMM models are trained using error sequences. The main goal of training is that failure characteristics are generated from the error sequences. Once the models are trained, new upcoming failure is evaluated from the error sequences. To predict upcoming failure, sequence likelihood is computed from HMM models.

In this paper, there is a sequence of log data over timestamp, which we needed to train our HMM model using a set of the sequence of log output as observation $O = (\text{info, warn, error, fatal})$. $N = 4$ for HMM model is denoting the stages in time that are allowed in different transitions in the HMM training. Each error

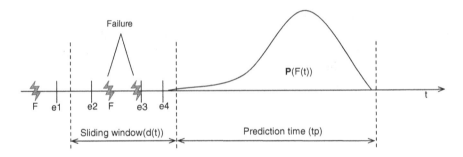

Fig. 3. Failure prediction, where e1, e2, e3 and e4 are error sequences, and F is the failure in the system.

sequence may result in failure-prone or non-failure system. The Failure-prone system has a similar pattern of errors, which result in failure. The probabilityof log data are computed using Forward-Backward Algorithm as proposed in [8].

Training the Model: First, from the log template sequence of an error message is extracted an output sequence composed of 1, 2, 3, 4, 5 and 6's, one number for each time step by rounding the timestamp of logs of a job to the nearest integer. Thus, the state transition in the HMM takes place every time step until the absorption state is reached. The choice of time step determines the speed of learning and its accuracy. A log sequence of a job always starts from the state x_1 and ends at x_2, and the initial probabilities for π are fixed to be 0.5. With the output sequence as described, we compute the most likely hidden state transition sequence and the model parameters $\lambda = (A, B, \pi)$. During training, the HMM parameters π, A, B are optimized. These parameters are maximized in order to maximize sequence likelihood. For initial steps, number of states, number of observation, transition probability and emission probability are pre-specified. In this experiment, initial parameters are calculated from the past observation, such that the model can predict accurately from the initial phase. As training of model progress, the parameter value gets closer to the actual value. Training in HMM is done using Expectation-Maximization algorithm, where backward variable β and forward variable α are evaluated. This algorithm helps to maximize the model parameters based on maximum likelihood. If the model started randomly from a pre-specified HMM parameter, it will take several iterations to get superior parameters, which best fit, the model for prediction. The goal of training datasets is to adjust the HMM parameters such that error sequences are best represented and that the model transits to a failure state each time a failure occurs in the training datasets.

There are a few existing methods such as; Baum-Welch algorithm and gradient-descent techniques, uses iterative procedures to get the locally maximized likelihood. However, this iterative procedure might be significantly slow if the observed sequence is large. In this paper, we proposed a slightly different algorithm to train data, which is significantly faster than the traditional method. The idea is to formulate the probability of the observation sequence

Algorithm 1. Failure State Prediction Algorithm

1: Initialized $O = \{o_1, o_2, ..o_6\}$ ▷ different types of error as observation
2: $S = \{Healthy, Failure\}$ ▷ two hidden state
3: $m = 2$ ▷ m is number of hidden state
4: $n = 6$ ▷ n is number of different types of errors
5: Initialized A_{ij}, B_{ij} ▷ emission matrix B_{ij} stores the probability of observable
     sequences o_j from state s_i ▷ transition matrix A_{ij} stores the transition
     probability of transiting from state s_i to state s_j
6: Initialized Π ▷ an array of initial probabilities
7: $Y = \{y_1, y_2...y_k\}$ ▷ an error sequence of observation
8: **Map:**
9: Initialized $StatePathProb$
10: Update A_{ij}, B_{ij}
11: $PathProb = StatePathProb * B_{ij}$
12: **for** each state s_j **do**
13:      $StatePathProb[j, i] \leftarrow \max_k (StatePathProb[k, i - 1] \cdot A_{kj} \cdot B_{jy_i})$
14:      $PathProb[j, i] \leftarrow \arg\max_k (PathProb[k, i - 1] \cdot A_{kj} \cdot B_{jy_i})$
15: **end for**
16: $z_i \leftarrow \arg\max_k (StatePathProb)$
17: $x_i \leftarrow S_i$
18: **for** $i \leftarrow T - 1, ..., 1$ **do**   ▷ T is length of observable sequence
19:      $z_i \leftarrow PathProb[z_i, i]$
20:      $x_i \leftarrow zi$
21: **end for**
22: $emit(timestamp, x)$

O_t, O_{t+1} pairs and then to use Expectation-Maximization algorithm to learn for this model λ.

In order to train the model, there is a need to find the repetitive error sequence in the data. To do so first, we need to compute the likelihood of raw data in the desired range. This problem is computed using EM algorithm. The EM consists of two steps: an expectation (E) step, which creates a function for calculating log-likelihood from the current estimate, and a maximization (M) step, which computes parameters maximizing the expected log-likelihood calculated on the E step. This EM algorithm is carried on a map and reduce task.

Prediction: The Hidden state is calculated using Viterbi algorithm. There is a sequence of observations $0 = O_1, O_2....O_n$ with given model $\lambda = (A, B, \pi)$. The aim of Viterbi algorithm is to find optimal state sequence for the underlying Markov chain, and thus, reveal the hidden part of the HMM λ. The final goal of Viterbi is to calculate the sequence of states (i.e. $S = \{S_1, S_2, ...S_n\}$), such that

$$S = argmax_s P(S; O, \lambda) \tag{2}$$

Viterbi algorithm returns an optimal state sequence of S. At each step t, the algorithm allow S to retain all optimal paths that finish at the N states. At t+1, the N optimal paths get updated and S continues to grow in this manner. Figure 4 shows details architecture of Viterbi algorithm implementation.

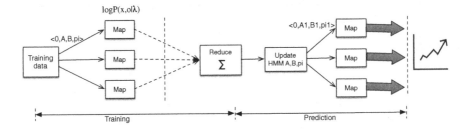

Fig. 4. Architecture of an algorithm predicting failure state using MapReduce programming framework.

The goal is to predict hidden state from the given observation $0 = \{O_1, O_2....O_n\}$. No reducer is used and on each mapper, a local maximum is calculated and state path based on maxima are observed.

4   Result

Setup: Our cluster is comprised of 11 nodes with CentOS Linux distro, one node each for Namenode, Secondary Namenode, HBase Master, Job Tracker, and Zookeeper. The remaining 6 nodes act as Data Nodes, Regional Severs, and Task Trackers. All nodes have an AMD Opteron (TM) 4180 six-core 2.6 GHz processor, 16 GB of ECC DDR-2 RAM, 3×3 TB secondary storage and HP ProCurve 2650 switch. Experiments were conducted using RHIPE, Hadoop-0.20, Hbase-0.90 Apache releases. Our default HDFS configuration had a block size of 64 MB and the replication factor of 3.

The prediction techniques presented in this paper have been applied to the data generated while performing operations in the Hadoop Cluster. With 1 month of Hadoop log data, we trained HMM model using sliding windows varying from 1 to 2 h in length. A Large amount of Hadoop log data was generated using SWIM [26], a tool to generate arbitrary Hadoop jobs that emulate the behaviors of true production jobs of Facebook. We used AnarchyApe [6] to create different types of failure scenarios in Hadoop cluster. AnarchyApe is an open-source project, created by Yahoo! developed to inject failures in Hadoop cluster [10].

4.1   Types of Error

Errors such as operational, software, configuration, resource and hardware are present in Hadoop cluster. In this analysis, hardware failure was not considered. Operational, software and resource errors are taken into consideration to detect a failure in the software level of Hadoop cluster. Operation errors include missing operations and incorrect operations. Operation errors are easily identified in log messages by operations: $HDFS\_READ$, $HDFS\_WRITE$, etc. Resource errors (memory overflow, node failure) refer to resource unavailability occurring in the computing environment. Software errors (Java I/O errors, unexceptional) refer to software bugs and incompatibility. These types of error are detected on different DataNodes. Log messages are classified into six different types of error: Network connection, Memory overflow, Security setting, Unknown, Java I/O error, NameNode failure as shown in Fig. 5. Errors like network connection and security setting are most occurring errors in the Hadoop cluster. In the pre-processing step, each log message is tagged with certain error ID and using the clustering algorithm like k-means, different types of error are analyzed.

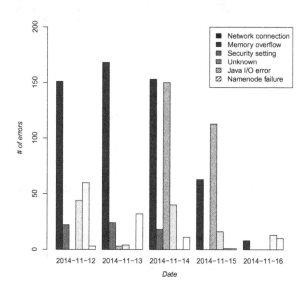

Fig. 5. Different types of error in Hadoop cluster during 5 days interval.

4.2   Predicting Failure State in Hadoop Cluster

Different types of error are used as input observation in our model i.e. $O = \{O_1, O_2....O_n\}$. O_1 to O_6 are error sequences, and O7 is a non-error sequence from the log template. This observation is used in the model $\lambda = (\pi, A, B)$ to detect $S = \{Healthy, Failure\}$. Based on the error sequence in the HMM model, with the help of Viterbi algorithm, hidden state sequences are generated which are shown in gray and red line in Fig. 6. The red line indicates failure state and gray indicates non-failure state. Similarly, black and gray line shows actual failure state. Based on the probability of the previous state and HMM parameters, the failure, and non-failure states are determined. Error sequences are predicted using EM algorithm, and based on predicted error sequences, hidden states (failure or non-failure) are predicted. Error in prediction is calculated by differencing the actual and predicted value as shown in the graph. At first step, our model is trained from the previous record. As the time passes, the model gets more

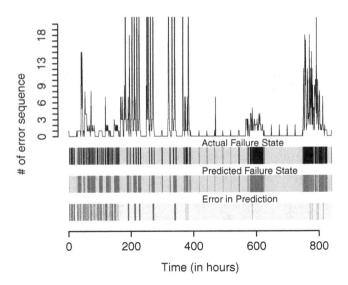

Fig. 6. Using HMM to predict failure and normal state. Error sequences are the observation or observable state of HMM and gray-black bar indicate the hidden state, gray line indicate non-failure state and black indicate failure state. Similarly, red line indicates predicted failure state. And blue line indicates prediction error (Color figure online).

accurate. The training of the model depends solely on the initial parameter. For this example, initial parameters are calculated from the past record. That is why this model has similar behavior from the initial point, but not an accurate prediction.

The models' ability to predict failure precisely is evaluated by four metrics: precision, recall, false positive rate and F-measure. These metrics are frequently used for prediction evaluation and have been used in many relevant researches as in e.g. [25]. Four metrics are defined in Table 1.

From the above observation, we used log entries of 800 h out of which; first 650 h is used for training and last 150 h is used for prediction. In total, we have 24000 observations for 150 h of prediction time. Different cases for prediction is shown in Table 2. The accuracy of the model is 91.25 % (9000 + 12900/24000). And the precision and recall is 0.93 and 0.91 respectively.

Table 1. Definition of metrics.

Metric	Definition	Calculation
Precision	$p = \frac{TP}{TP+FP}$	0.93
Recall	$r = \frac{TP}{TP+FN}$	0.91
False positive rate	$fpr = \frac{FP}{FP+TN}$	0.091
F-measure	$F = \frac{2pr}{p+r}$	0.92
Accuracy	$accuracy = \frac{TP+TN}{TP+FP+FN+TN}$	0.91

Table 2. Observation for different cases.

Predicted failure state	Predicted non-failure state
TN: 9000	FP: 900
FN: 1200	TP: 12900

Higher precision ensure fewer false positive errors, while a model with high recall ensures lesser false negative errors. Ideal failure prediction model would achieve higher precision and recall value, i.e. precision = recall = 1. However, both high recall and precision are difficult to obtain at the same time. They are often inversely proportional to each other. Improving recall in most cases lowers the precision and vice-versa. F-measure ensures that the model is accurate or not. It provides both precision and recall are reasonably high. In HMM method, a threshold value allows the control of true positive rate and false positive rate. This is one of the big advantages of HMM, method over other techniques.

4.3   Scalability

To test the implementation of MapReduce HMM model in the cluster, we fixed the number of nodes in the cluster to be 6. And, then tested HMM by varying the number of data size from 1 GB (85 million error sequences) to 7 GB (571 million error sequences). Figure 7a demonstrates the scalability of the algorithm. It shows a steady increase in execution time with the growth in data size. The brown and black lines in the graph represent parallel and sequential execution of map task. It is obvious that parallel execution outperform.

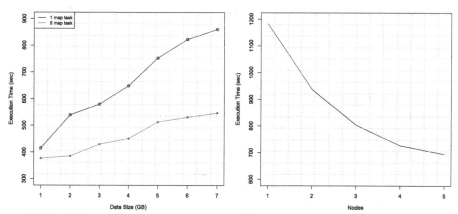

(a) Scalability with increases in data size (b) Scalability with increases in cluster size.

Fig. 7. Scalability of failure prediction algorithm

In the Fig. 7(b), we set the number of nodes participating in the MapReduce calculations to 1, 2, 3, 4, or 5. The algorithm was then tested on the dataset of size 5 GB (138 million error sequences). The experimental result shows that the execution time improves with an increase in the number of nodes. This increase can significantly improve the system processing capacity for the same scale of data. By adding more nodes to the system, the performance improves and computation can be distributed across the nodes [20]. The ideal scalability behavior would illustrate a linear line in the graph. However, it is impossible to realize this ideal behavior due to many factors such as network overheads.

5   Conclusion

As failures in cluster systems are more prevalent, the ability to predict failures is becoming a critical need. To address this need, we collected Hadoop logs from Hadoop cluster and developed our algorithm on the log messages. The messages in the logs contain error and non-error information. The messages in the log were represented using error IDs, which indicate message criticality. This paper introduced a novel failure prediction method using distributed HMM method over distributed computation. The idea behind this model is to identify the error pattern that indicates an upcoming failure. A machine learning approach like HMM has been proposed here, where the model is trained first using previously pre-processed log files and then it is used to predict the failures. Every log entry is split into equal intervals, defined by sliding window. These entries are separated into error sequence and non-error sequence. Training of the model is done using past observation. Viterbi's algorithm does the prediction of hidden state. Experimental results using Hadoop log files provide an accuracy of 91 % and F-measure of 92 % for 2 days of prediction time. These results indicate that it is promising to use the HMM method along with MapReduce to predict failure.

References

1. Apache: Apache flume (2010). https://flume.apache.org/FlumeUserGuide.html
2. Baum, L.E., Eagon, J., et al.: An inequality with applications to statistical esti-mation for probabilistic functions of markov processes and to a model for ecology. Bull. Am. Math. Soc. **73**(3), 360–363 (1967)
3. Box, G.E., Jenkins, G.M., Reinsel, G.C.: Time Series Analysis: Forecasting and Control. Wiley, New York (2013)
4. Chang, H., Kodialam, M., Kompella, R.R., Lakshman, T., Lee, M., Mukherjee, S.: Scheduling in mapreduce-like systems for fast completion time. In: 2011 Proceed-ings IEEE INFOCOM, pp. 3074–3082. IEEE (2011)
5. Daidone, A., Di Giandomenico, F., Bondavalli, A., Chiaradonna, S.: Hidden markov models as a support for diagnosis: formalization of the problem and synthesis of the solution. In: 25th IEEE Symposium on Reliable Distributed Systems, SRDS 2006, pp. 245–256. IEEE (2006)
6. David: anarchyape (2013). https://github.com/david78k/anarchyape

7. Dean, J., Ghemawat, S.: Mapreduce: simplified data processing on large clusters. In: Proceedings of the 6th Conference on Symposium on Opearting Systems Design & Implementation, OSDI 2004, vol. 6, p. 10. USENIX Association, Berkeley (2004). http://dl.acm.org/citation.cfm?id=1251254.1251264

8. Dempster, A.P., Laird, N.M., Rubin, D.B.: Maximum likelihood from incomplete data via the em algorithm. J. R. Stat. Soc. Ser. B (Methodol.) **39**, 1–38 (1977)

9. Durbin, R.: Biological Sequence Analysis: Probabilistic Models of Proteins and Nucleic Acids. Cambridge University Press, Cambridge (1998)

10. Faghri, F., Bazarbayev, S., Overholt, M., Farivar, R., Campbell, R.H., Sanders, W.H.: Failure scenario as a service (fsaas) for hadoop clusters. In: Proceedings of the Workshop on Secure and Dependable Middleware for Cloud Monitoring and Management, p. 5. ACM (2012)

11. Fahad, A., Alshatri, N., Tari, Z., ALAmri, A., Y Zomaya, A., Khalil, I., Foufou, S., Bouras, A.: A survey of clustering algorithms for big data: taxonomy & empirical analysis (2014)

12. Fonseca, R.: X-trace (2010). https://github.com/rfonseca/X-Trace

13. Fulp, E.W., Fink, G.A., Haack, J.N.: Predicting computer system failures using support vector machines. In: Proceedings of the First USENIX Conference on Analysis of System Logs, WASL 2008, p. 5. USENIX Association, Berkeley (2008). http://dl.acm.org/citation.cfm?id=1855886.1855891

14. Hassan, M.R., Nath, B., Kirley, M.: A fusion model of hmm, ann and ga for stock market forecasting. Expert Syst. Appl. **33**(1), 171–180 (2007)

15. Huang, X., Acero, A., Hon, H.W., Foreword By-Reddy, R.: Spoken Language Processing: A Guide to Theory, Algorithm, and System Development. Prentice Hall PTR, Upper Saddle River (2001)

16. Kanungo, T., Mount, D.M., Netanyahu, N.S., Piatko, C.D., Silverman, R., Wu, A.Y.: An efficient k-means clustering algorithm: analysis and implementation. IEEE Trans. Pattern Anal. Mach. Intell. **24**(7), 881–892 (2002). http://dx.doi.org/10.1109/TPAMI.2002.1017616

17. Konwinski, A., Zaharia, M., Katz, R., Stoica, I.: X-tracing hadoop (2008)

18. Liang, Y., Zhang, Y., Sivasubramaniam, A., Sahoo, R.K., Moreira, J., Gupta, M.: Filtering failure logs for a bluegene/l prototype. In: Proceedings of the International Conference on Dependable Systems and Networks, DSN 2005, pp. 476–485. IEEE (2005)

19. de Botelho Marcos, P.: Maresia: an approach to deal with the single points of failure of the mapreduce model (2013)

20. Mccreadie, R., Macdonald, C., Ounis, I.: Mapreduce indexing strategies: studying scalability and efficiency. Inf. Process. Manage. **48**(5), 873–888 (2012). http://dx.doi.org/10.1016/j.ipm.2010.12.003

21. Ng, F.: Analysis of hadoops performance under failures. Rice University

22. Plötz, T., Fink, G.A.: Markov Models for Handwriting Recognition. Springer, Heidelberg (2011)

23. Rabiner, L.: A tutorial on hidden markov models and selected applications in speech recognition. Proc. IEEE **77**(2), 257–286 (1989)

24. Sahoo, R.K., Sivasubramaniam, A., Squillante, M.S., Zhang, Y.: Failure data analysis of a large-scale heterogeneous server environment. In: 2013 43rd Annual IEEE/IFIP International Conference on Dependable Systems and Networks (DSN), p. 772 (2004)

25. Salfner, F., Malek, M.: Using hidden semi-markov models for effective online failure prediction. In: 26th IEEE International Symposium on Reliable Distributed Systems, SRDS 2007, pp. 161–174. IEEE (2007)

26. SWIMProjectUCB: Swimprojectucb/swim (2012). https://github.com/SWIMProjectUCB/SWIM

27. Tai, A.H., Ching, W.K., Chan, L.Y.: Detection of machine failure: hidden markov model approach. Comput. Ind. Eng. **57**(2), 608–619 (2009)

28. Tan, J., Pan, X., Kavulya, S., Gandhi, R., Narasimhan, P.: Salsa: analyzing logs as state machines. WASL **8**, 6–6 (2008)

29. Teoh, T.T., Cho, S.Y., Nguwi, Y.Y.: Hidden markov model for hard-drive failure detection. In: 2012 7th International Conference on Computer Science & Education (ICCSE), pp. 3–8. IEEE (2012)

30. Viterbi, A.J.: Error bounds for convolutional codes and an asymptotically optimum decoding algorithm. IEEE Trans. Inf. Theory **13**(2), 260–269 (1967)

31. Wang, F., Qiu, J., Yang, J., Dong, B., Li, X., Li, Y.: Hadoop high availability through metadata replication. In: Proceedings of the First International Workshop on Cloud Data Management, pp. 37–44. ACM (2009)

32. White, T.: Hadoop: The Definitive Guide. O'Reilly Media, Inc., Sebastopol (2012)

33. Wilson, A.D., Bobick, A.F.: Parametric hidden Markov models for gesture recognition. IEEE Trans. Pattern Anal. Mach. Intell. **21**(9), 884–900 (1999)

34. Zawawy, H., Kontogiannis, K., Mylopoulos, J.: Log filtering and interpretation for root cause analysis. In: ICSM, pp. 1–5. IEEE Computer Society (2010). http://dblp.uni-trier.de/db/conf/icsm/icsm2010.html#ZawawyKM10

Emerging Pragmatic Patterns in Large-Scale RDF Data

Weiyi Ge[1]([✉]), Wei Hu[2], Chenglong He[1], and Shiqiang Zong[1]

[1] Science and Technology on Information Systems Engineering Laboratory,
Nanjing 210007, China
geweiyi@gmail.com, hechenglong@nuaa.edu.cn, sqzong@yeah.net
[2] State Key Laboratory for Novel Software Technology, Nanjing University,
Nanjing, China
whu@nju.edu.cn

Abstract. With the development of the Linked Data, an increasing number of RDF data sets are published in many application domains. To understand the underlying meaning and characteristics of large RDF data, and to reuse popular domain terms when publishing data, capturing emerging pragmatic patterns is critical. In this paper, we propose the notion of term co-instantiation graph (TIG) and a method to build a TIG for a given RDF dataset. We also describe a clustering-based approach to distill a set of pragmatic patterns from a TIG, which reveal the pragmatic custom of highly-correlated terms. Through extensive experiments on a real big dataset containing 21 M RDF documents, we analyze the macroscopic structure of the term co-instantiation graph and pragmatic patterns from the complex network point of view, and demonstrate our approach can not only give an elaborated ontology partitioning from the pragmatic perspective to ease the ontology reuse, but also provide a new way to explore the Linked Data.

Keywords: Pragmatic pattern · Term co-instantiation graph · Clustering · Complex network analysis · Linked data

1 Introduction

The Semantic Web vision and its related technology stack have brought out the development of a Web of data, or the Linked Data. In recent years, a considerable amount of RDF documents have been published on the Web by various parties. Especially, as a consequence of Linking Open Data project (http://linkeddata. org) launched by Semantic Web community, lots of RDF datasets in a wide range of domains are introduced. In order to make the Linked Data step forward quickly and steadily, RDF data producers are encouraged to reuse popular domain terms when publishing their own data. Those terms which are often used together constitute a pragmatic pattern.

It should be stressed that capturing emerging pragmatic patterns in large-scale RDF data is a fundamental issue. Firstly, investigation [1] shows that most

© Springer International Publishing Switzerland 2015
W. Qiang et al. (Eds.): CloudCom-Asia 2015, LNCS 9106, pp. 247–260, 2015.
DOI: 10.1007/978-3-319-28430-9_19

Semantic Web terms (95.1 %) have no instances, and terms from different vocabularies are frequently co-instantiated in common RDF documents, so it is necessary to take into account the relatedness among terms from the viewpoint of pragmatics or usage for ontology partitioning or modularization. Secondly, salient pragmatic patterns of co-instantiated terms should be listed explicitly to recover topics or meanings of RDF data for RDF summarization or exploration. In addition, pragmatic patterns can also be used in other scenarios, such as SPARQL query [2], query expansion [3], and relation exploration and search [4].

To capture emerging pragmatic patterns in large-scale RDF data, we firstly establish a matrix, called the TIM matrix, to represent the information about terms instantiation in all RDF documents. Then we propose the notion of *term co-instantiation graph* (TIG), which is a weighted undirected graph, and a way to build a TIG from TIM by measuring the relatedness between terms. We describe an agglomerative clustering algorithm on TIG to distill a set of *pragmatic patterns*, each of them is a subgraph of the term co-instantiation graph deduced by several highly-correlated terms. We perform our experiments on a large dataset from the Falcons search engine [5], which covers a huge number of RDF data on the Linked Data. We analyze the TIG from the complex network point of view, and report the pragmatic patterns captured by our clustering-based approach. Experiments show that our approach can not only give an elaborated ontology partitioning from the pragmatic perspective, but also provide a new way to explore the Linked Data.

The rest of this paper is structured as follows. We start by giving the preliminaries and notations used throughout the paper. Section 3 presents a method to build TIG for a given RDF dataset. Section 4 proposes a clustering-based algorithm to distill pragmatic patterns. In Sect. 5, we report our experimental results on a large dataset from Falcons. Section 6 discusses related work. Finally, Sect. 7 concludes this paper with future work.

2   Preliminaries

Given a set of URIs U, blank nodes B, and literals L, an *RDF triple* is a triple $\langle s, p, o \rangle \in (U \cup B) \times U \times (U \cup B \cup L)$. An *RDF graph* is a set of RDF triples, and an *RDF document*, denoted by d, is a serialization of an RDF graph.

In this paper, we assume that classes and properties are disjoint. We design a discriminator, which can tell whether a URI is a class, a property or neither. This discriminator is based on a heuristic approach which using some RDF(S)[1] and OWL DL[2] entailment rules. A URI u is a *class* (*property*) iff entailing an RDF triple $\langle u, \mathtt{rdf:type}, \mathtt{rdfs:Class} \rangle$ ($\langle u, \mathtt{rdf:type}, \mathtt{rdf:Property} \rangle$). For instance, a URI u may be a class if there is an RDF triple whose subject is u and predicate is $\mathtt{rdfs:subClassOf}$; u is recognized to be a property if there is an RDF triple whose predicate is $\mathtt{owl:onProperty}$ and object is u.

---

[1] http://www.w3.org/TR/2004/REC-rdf-mt-20040210/.

[2] http://www.w3.org/TR/owl-absyn/.

A *vocabulary* is a non-empty set of URIs (named the constituent terms) that denote classes/properties with a common *URI namespace*[3]. For convenience, *qualified names* are used for the URIs with famous namespaces in the paper, e.g. foaf:Person for http://xmlns.com/foaf/0.1/Person.

Unambiguously, classes and properties are uniformly referred to as *terms* throughout this paper. For notations, we use t to denote a term, and d to denote an RDF document. We define that d *instantiates* t as a class if there is an RDF triple $\langle s, \text{rdf:type}, t \rangle$ in d. Similarly, t is instantiated as a property in d if there exists an RDF triple $\langle s, t, o \rangle$ in d. The number of times that d instantiates t is given by counting distinct RDF triples that satisfying the above requirement. If an instantiation of a term in an RDF document is inconsistent with the type of the term (class or property) recognized from the discriminator, we discard the instantiation. We find that there indeed exist some classes (properties) being instantiated as properties (classes) in some RDF documents.

Because an RDF document can instantiate a set of terms, and a term may be instantiated in a set of RDF documents, we use a matrix to represent the information about term instantiation in RDF documents.

Definition 1 (Term Instantiation Matrix (TIM)). *TIM is a $n \times m$ matrix:*

- *rows are a set of terms $T = \{t_1, t_2, \ldots, t_m\}$;*
- *columns are a set of RDF documents $D = \{d_1, d_2, \ldots, d_n\}$;*
- *entry a_{ij} denotes the number of times (#RDF triples) that d_j instantiates t_i, in which a zero entry indicates non-instantiation.*

Term co-instantiations in the same RDF documents reveal their relatedness in pragmatics. We use a graph to represent this information.

Definition 2 (Term Co-instantiation Graph (TIG)). *TIG $= (T, E, W_T, W_E)$ is a weighted undirected graph:*

- *T, a set of nodes, where each node is an instantiated term;*
- *$E \subset T \times T$, a set of undirected edges, where $(t_1, t_2) \in E$ if and only if t_1 and t_2 are instantiated by at least one of the same RDF documents;*
- *$W_T : T \to \mathbb{N}$ is a weighting function that maps each term to a natural number w_i, and the weight indicates the instantiation times of the term in the dataset;*
- *$W_E : E \to \mathbb{R}$ is a weighting function that maps each edge (t_i, t_j) to a non-negative real number w_{ij}, and the weight indicates the relatedness between t_i and t_j.*

The co-instantiation relation between terms corresponds to a *partitioning* on TIG, which separates TIG into a set of pragmatic patterns $\Delta = \{\delta_1, \delta_2, \ldots, \delta_l\}$.

Definition 3 (Pragmatic Pattern). *A pragmatic pattern, denoted by $\delta = TIG(T')$, is a connected subgraph of TIG induced by some terms T', where T' is a non-empty subset of T.*

---

[3] http://www.w3.org/TR/swbp-vocab-pub/.

Note that it is not required that T' is contained in a single vocabulary. We can extend the definition by adding the axioms describing the terms in a pattern, which come from their dereference documents and can be considered as the authoritative descriptions of these terms. In this paper, we only adopt a simple definition. A *partitioning* Δ on TIG separates TIG into a set of pragmatic patterns $\Delta = \{\delta_1, \delta_2, \ldots, \delta_l\}$.

3   Term Co-Instantiation

Given an RDF document set, we will firstly introduce the establishment of TIM, and then present the computation of the relatedness between terms to build TIG.

3.1   Establishing TIM

Parallel parsing RDF documents to collect the instantiated terms and gain the instantiation times of each term in each RDF document is straightforward to establish a raw TIM matrix. More specifically, we assume that there is an external discriminator that can tell whether a URI is a class, a property or neither. For every RDF document in the RDF document set, we firstly obtain a set of candidate classes and properties instantiated in this RDF document. Then, we submit every term to the discriminator, get its suggested type and compare with the recognized instantiation type. If they are inconsistent, we discard the term since it is probably misusing. Please note that a lot of current search engines, such as Falcons, can provide such functionality. Next, we refine this raw TIM matrix by the following three heuristic rules:

1. Remove all the terms from the RDF, RDFS, OWL, DAML+OIL, DC and SKOS ontologies;
2. Remove all the terms having no description in their dereference documents;
3. Remove all the documents that do not instantiate any term after applying the above two rules.

The first rule is intended to remove the meta-level terms which are instantiated in nearly every RDF document and widely used to define general terms. While the second one is to eliminate the spurious or misspelt terms. As an example, the FOAF vocabulary does not have `foaf:city`. As another example, http://xmlns.com/foaf/0.1/givenName is a misspelling of http://xmlns.com/foaf/0.1/givenname. Results are stored in a database, where term and RDF document are two keys of the table to locate an entry for recording the instantiation times in the matrix.

3.2   Building TIG

After executing the heuristics aforementioned, TIM is used for building a term co-instantiation graph (TIG) by measuring the relatedness between terms. For every entry a_{ij} in TIM, we firstly normalize its value to the $[0,1]$ range based on the traditional TF [6] technique, in order to indicate its importance or frequency. More specifically, let $\widetilde{a_{ij}}$ be the normalized value of an entry a_{ij}. We compute $\widetilde{a_{ij}} = a_{ij}/\sum_{i=1}^{|T|} a_{ij}$. We use $\widetilde{\text{TIM}}$ to represent the normalized matrix from TIM.

The relatedness between any two terms t_i and t_j in T is defined as the cosine similarity between the corresponding normalized row vectors in $\widetilde{\text{TIM}}$:

$$rel(t_i, t_j) = \frac{\sum_{l=1}^{|D|} \widetilde{a_{il}}\widetilde{a_{jl}}}{\sqrt{\sum_{l=1}^{|D|} \widetilde{a_{il}}^2} \cdot \sqrt{\sum_{l=1}^{|D|} \widetilde{a_{jl}}^2}}. \tag{1}$$

In practice, TIM is very sparse, so it is not necessary to calculate the relatedness between every pair of terms in T because most of them are orthogonal. It is observed that we only need to compute the relatedness between two terms if they are co-instantiated in at least one RDF document in D. Hence, we first generate the term pairs having the relatedness, and then measure the relatedness between terms in each term pair.

There is an optimization for obtaining the term pairs. The principle is that, if the terms instantiated in an RDF document d_i is a superset of those in another d_j, there is no need to generate the term pairs from d_j, since they must be a subset of the term pairs from d_i. An algorithm to generate possibly related term pairs is shown in Algorithm 1. Firstly, we sort all RDF documents in D by the sizes of their instantiated terms in Line 2. Then, we scan every RDF document d_j from the one with the largest term size to the smallest one. If we detect that terms from d_j is not subset of any other ones from d_i (Line 4 to 9), we consider d_j for generating term pairs (Line 10 to 18).

For each possibly related term pair $(t_i, t_j) \in \mathcal{T}$, we measure their relatedness $rel(t_i, t_j)$ by Eq. 1. We build a term co-instantiation graph TIG by assigning the accumulated instantiation times $\sum_{j=1}^{|D|} a_{ij}$ to $W_T(i)$ for each term t_i, and generating an undirected edge (t_i, t_j) for any two terms t_i, t_j and setting $W_E(i, j) = W_E(j, i) = rel(t_i, t_j)$.

4 Pragmatic Patterns

4.1 Partitioning

The objective of our partitioning algorithm is to divide a term co-instantiation graph TIG $= (T, E, W_T, W_E)$ into a set of pragmatic patterns $\Delta = \{\delta_1, \delta_2, \ldots, \delta_l\}$, where, by certain measure, the cohesiveness among the terms in any δ_i is high; while the coupling between different δ_i, δ_j is low. The pragmatic patterns also satisfy that: (i) $\forall \delta_i = \text{TIG}(T_i), \delta_j = \text{TIG}(T_j) \in \Delta, \delta_i \neq \delta_j \mid T_i \cap T_j = \emptyset$; and (ii) $\cup_{i=1}^{l} T_i = T$.

The proposed partitioning algorithm is a hierarchical agglomerative clustering (HAC) algorithm principally inspired by ROCK [7], which is a very scalable algorithm in the field of data mining. A principle difference between ROCK and our algorithm is that we use the $cut()$ function as the criterion function in order

---

Algorithm 1. An algorithm to generate term pairs having the relatedness

---

Input: a term co-instantiation matrix TIM $= [a_{ij}]_{m \times n}$
Output: a set of term-pairs $\mathcal{T}$
1 $T(d_j) = \{t_i | a_{ij} \neq 0\}$;   // a set of terms instantiated in d_j
2 Sort $D = \{d_1, d_2, \cdots, d_n\}$ by $|T(d)|$ desc;
3 **foreach** $d_j \in D$ **do**
4   **for** $i = 1$ **to** $j - 1$ **do**
5   **if** $T(d_j) \subseteq T(d_i)$ **then**
6        $isSubset = true$;
7   break;
8   **end**
9   **end**
10   **if** $not\ isSubset$ **then**
11   **for** $i = 1$ **to** m **do**
12   **for** $k = i + 1$ **to** m **do**
13   **if** $a_{ij} \neq 0$ *and* $a_{kj} \neq 0$ **then**
14   Add (t_i, t_k) into $\mathcal{T}$;
15   **end**
16   **end**
17   **end**
18   **end**
19 **end**
20 **return** $\mathcal{T}$;

---

to improve the efficiency. Another important difference is that we adopt floating point values (relatedness) between terms instead of binary values.

As the criterion function, $cut()$ is to calculate both the cohesiveness and the coupling, which measures the distance between two pragmatic patterns by accumulating the aggregated inter-connectivity of them. Let δ_i, δ_j be two pragmatic patterns. $cut()$ between δ_i and δ_j is defined as follows:

$$cut(\delta_i, \delta_j) = \frac{\sum_{t_i \in \delta_i} \sum_{t_j \in \delta_j} rel(t_i, t_j)}{|\delta_i| \cdot |\delta_j|}, \qquad (2)$$

where $|\delta_i|$ gets the number of terms in δ_i. When δ_i and δ_j are identical, it computes the cohesiveness of that pragmatic pattern,

$$cohes(\delta_i) = cut(\delta_i, \delta_i). \qquad (3)$$

When δ_i and δ_j are different, it computes the coupling between them,

$$coupl(\delta_i, \delta_j) = cut(\delta_i, \delta_j). \qquad (4)$$

Similar $cut()$ functions are often seen in spectral clustering and information retrieval [6,8,9]. The advantage of our function is that it generates almost equally-sized pragmatic patterns, which means that it would not suffer from the heavily-imbalanced pragmatic patterns (called a skewed partitioning). In addition, using a uniform function to calculate both cohesiveness and coupling is firstly proposed in [10], called silhouette coefficient.

Our partitioning algorithm is shown in Algorithm 2, which follows a traditional framework of agglomerative clustering [11]. It accepts input as a term co-instantiation graph, TIG, to be partitioned. Initially, it creates a pragmatic pattern for every term. The cohesiveness of a term is calculated by a variation of its instantiation times (Line 6). As a result, the more popular terms can be selected earlier for merging. We propose the ln() function, because the instantiation times scale largely, and we want to decrease their impacts on further selection. However, we guarantee that the cohesiveness of any term is greater

---

Algorithm 2. A hierarchical agglomerative clustering algorithm

---

Input: a term co-instantiation graph $TIG = (T, E, W_T, W_E)$, and a parameter ϵ limiting the maximum number of terms in each pragmatic pattern $(\epsilon \ll |T|)$

Output: a set of pragmatic patterns $\Delta = \{\delta_1, \delta_2, \ldots, \delta_l\}$

```
 1  foreach t_i ∈ T do                                    // Initialization
 2  │   δ_i := create(t_i);
 3  │   Δ := Δ ∪ {δ_i};
 4  end
 5  foreach δ_i ∈ Δ do
 6  │   cohes(δ_i) := ln(e + W_T(i));                      // Initial cohesiveness
 7  │   foreach δ_j ∈ Δ satisfying j ≠ i do
 8  │   │   coupl(δ_i, δ_j) := W_E(i, j);                  // Initial coupling
 9  │   end
10  end
11  while true do                                         // Clustering
12  │   δ_s := arg max(cohes(δ_i));
13  │   δ_t := arg max(coupl(δ_s, δ_j));                   // j ≠ s
14  │   if |δ_s| + |δ_t| > ε or cohes(δ_s) = 0 then
15  │   │   return Δ;                                      // Termination
16  │   end
17  │   else if coupl(δ_s, δ_t) = 0 then                   // Isolated cluster
18  │   │   cohes(δ_s) := 0;
19  │   end
20  │   else                                              // Merging
21  │   │   δ_p := δ_s ∪ δ_t;
22  │   │   cohes(δ_p) := cohes(δ_s) + cohes(δ_t) + coupl(δ_s, δ_t);
23  │   │   foreach δ_i ∈ Δ satisfying i ≠ p, s, t do
24  │   │   │   coupl(δ_p, δ_i) := coupl(δ_s, δ_i) + coupl(δ_t, δ_i);
25  │   │   │   coupl(δ_i, δ_p) := coupl(δ_p, δ_i);
26  │   │   end
27  │   │   Δ := Δ ∪ {δ_p} \ {δ_s, δ_t};
28  │   end
29  end
```

than 1.0, which would give every term an opportunity to participate in selecting another term/pattern for merging, since the value of $cut(\delta_i, \delta_j)$ is usually less than 1.

During each iteration, it selects a pragmatic pattern δ_s with the maximum cohesiveness (Line 12) and searches for a pragmatic pattern δ_t that has the maximum coupling with δ_s (Line 13). After merging δ_s and δ_t to create a new δ_p (Line 21), it updates the cohesiveness of δ_p as well as its coupling to other ones (Line 22 to 26). If δ_s is fully isolated (Line 17), we set its cohesiveness to zero (a trick in implementation), which implies that selecting it to further merge provides no benefit. The algorithm terminates (Line 15) when the maximum number of the terms in any pragmatic pattern exceeds ϵ or there is no pragmatic pattern whose cohesiveness is larger than zero (implying that it is unnecessary to continue merging, since every pragmatic pattern is completely separated). In practice, ϵ is determined based on experience, e.g. $\epsilon = 9$.

One may think that it is costly to update the cohesiveness and coupling when a new pragmatic pattern is created. In fact, it is not necessary to re-sum the relatedness between those terms in the new pragmatic pattern. For instance, let δ_p be the new pragmatic pattern merged from δ_s and δ_t, the sum of the relatedness in δ_p can be computed as follows:

$$
\sum_{t_i \in \delta_p} \sum_{t_j \in \delta_p} rel(t_i, t_j) = \sum_{t_i \in \delta_s} \sum_{t_j \in \delta_s} rel(t_i, t_j)
$$
$$
+ \sum_{t_i \in \delta_t} \sum_{t_j \in \delta_t} rel(t_i, t_j)
$$
$$
+ \sum_{t_i \in \delta_s} \sum_{t_j \in \delta_t} rel(t_i, t_j). \tag{5}
$$

Recall Formula (2), the sum of the relatedness in δ_s (or δ_t) could be computed by multiplying the cohesiveness of δ_s (or δ_t) to $(|\delta_s| \cdot |\delta_s|)$ (or $(|\delta_t| \cdot |\delta_t|)$), while the sum of the relatedness between δ_s and δ_t can be got by multiplying the coupling between δ_s and δ_t to $(|\delta_s| \cdot |\delta_t|)$. In the past steps, the cohesiveness and coupling have already been computed. In our algorithm shown in Algorithm 2, we use $+$ to represent this optimization (Line 22 and 24).

Compared to some other data clustering approaches, our algorithm is efficient. The time complexity of our algorithm is $O(n^2)$, where n is the number of terms ($n = |T|$). The maximum times of iterations for partitioning is n. In each iteration, the most time-consuming step is to update the coupling of δ_p with others (Line 23 to 26), which takes at most k times, where k is the number of pragmatic patterns in the iteration ($k \leq n$). Assume that we do not sort pragmatic patterns by the cohesiveness (or coupling), thus selecting δ_s (or δ_t) spends no more than k times. Totally, the time complexity is $O(n^2)$. In general, the time complexity of agglomerative partitioning is at least $O(n^2)$ [11], hence our algorithm has already achieved such lower bound. The time complexity of spectral clustering [8,9] (a kind of divisive partitioning) is much higher due to the high computational cost ($O(n^3)$) of singular value decomposition.

5 Experiments

In this section, we will report the results of an experimental study on a large dataset[4] collected by the Falcons search engine until September 2009.

5.1 Statistics of the Dataset

The dataset consists of 21.6 M RDF documents coming from 21 K domains, with 1 k RDF documents in a domain averagely. RDF documents from several domains, e.g. `bio2rdf.org`, `dbpedia.org`, `hi5.com`, `geonames.org`, `opiumfield.com`, `l3s.de`, `liveinternet.ru`, `fu-berlin.de`, `dbtune.org`, constitute the main part of our dataset. Each RDF document consists of 134 RDF triples in average, and only nine RDF documents are composed of more than one million RDF triples.

From these RDF documents, Falcons identifies 2,868,214 classes and 264,315 properties. All the terms are constituted in 12,467 ontologies, implying that there are 251.27 terms in each vocabulary in average.

5.2 Analysis of TIM and TIG

With this dataset, we get a TIM matrix with 91,753 rows and 18,739,990 columns[5]. All those 91,753 terms are distributed in 3,538 ontologies. The number of instantiated terms within each vocabulary is 25.93 in average, which is much smaller than 251.27, the average number of defined terms in each vocabulary. It demonstrates that a large amount of terms are defined but not instantiated in any RDF document.

We establish the TIM matrix and find that only 0.84 % entries (=144,618, 460) in TIM are non-zeros. By traversing the matrix, we find that the maximum term instantiation times in a RDF document is 924 and the average is 7.72, which provide a clue to determine ϵ for clustering.

We measure the relatedness among terms for building TIG. In practice, TIM is very sparse, so it is not necessary to calculate the relatedness between every pair of terms in T since most of them are orthogonal. We only need to compute the relatedness between two terms if they are co-instantiated in at least one RDF document in D. Hence, we first generate the term pairs having the relatedness, and then measure the relatedness between terms in each term pair.

5.3 Analysis of Pragmatic Patterns

We perform our partitioning algorithm in Algorithm 2 to construct a set of pragmatic patterns. The maximum number of terms in each pragmatic pattern (ϵ) is set to 9, which is a bit greater than 7.72, the average number of instantiated terms in each RDF documents, since we expect to keep the naturally well-organized

---

[4] http://ws.nju.edu.cn/olg/.

[5] It is refined from the raw TIM matrix by using heuristic rules, c.f. Sect. 3.1.

terms being clustered together. Besides, from our experience, humans are getting easy to understand a pragmatic pattern with less than ten terms.

It takes nearly seven days to generate the relatedness between terms in 156, 158 term pairs by parallel computing with a ten-node cluster which uses Spark framework to accelerate the processing, and only spends around one hour to divide the terms in 26,103 pragmatic patterns, which demonstrates the efficiency of our clustering algorithm. There are 6,581 pragmatic patterns with nine terms respectively (64.55 % terms), which demonstrates that our partitioning algorithm is capable of generating nearly equally-sized pragmatic patterns.

We pick up the top-100 pragmatic patterns having maximum instantiation times. The instantiation times of a pragmatic pattern is equal to the accumulation of the instantiation times of all the terms in that pragmatic pattern. The relatedness between any two pragmatic patterns is calculated by reusing the $cut()$ function in Formula (2). We depict the macroscopic network of the top-100 instantiated pragmatic patterns in Fig. 1 by Pajek [12], in which each node denotes a pragmatic pattern, and each edge denotes the relatedness between two pragmatic patterns satisfying $cut() > 0.001$. There are 295 edges in the figure. We manually add some tags to help understanding.

According to the figure, we can observe that the top-100 pragmatic patterns perform a clustering behavior. For instance, on the right of this figure, a lot of pragmatic patterns strongly connect to each other since the terms in them all belong to the Bio2RDF vocabulary. Moreover, four groups of terms spaces, i.e. DBpedia, Person, Music and Publication, have both many internal and external edges. This tells that the terms in the ontologies, such as FOAF, DBLP, are frequently instantiated together on the Linked Data. Besides, this figure demonstrates that pragmatic patterns can provide a new way to observe the Linked Data other than the one provided by the Linking Open Data dataset cloud.[6]

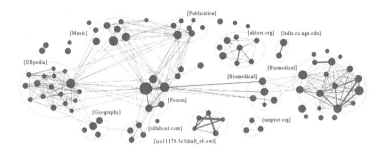

Fig. 1. Top-100 instantiated pragmatic patterns

---

[6] http://lod-cloud.net/.

5.4   Effectiveness

To help readers to understand and evaluate the effectiveness further, we illustrate an example about partitioning terms from the FOAF vocabulary 0.96[7], as FOAF is a well-known vocabulary containing many popular terms. FOAF 0.96 defines 66 terms, which are generally grouped in five broad categories: FOAF Basics, Personal Info., Online Accounts/IM, Projects & Groups, and Documents & Images.

We find that four of the 66 terms are not instantiated in any RDF document, so they are not considered to produce pragmatic patterns. By measuring the relatedness between the rest 62 terms and performing clustering, 26 pragmatic patterns are constructed. All these pragmatic patterns may contain terms from several other ontologies. In fact, the 26 pragmatic patterns contain 155 terms other than FOAF. For example, foaf:made is grouped with http://purl. org/ontology/mo/MusicGroup and other seven terms in the Music vocabulary. We list 12 pragmatic patterns having more than two terms in FOAF in Fig. 2 (omitting the namespace and the *cut*() values between pragmatic patterns).

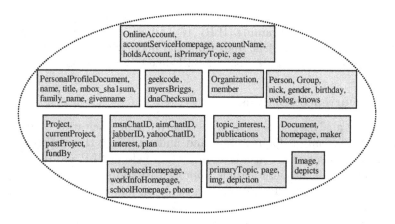

Fig. 2. An example about FOAF

In the figure, we can observe that, although the pragmatic patterns are distilled from the viewpoint of pragmatics, they still clearly follow the five general categories. For instance, foaf:name, foaf:title are grouped together for describing foaf:PersonalProfileDocument; while foaf:fundBy, foaf:currentProject and foaf:pastProject are clustered with foaf:Project.

More importantly, we can find that these pragmatic patterns illustrate an interesting subdivision further. As an example, foaf:publications, foaf:knows and foaf:currentProject describe different facets of the information regarding a person. In our clustering result, they have been well-separated

---

[7] http://xmlns.com/foaf/spec/20091215.html. Our experimental dataset is crawled in 2009. By then, the FOAF's version is 0.96.

into different pragmatic patterns. If a data producer would like to describe a person's academic interests, he/she would find that `foaf:topic_interest` and `foaf:publications` are two relevant terms.

In summary, the pragmatic patterns constructed by our approach can give an elaborated partitioning upon ontologies from the pragmatic perspective, and facilitate term reuse and RDF data exploration on the Linked Data.

6 Related Work

6.1 Pattern Mining in the Semantic Web

A notion of pattern for ontology design [13] are introduced as a "cookbook recipe" provided for ontologists to model in a rich and rigorous manner. It relies on crowdsourcing by experts to generate patterns, which requires a lot of manual efforts, and can not identify emerging patterns in time. Frequent pattern mining [14] and clustering [15] are introduced to automatically discover frequent patterns from OWL DLP to represent domain knowledge. But it only focus on the schema level of the Semantic Web, we argue that it is necessary to take into account the relatedness among terms from the viewpoint of pragmatics or usage. Inductive logic programing (ILP) [16] and association rule mining [17,18] are introduced to learn logical rules from the underlying data. Although with sophisticated results, both are easily overwhelmed by the amount of data.

6.2 Co-occurrence Analysis

Co-occurrence analysis, such as term co-occurrence or co-citation analysis, provides useful information for mapping and understanding the structures of the underlying document sets, which is an important field in NLP and IR. For example, Chen et al. [19] propose a topic approach to address the vocabulary (difference) problem in scientific information retrieval, taking the molecular biology domain as an example. Dhillon [20] gives an idea of modeling a set of documents as a bipartite graph between documents and terms, and introduces a spectral co-clustering algorithm to simultaneously separate documents and terms into different term/document spaces.

7 Conclusion

In summary, the main contributions of the paper is listed as follows.

- We have proposed the notion of term co-instantiation graph for a RDF dataset, and described a method to collect term instantiation information in RDF documents and measure the relatedness between terms for building a term co-instantiation graph.
- We have presented a clustering-based algorithm to distill a set of pragmatic patterns from the term co-instantiation graph, which reveal the pragmatic custom of highly-correlated terms populated in the Linked Data.

– We have analyzed the macroscopic structure of the term co-instantiation graph and pragmatic patterns from the complex network point of view. The experimental results demonstrates our approach can not only give an elaborated partitioning upon vocabularies from the pragmatic perspective, but also provide a new way to explore the Linked Data other than the one provided by the LOD cloud.

The spirit of the Linked Data desires a new way to search and organize RDF data, other than the current one for the traditional Web. The work reported in this paper is a first step towards exploring the Web of data by pragmatic patterns, and many issues still need to be addressed. In future work, we look forward to proposing methods to generate more sophisticated pragmatic patterns (e.g. revealing explicit term relations). We also plan to evaluate the effectiveness of pragmatic patterns in semantic search. Besides, combining semantics and pragmatics for ontology partitioning is another interesting topic to be studied.

Acknowledgments. This work is supported by the National Natural Science Foundation of China (NSFC) under Grants 61402426 and partially supported by Collaborative Innovation Center of Novel Software Technology and Industrialization.

References

1. Ding, L., Finin, T., Joshi, A.: Analyzing social networks on the semantic web. IEEE Intell. Syst. **9**(1), 451–458 (2005)
2. Campinas, S., Perry, T.E., Ceccarelli, D., Delbru, R., Tummarello, G.: Introducing rdf graph summary with application to assisted sparql formulation. In: 2012 23rd International Workshop on Database and Expert Systems Applications (DEXA), pp. 261–266. IEEE (2012)
3. Zhang, Z., Gentile, A.L., Blomqvist, E., Augenstein, I., Ciravegna, F.: Statistical knowledge patterns: identifying synonymous relations in large linked datasets. In: Alani, H., et al. (eds.) ISWC 2013, Part I. LNCS, vol. 8218, pp. 703–719. Springer, Heidelberg (2013)
4. Cheng, G., Zhang, Y., Qu, Y.: Explass: exploring associations between entities via top-K ontological patterns and facets. In: Mika, P., et al. (eds.) ISWC 2014, Part II. LNCS, vol. 8797, pp. 422–437. Springer, Heidelberg (2014)
5. Cheng, G., Ge, W., Qu, Y.: Falcons: searching and browsing entities on the semantic web. In: Proceedings of WWW, pp. 1101–1102 (2008)
6. Salton, G., McGill, M.H.: Introduction to Modern Information Retrieval. McGraw-Hill, New York (1983)
7. Guha, S., Rastogi, R., Shim, K.: Rock: a robust clustering algorithm for categorical attributes. In: Proceedings of ICDE, pp. 512–521 (1999)
8. Kannan, R., Vempala, S., Vetta, A.: On clustering: good, bad and spectral. J. ACM **51**(3), 497–515 (2004)
9. Shi, J., Malik, J.: Normalized cuts and image segmentation. IEEE Trans. Pattern Anal. Mach. Intell. **22**(8), 888–905 (2000)
10. Rousseeuw, P.: Silhouettes: a graphical aid to the interpretation and validation of cluster analysis. J. Comput. Appl. Math. **20**(1), 53–65 (1987)

11. Han, J., Kamber, M.: Data Mining: Concepts and Techniques, 2nd edn. Morgan Kaufmann Publishers Inc., San Francisco (2005)
12. de Nooy, W., Mrvar, A., Batagelj, V.: Exploratory Social Network Analysis with Pajek. Cambridge University Press, Cambridge (2005)
13. Gangemi, A.: Ontology design patterns for semantic web content. In: Gil, Y., Motta, E., Benjamins, V.R., Musen, M.A. (eds.) ISWC 2005. LNCS, vol. 3729, pp. 262–276. Springer, Heidelberg (2005)
14. Józefowska, J., Lawrynowicz, A., Lukaszewski, T.: Faster frequent pattern mining from the semantic web. In: Intelligent Information Processing and Web Mining. Advances in Soft Computing, vol. 35, pp. 121–130. Springer, Heidelberg (2006)
15. Fanizzi, N., dAmato, C., Esposito, F.: Metric-based stochastic conceptual clustering for ontologies. Inf. Syst. **34**(8), 792–806 (2009)
16. Lisi, F.A., Esposito, F.: Mining the semantic web: a logic-based methodology. In: Hacid, M.-S., Murray, N.V., Raś, Z.W., Tsumoto, S. (eds.) ISMIS 2005. LNCS (LNAI), vol. 3488, pp. 102–111. Springer, Heidelberg (2005)
17. Galárraga, L.A., Teflioudi, C., Hose, K., Suchanek, F.: Amie: association rule mining under incomplete evidence in ontological knowledge bases. In: Proceedings of the 22nd international conference on World Wide Web, pp. 413–422. International World Wide Web Conferences Steering Committee (2013)
18. Nebot, V., Berlanga, R.: Finding association rules in semantic web data. Knowl.-Based Syst. **25**(1), 51–62 (2012)
19. Chen, H., Ng, T.D., Martinez, J., Schatz, B.R.: A concept space approach to addressing the vocabulary problem in scientific information retrieval: an experiment on the worm community system. J. Am. Soc. Inform. Sci. **48**(1), 17–31 (1997)
20. Dhillon, I.S.: Co-clustering documents and words using bipartite spectral graph partitioning. In: Proceedings of KDD, pp. 269–274 (2001)

Big Data and Social Network

Cross-Correlation as Tool to Determine the Similarity of Series of Measurements for Big-Data Analysis Tasks

Marcus Hilbrich[1]([⊠]) and Ralph Müller-Pfefferkorn[2]

[1] Software Quality Lab (s-lab), Universität Paderborn, Paderborn, Germany
marcus.hilbrich@uni-paderborn.de
[2] Center for Information Services and High Performance Computing (ZIH),
Technische Universität Dresden, Dresden, Germany

Abstract. One aspect of the so called Big Data challenge is the rising quantity of data in almost all scientific, social, governmental and commercial disciplines. As a result there are many ongoing developments of analysis techniques to substitute manual processes with automatic or semi-automatic algorithms. This means the knowledge of data analysts has to be transferred to algorithms which can be executed simultaneously on many data sets. Such, the rising amount of data can be analysed in an constant quality and in a shorter time. Even if the number of existing algorithms is enormous, a ready to use solution for each problem doesn't exist. Especially for analysing and comparing series of measurements, e.g. for analysing data of activity trackers or to monitor service execution infrastructures, we discovered a lack of options. Thus we explain the basics of an algorithm using the cross-correlation function to determine a meaningful value of similarity for two or more series of measurements. We used the new method to analyse and categorise job centric monitoring data.

Keywords: Cross-correlation · Monitoring · Similarity · Big-data · Analysis · Series of measurements · Jobs · Cloud · Grid · HPC · Smart-data · Normalisation

1 Introduction

Terms like Big Data or Smart Data mark todays hot topics. One of the challenges in this complex research field is the analysis of huge amounts of data. Due to the massive data quantity a manual handling is no longer feasible. Therefore many different strategies for automatic and semi-automatic analysis strategies are developed.

As already shown and discussed before [18] we see a lack of methods to analyse and compare series of different measurements. An example use case is job centric monitoring which we will introduce in this paper, but also other fields of monitoring applications, analysis of informations of home automation systems

© Springer International Publishing Switzerland 2015
W. Qiang et al. (Eds.): CloudCom-Asia 2015, LNCS 9106, pp. 263–282, 2015.
DOI: 10.1007/978-3-319-28430-9_20

and wearables, or for comparing social behaviour/activities can benefit from the introduced method. In this paper we show a new semi-automatic strategy to analyse huge quantities of such measurement series using cross-correlation as basis.

In the scenario, which is described in the following section, we use a so called reference. A reference is a series of measurements with an already known behaviour. To find deviant behaviour we have to compare all series of measurements to the reference, preferable (semi-)automatically. In addition a method to evaluate the similarity is needed, which we will introduce after a presentation of related work. We will also show how we can improve the method in two steps to achieve more meaningful results. Based on the comparison of the measurements with the reference, it can be automatically decided whether the series shows known behaviour or an additional manual analysis is needed. This manual analysis can lead to an additional reference so that future series with the same behaviour can be automatically handled from now on. To use the method for Big Data challenges, the computing time for the comparison has to be of low computational complexity and low runtime. This and additional details are covered in the last part of the paper.

2   Use Case

The analysis method described in this paper is a general one and can be adapted to various scenarios where series of measurements are used. Originally, the strategies were developed for job centric monitoring [29].

Job centric monitoring in short, is an observation strategy for program and application execution or service provision (this is what we call jobs) which are processed on local, remote or distributed computing systems. It collects job specific performance data (like CPU or memory usage) over the life time of a job. It can be done on HPC systems or clusters as well as on conceptional strategies like portals for job execution or Grid and Cloud environments. Based on the still increasing capabilities of such systems the number of jobs executed by a user ca be enormous. The observation capabilities of the jobs are hindered at the same time by additional management layers introduced by Grid, Cloud or portal services which often prevent a direct job observation for users. As a result the job execution is a mostly unobserved task where errors, misbehaviour of jobs or computing systems as well as optimisation potentials can't be identified. Based on a large number of similar jobs (like analyses of different data sets or parametrized simulations) and the limited knowledge of users about details of the job observation a job centric monitoring infrastructure had been introduced [16,17] which was enhanced to semi-automatically find erroneous and unwanted behaviour of jobs. This was needed because traditional analysis strategies based on visualisation and manual analysis can no longer keep up with the rising number of jobs - even with the introduction of more compact visual representations like colour coding as shown in Fig. 1. Such, the analyses and comparison of series of measurements from job centric monitoring of similar jobs is a big data challenge.

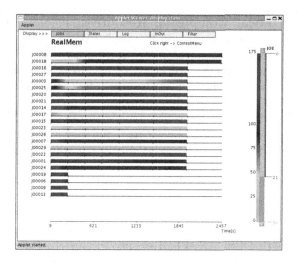

Fig. 1. Screenshot of the AMon-GUI showing monitoring data of multiple jobs (one bar per job) with a color-coded representation of the consumed memory of the individual jobs

The classical analysis strategy for job centric monitoring data is to compare a jobs data with the data from normal, exceptional, and faulty runs. An exception or fault depends on the already determined standard job behaviour. Examples are job sets that use the same program or service with analogous data, which is common in physics, chemistry or life science. Fealty behaviour is not easy to describe and even if possible it is not easy to find an generally applicable automatic detection strategy [8]. Thus it depends on the analysis expert which errors and discrepancies are detected.

For semi-automatic analysis we adapt this strategy. The first job of a new job set has to be analysed manually because there is no reference yet. This can be done with the help of the application expert, e.g. when introducing a new service. Additional jobs can be compared to the first one which is taken as reference and be categorised as similar/faultless or unalike/differing. A similar job presents already known behaviour and doesn't need any future attention, while a differing one has to be manually categorised as faulty or error free. Based on the decision this job can be used as an additional reference. In this way, enormous numbers of jobs can be automatically analysed with minimal user interaction. The missing link is an algorithm which can categorise such measurement series of jobs as similar or unalike. The basics of such an algorithm are described in this paper.

3   Related Work

A lot of scientific work has been done on monitoring in general and in similar fields of research.

Monitoring on Local Computing Resources: For local monitoring on UNIX or Linux systems command line tools like ps, top or free[1] or graphical ones like Gnome System Monitor[2] can be used. On clusters Ganglia[3] gives information about the utilisation of resources. The impact of a specific user or job can not be identified directly by these tools and automatic categorisation of the data is only rudimentary if present at all.

Tracing and Profiling: Profiling like done with GNU gprof[4] gives information which functions of a program are used and how long they are used. This information is often presented as statistical evaluation. Even more detailed information are given by tracing tools like Vampir [4]. But such analyses significantly slowdown the application. Thus they are not valuable for monitoring many jobs at once. Data analyses for such systems are currently under development and use e.g. k-means [15,28] based clustering, sometimes in combination with silhouette [35,36] or other clustering algorithms [22] like k-medoids [23] or DBSCAN [10]. To apply an clustering algorithm to series of measurements, an lossy transformation is needed in additional. Such algorithms are e.g. based on Fourier-coefficients [11], wavelets [21,39] or stepwise approximation [24]. Sequence comparison [14,19,30,31] is an additional family of algorithms to analyse tracing and profiling data.

Accounting: Accounting is used to measure the utilisation of computing systems and for billing of the use of resources. Examples are SGAS [9] and DGAS [33]. Based on the fact that only basic summary information of a job have to be recorded the amount of data to be handled is low and the capabilities to extract information about the job behaviour are limited.

Resource Monitoring: The task of resource monitoring is to record information about computing resources or components in distributed infrastructures like Grids. Collected are information about hardware, middlewares, the offered services, known outages, planed maintenances and utilisation or free resources. This allows to create statistics of reliability and utilisation of resources and services or to assign jobs to available resources. Examples are D-Mon [2], CMS Dashboard [1] and Ganga [38]. The information evaluation is mostly limited to usage statistics.

The specific needs for handling and analysing series of measurements of job centric monitoring data are not properly considered by any of these tools or research topics. Thus we investigated additional algorithms e.g. from the field of genetic algorithms [3,12,20,37], pattern matching for intrusion detection [7,27,32,34], statistical analysis of events [6], machine learning [5,25,26] and some more [18]. Finally we came up to develop a method which is based on the cross-correlation function.

---

[1] http://procps.sourceforge.net/index.html

[2] http://library.gnome.org/users/gnome-system-monitor/stable/index.html

[3] http://ganglia.info/

[4] http://sourceware.org/binutils/docs/gprof/

4   Cross-Correlation

Correlation functions, known from the field of signal analysis [13], calculate a so called similarity of two functions. The cross-correlation function is defined as follows:

$$(f_1 \star f_2)(\tau) = \int f_1(t) \cdot f_2(t + \tau) \, dt \tag{1}$$

τ denotes the offset between the functions f_1 and f_2. There are also adaptations for discrete signals available and it is possible to adapt the continuous function to non-discrete sequences[5] of measurements by using interpolation.

A typical application for cross-correlation is to determine the time-based shift of two signals. An example is shown in Fig. 2. $x_1(t)$ is the transmitted signal while $x_2(t)$ is the received signal. To determine the time shift between both signals (e.g., to calculate the distance the signal travelled in radar systems) the cross-correlation is calculated. The correlation function on the right side of Fig. 2 shows the similarity of both signals depending on an constant time shift (τ). The maximum value of the cross-correlation indicates the corresponding value of τ, which is the time shift between f_1 and f_2.

The classical usage of cross-correlation allows to align two functions with an arbitrary offset. For more complex scenarios containing dynamic changes (e.g. caused by system noise, different CPU-speed or memory bandwidth of different executing systems or a variable performance of an network connection) of the offset and extra execution loops with repeating sequences a transformation of one of the functions is needed to compensate the variable time drifts. This transformation can also realise a defined fixed offset like expressed by τ. Consequently, the offset in the cross-correlation can be removed by defining $\tau = 0$.

The optimisation problem to fined an adequate transformation is future work and not covered by this paper. In short, the cross-correlation has to be maximized by finding a transformation respecting the dynamic time shift. Without this transformation the cross-correlation result would be useless as similarity metric

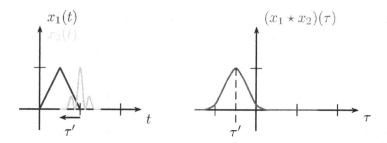

Fig. 2. Example of two similar but non identical signals $x_1(t)$ and $x_2(t)$ (left figure). Between both signals is a time lag τ'. Also shown is the cross-correlation $(x_1 \star x_2)(\tau)$ of both functions (right figure) which has its maximum at τ'.

---

[5] Non-discrete sequences are series of measurements which are not guaranteed to be taken at constant time intervals.

in a contexts like job centric monitoring or other scenarios with a dynamic change of measurement times. One strategy to determine the dynamic time shift of a complete time series is educated guessing. Educated guessing uses characteristic points like fast rising CPU usage, starting of jobs, or writing output to a file in the case of job centric monitoring. It matches corresponding points in both series for a coarse alignment and recalculates the cross-correlation for each pair. Depending on the cross-correlation result the respective pair of corresponding points is retained or discarded. Additional optimization strategies are genetic algorithms or machine learning.

But before we can solve this optimisation problem we have to deal with the effects of the cross-correlation itself and have to check whether the kind of similarity the function delivers corresponds to our demands on an measurement for similarity.

5   Calculating the Cross-Correlation for Series of Measurements

The cross-correlation gives a function (depending on τ) calculated from two functions. We have to adapt it to series of measurements as input and to give a value (not an function) which represents the similarity of both series of measurements instead of an function.

If we accept that a time-based alignment has to be applied to the series of measurements to arrange corresponding parts of the execution process of both jobs to the same time index we can determine the value of $\tau = 0$. This means we calculate a cross-correlation coefficient which is an value.

To calculate the cross-correlation coefficient, it is not needed to calculate approximation functions of the series of measurements. The coefficient can be calculated step wise. For each interval between two measurements a local coefficient can be calculated and the local coefficient can be summed up to get the result. This is possible due to the additivity of integration on intervals.

The values in-between two measurements of a measurement series are not defined. To calculate the integral over the time, it is needed to interpolate the values in between. In the following we use a linear interpolation as a first approximation. More complex interpolations are not considered as we do not expect any issues with accuracy at this point.

Another difficulty of series of measurements is the measurement time, which is not necessarily synchronised for both series. To realise a step wise integration, additional data points are therefore interpolated. More precisely, for each measurement time of one series a data point is calculated for the other series of measurements.

Now, both series of measurements have data points to the same times. Thus the cross-correlation coefficient for each interval of two contiguous data points can be calculated.

To simplify the calculations for a single interval we adapt the measurement time. The first measurement time of the value is transformed to zero, the second time has to be adapted according to $t_2 - t_1$. This transformation represents a moving of the coordinate plane illustrated by Fig. 3. The time of the interval remains the same and dose not alter the calculation.

Based on the linear interpolation and the time transformation the continuous functions for the measurements $f_1(t)$ and $f_2(t)$ are the following:

$$f_1(t) = \frac{t \cdot (p_{1,2} - p_{1,1})}{t_2 - t_1} + p_{1,1}$$

$$f_2(t) = \frac{t \cdot (p_{2,2} - p_{2,1})}{t_2 - t_1} + p_{2,1}$$

(2)

Both formulas have to be applied to the cross-correlation (formula 1) for $\tau = 0$ to calculate a local cross-correlation coefficient c. This can be solved to an easy to apply calculation:

$$c = \int_0^{t_2-t_1} \left(p_{1,1} + t \cdot \frac{p_{1,2} - p_{1,1}}{t_2 - t_1} \right) \cdot \left(p_{2,1} + t \cdot \frac{p_{2,2} - p_{2,1}}{t_2 - t_1} \right) dt$$

(3)

$$= \int_0^{t_2-t_1} p_{1,1} \cdot p_{2,1} + t \cdot \frac{p_{1,1} \cdot (p_{2,2} - p_{2,1})}{t_2 - t_1} + t \cdot \frac{p_{2,1} \cdot (p_{1,2} - p_{1,1})}{t_2 - t_1}$$

$$+ t^2 \cdot \frac{(p_{1,2} - p_{1,1}) \cdot (p_{2,2} - p_{2,1})}{(t_2 - t_1)^2} \, dt$$

$$= p_{1,1} \cdot p_{2,1} \cdot t \Big|_0^{t_2-t_1} + \frac{t^2}{2} \cdot \frac{p_{1,1} \cdot (p_{2,2} - p_{2,1})}{t_2 - t_1} \Big|_0^{t_2-t_1}$$

$$+ \frac{t^2}{2} \cdot \frac{p_{2,1} \cdot (p_{1,2} - p_{1,1})}{t_2 - t_1} \Big|_0^{t_2-t_1} + \frac{t^3}{3} \cdot \frac{(p_{1,2} - p_{1,1}) \cdot (p_{2,2} - p_{2,1})}{(t_2 - t_1)^2} \Big|_0^{t_2-t_1}$$

$$= t \cdot \left(\frac{t^2}{3} \cdot \frac{(p_{1,2} - p_{1,1}) \cdot (p_{2,2} - p_{2,1})}{(t_2 - t_1)^2} \right.$$

$$\left. + \frac{t}{2} \left(\frac{p_{1,1} \cdot (p_{2,2} - p_{2,1})}{(t_2 - t_1)} + \frac{p_{2,1} \cdot (p_{1,2} - p_{1,1})}{(t_2 - t_1)} \right) + p_{1,1} \cdot p_{2,1} \right) \Big|_0^{t_2-t_1}$$

$$= t \cdot \left(t \cdot \left(\frac{t}{3} \cdot \frac{(p_{1,2} - p_{1,1}) \cdot (p_{2,2} - p_{2,1})}{(t_2 - t_1)^2} \right. \right.$$

$$\left. \left. + \frac{p_{1,1} \cdot (p_{2,2} - p_{2,1}) + p_{2,1} \cdot (p_{1,2} - p_{1,1})}{2 \cdot (t_2 - t_1)} \right) + p_{1,1} \cdot p_{2,1} \right) \Big|_0^{t_2-t_1}$$

$$c = (t_2 - t_1) \cdot \left(\frac{(p_{1,2} - p_{1,1}) \cdot (p_{2,2} - p_{2,1})}{3} \right.$$

$$\left. + \frac{p_{1,1} \cdot (p_{2,2} - p_{2,1}) + p_{2,1} \cdot (p_{1,2} - p_{1,1})}{2} + p_{1,1} \cdot p_{2,1} \right)$$

(4)

A program can calculate all the local cross-correlation coefficients and sum them up to the cross-correlation coefficients of both series of measurements C.

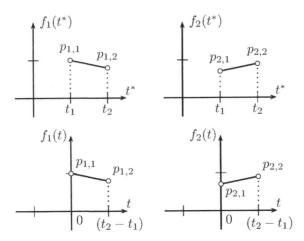

Fig. 3. Moving the coordinate plane for future calculation of the cross-correlation coefficients. The time of the first measurement t_1 (upper diagrams) is defined as 0 (lower diagrams), t_2 is adapted according to the movement.

From the mathematical and programming point of view, the cross-correlation coefficients can be calculated and compared with other ones. As already described, one of the series of measurements is a reference, the other one is the data of a job to test. To interpret the number of a calculated cross-correlation coefficient, it can be compared with the self-correlation coefficient of the reference.

The self-correlation coefficient is calculated similar to the cross-correlation coefficient but for both input functions the data of the reference are used. The self-correlation coefficient C_S is the ideal value of the cross-correlation coefficient because it represents the similarity of the reference to itself. Such a normalised cross-correlation coefficient $C_n = \frac{C}{C_S}$ is introduced. This is called the global normalisation to distinct it from later introduced normalisations. To the method we refer in following as the naive implementation.

The ideal value for the normalised cross-correlation coefficient C_n is 1 and each difference to this value is a discrepancy. Based on such discrepancies, the historical data of the analysis process and the users feedback the job can now be categorised as similar or differing to the reference (normal program run or a known error) - depending on the reference used. If the jobs monitoring data are not similar to a reference we found an unknown or not yet categorised behaviour which has to be analysed further and will probably give an additional reference.

6 Example Using the Naive Implementation of the Cross-Correlation

For a better understanding of the method in the following we will use a simple and artificial job behaviour and look only at one measurement category. In this way we can also better demonstrate the effects we have found by analysing more complex jobs.

The monitoring data of the basic job 1 we are using is shown in Fig. 4. Shown is a measurement of the CPU load over a runtime of about eight hours. At the beginning of the job the CPU usage increases from 0 to 1. This is the starting phase of the job. Afterwards a working phase follows with a constant value of 1 which means that one CPU is used to 100 %. In the last phase the job ends with decreasing CPU-usage. There is no measurement with zero percent CPU usage at the end as the resources of the job got deallocated before such a measurement was taken. An example for such a behaviour is an application that reads in data at the beginning, does some extensive calculations and write output data at the end.

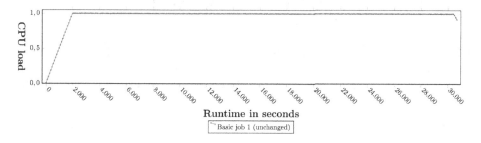

Fig. 4. Plot of the used monitoring data (CPU load) of basic job 1.

To demonstrate the use of the cross-correlation we constructed additional jobs based on the basic job. In the following so called gaps are applied. A gap changes the monitored values in a defined time interval. In our example the value is varied by 1 over 10 % of the runtime of the job, either as an increase $(<+<)$ or a decrease $(<-<)$ (Fig. 5).

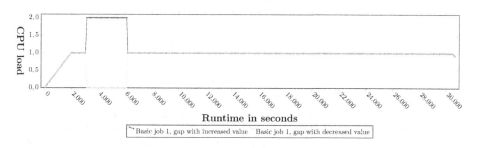

Fig. 5. Adaptations of basic job 1 with gaps $(<+<)$ and $(<-<)$.

By using basic job 1 as reference the normalised cross-correlation coefficient can be calculated using the naive method:

	no gap	$(<+<)$	$(<-<)$
naive method	1.00	1.10	0.90

If the basic job 1 is compared with itself (no gap, self-correlation) it is clear that the ideal value for similarity is reached. Also shown is that a change of 10 % of the basic job results in a 10 % decrease or increase of the cross-correlation coefficient.

Here we have seen the first phenomena which can be called unusual. A job with a discrepancy to the reference can have a higher cross-correlation coefficient than the reference. This means a higher value does not correspond to a better similarity in all cases. The difference to the ideal value (self-correlation) has to be considered.

7   Compensation of Locally Increased and Decreased Cross-Correlation Coefficients

By evaluating a gap $(<+-<)$, which increases and decreases the measurement values of the reference (see Fig. 6), it can be shown that the corresponding increase and decrease of the cross-correlation coefficient can compensate each other.

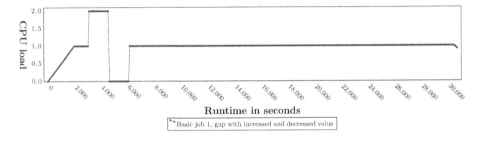

Fig. 6. Adaptations of basic job 1 with gap $(<+-<)$.

As result a job with a clear anomaly has an ideal value of the cross-correlation:

	no gap	$(<+<)$	$(<-<)$	$(<+-<)$
naive method	1.00	1.10	0.90	1.00

To avoid such an unwanted compensation and to get a clear signal for a differing job behaviour all discrepancies between job and reference should clearly show either an increase or a decrease of the cross-correlation coefficient! We decided to always take a decrease as signal. As result the ideal normalised cross-correlation coefficient 1 will be the highest possible result. Each variation of the basic behaviour results in lowering[6] of the coefficient. If we just consider measurement series of positive values (like CPU or memory usage) the result

---

[6] This lowering has to respect the leading sign. A cross-correlation coefficient less then zero represents similar functions with reverse signs.

will be a cross-correlation coefficient between 0 for no similarity and 1 for ideal similarity. This can be realised by introducing a local normalisation.

Local normalisation can be done in various ways. The most intuitive one is to normalise the cross-correlation coefficient for each segment between two measurement points. The variant we have chosen for practical use is to transform the measurements of the job. Both variants are described in the following.

The first step for both ways is to check for which intervals or measurement points a normalisation has to be applied. In the case of normalising the local coefficient each interval of two contiguous points is checked separately. We have to consider three different cases:

1. The values at both points have lower values than the corresponding points of the reference. Then the (non-normalised) local coefficient is below the self-correlation coefficient and a local normalisation is not needed for this segment.
2. Both values of the job are larger then the corresponding values of the reference. So the local coefficient for this segment is higher then the self-correlation coefficient and normalisation has to be applied to transform the discrepancy to a lower value of the local coefficient.
3. One value of the job is below and the other point is above the corresponding value of the reference (see Fig. 7). This means that only a part of the interval has to be normalised. For our experiments we calculated the point of intersection and added it to the measurements. The resulting new sections can be handled according to the first or second case.

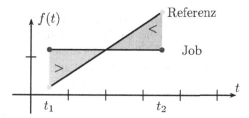

Fig. 7. In case of an intersection of job and reference the local cross-correlation coefficient has to be normalised only for one part of the segment.

By using a transformation of the measurements of the job a cross-correlation coefficient above the self-correlation coefficient can be avoided. When a value of the job is below the value of the reference (first case from above) no transformation is needed. Higher values result in high coefficients (second and last case from above). Consequently these values have to be lowered by normalisation. Thus, intersection points between the reference and the normalised points of the job are avoided and a fragmentation of intervals is not needed. This makes the value normalisation easier and due to viewer calculations faster in comparison to the normalisation of the coefficients.

After identifying which local cross-correlation coefficients or measurement values have to be normalised, in the second step the normalisation has to be applied. For both ways of normalisation a number has to be lowered below a reference. As normalisation function we used:

$$v_{normed} = \frac{v_{ref}}{1 + \left(\frac{v - v_{ref}}{v_{ref}}\right)} \tag{5}$$

For the normalisation of the local coefficients, v_{normed} is the normalised cross-correlation coefficient, v the cross-correlation coefficient to normalise, and v_{ref} the ideal value for the segment which is defined by the self-correlation coefficient of the reference. In case of normalisation of the measurement values, v_{normed} is the normalised value, v the value of the job to normalise, and v_{ref} the value of the reference.

After the local normalisation process the calculation of the cross-correlation coefficient for the combination of the job and reference proceeds by summing up the locally normalised coefficients of all intervals. In addition the already introduced global normalisation is applied, which makes 1 the ideal value for similarity.

Based on these methods, the coefficients for basic job 1 with the tree types of gaps are:

	no gap	$(<+<)$	$(<-<)$	$(<+-<)$
local normalisation (by coefficients)	1.00	0.95	0.90	0.92
local normalisation (by value)	1.00	0.95	0.90	0.92

If no gap or the gap $(<-<)$ is applied, no local normalisation is used and the result does not change in comparison to the naive method. A gap with increased value $(<+<)$ yields a coefficient below the ideal due to the normalisation. Most important is that the gap $(<-<)$ leads to a decreased coefficient because compensation of decreased and increased values can no more be compensated by each other, so even this kind of gap can be detected.

Not yet discussed are measurements with negative values. In our use case of job centric monitoring we always have data which are presented by positive values. So a local coefficient of an interval is always positive and only increases but never decreases the summed up coefficient for the complete comparison of a job and a reference. By allowing negative values a so called reverse similarity (the measurement series have some kind of similarity but opposite signs) has to be considered. As result, similarity and reverse similarity can compensate each other. How such an effect has to be considered depends on the concrete usage scenario and leads to a more complex local normalisation then the one described above.

8   Coefficient Dependence on the Values of the Reference

The expectation on the analysis algorithm is that a discrepancy to a reference can be detected independent of the form of the reference. The cross-correlation

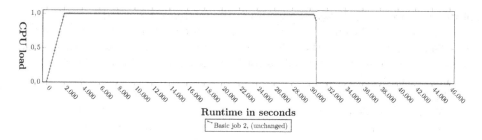

Fig. 8. Plot of the monitoring data (CPU load) of basic job 2.

coefficient does not fulfil this requirement directly. To demonstrate this, we introduce basic job 2 (Fig. 8), which adds an additional waiting phase to basic job 1 where e.g. data is transferred to a remote system and no CPU utilisation is caused.

Once again we applied the gap *(<-<)* at the running phase of the job. For an another job the gap *(>+>)* is applied, which increases the values during the waiting phase of the job. Both jobs are shown in Fig. 9.

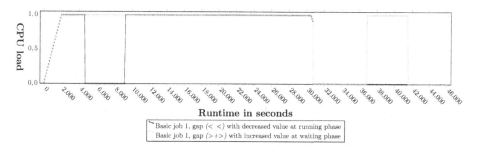

Fig. 9. Adaptations of basic job 2 with different gaps.

Calculating the cross-correlation coefficients with local normalisation (of the values) and global normalisation for basic job 2 and the gap-jobs leads to the following results:

	no gap	*(<-<)*	*(>+>)*
local normalisation (value normed)	1.00	0.84	1.00

The first two are as expected - a job matching the reference gives the ideal result 1 and a gap during the running phase of the job leads to a decreased value for the similarity. But the third one is clearly not wanted - a variation of the job during the waiting phase yielding the value for ideal similarity.

The cause for the unwanted behaviour is the way the cross-correlation is calculated. Formula 4 shows the calculation for an interval between two points of the measurements. When the values of the reference ($p_{1,1}$ and $p_{1,2}$) are both zero, the local coefficient is also zero. This is independent of the values of the job. Thus, intervals where the value of the reference is zero do not influence the cross-correlation at all. Small values for the reference have only a slight impact compared to intervals with higher values.

Our goal is to have an algorithm which does a detection over the whole reference, independent of its concrete values. To implement this behaviour we introduce an inverse cross-correlation coefficient in the following. This coefficient will have a high sensitivity when the values of the reference are low (maximal sensitivity is reached when the values of the reference are zero) and a low sensitivity for intervals with high values (for the maximum value of the reference no detection will be done). In short, the sensitivity of the reverse coefficient will be opposite to the cross-correlation coefficient. This allows to combine both coefficients to get meaningful results for the similarity of a job to the reference.

To calculate the inverse coefficient we do a value transformation for job and reference. We still want to compare the coefficients for different jobs calculated for the same reference, so it is needed to apply the same transformation to them. This means only the parameters of the reference can be used to find an adequate transformation.

The transformation has to change the sensitivity of the cross-correlation. Therefore low values have to be mapped to high values and high values to low once. As transformation we use:

$$x^s = -(x - x_{Max}) = x_{Max} - x \tag{6}$$

x^s is the shifted (transformed) value and x the original value of the measurement, either of the reference or the job. x_{Max} is always the maximum value of the reference. The maximum value of the job cannot be considered for the transformation based on the reasons given above.

For the reference the situation is quite simple. A value of null is transformed to the maximum value with maximal sensitivity and the maximal value is transformed to null with no sensitivity. Negative values are not caused by the transformation.

The values of the job can be higher than the maximum of the reference leading to negative values after the transformation. Because the negative values always show a discrepancy to the non-negative transformed values of the job and the negative, local inverse cross-correlation coefficient respects this behaviour we can accept negative values for the job. If preferred, negative values can also be avoided by applying the value normalisation from above before shifting the values. This avoids that values of the job are higher then the values of the reference.

A visualisation of the value transformation by shifting is shown in Fig. 10. The points of the job j_A, j_B and j_C can only be analysed by the cross-correlation coefficient while the points j_D, j_E and j_F can only be analysed by the inverse cross-correlation coefficient. Also shown is that the points (j_C and j_F) of the job can be transformed to negative values.

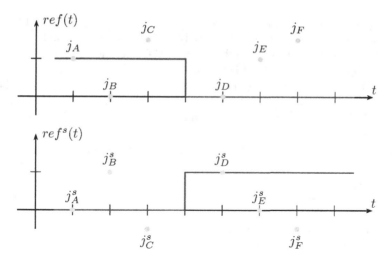

Fig. 10. Shifting of values for job and reference. The picture at top shows the values before, the lower one shows the values after the transformation. The reference is shown as a line (values interpolated), the job is shown by representative points.

Based on the inverse coefficient a new global normalisation to combine both coefficients is needed. The first version normalises both coefficients separately to 0.5 each and sums up both values, so that the ideal value remains 1. The formula for this combination cc_{normed_v1} is:

$$cc_{normed\_v1} = \frac{\frac{cc}{sc} + \frac{cc_i}{sc_i}}{2} \tag{7}$$

cc is the cross-correlation coefficient and cc_i the reversed one. sc and sc_i are the respective self-correlation coefficients.

The second version to calculate a normalised and combined coefficient cc_{normed_v2} is based on the same data as the first. It sums up both not yet globally normalised coefficients and applies the normalisation to an ideal value of 1 afterwards:

$$cc_{normed\_v2} = \frac{cc + cc_i}{sc + sc_i} \tag{8}$$

For basic job 2 with the gaps described above and the local normalisation introduced in Sect. 7 the results for all variants are:

	no gap	$(<-<)$	$(>+>)$
normalised cross-correlation coefficient	1.00	0.84	1.00
normalised, inverse cross-correlation coefficient	1.00	1.00	0.70
coefficient combination variant 1	1.00	0.92	0.85
coefficient combination variant 2	1.00	0.90	0.90

The values for the normalised cross-correlation coefficient were already presented above.

The inverted coefficient has an inverted sensitivity. Thus, the gap at the time of low values of the reference can be clearly detected with a value for similarity of 0.70. The gap at the time of high values of the reference can not be detected - an ideal similarity is signalised by a value of 1.00.

Both variants of combining the coefficient show ideal values when the job and the reference are the same (no gaps) and give reduced values for both gaps. We verified this, as well as the other results, with other basic jobs and various kinds of alterations.

For practical use we chose the second version of combining the coefficients. The first version weights both coefficients the same, independence of there sensitivity effects. For example, in basic job 2 the cross-correlation coefficient is responsible for about $\frac{2}{3}$ of the reference (mainly for the running phase), the inverse coefficient for about $\frac{1}{3}$ (the waiting phase). As result a gap or another discrepancy of the job to the reference during the waiting phase is over-represented. This is also shown by the values above. The gap $(<-<)$ changes the job value by 10 % and gives a combined coefficient of 0.92. The gap $(>+>)$ also changes the job value by 10 % but gives a stronger decrease of the combined coefficient to 0.85 because it is placed in a region where the inverse coefficient responsible for the detection.

The effect, that a discrepancy between reference and job leads to different reduction of the first version of the combined coefficient depending on which coefficient was used to detected it can become more powerful. A basic job with a shorter waiting time would yield a much stronger over-representation of the inverse coefficient. So the first version of combining coefficients is not feasible for real world applications even if the results from above look acceptable.

The second version weights according to the self-correlation coefficient and its inverse variant. The not normalised self-correlation coefficient is an very good metric for the sensitivity of the cross-correlation coefficient because it depends on the values of the reference and the time these values are present. In the example above this combination and normalisation of both coefficients gives ideal results. Both gaps alter the job values by 10 % and each yields a 10 % decrease for similarity represented by a coefficient of 0.90.

9   Conclusions and Future Work

The number of methods for semi-automatic data analysis is rising due to an ongoing demand for such algorithms and the missing of universal strategies. We already discussed this in an earlier publication [18] which also pointed out a lack of methods to analyse series of measurements. With this paper we contribute to the pool of analysis strategies and set a clear focus on data which is organised as series of measurements. This kind of data is not only common for job centric monitoring, the use case we showed, but also other fields of monitoring applications, analysing information of home automation systems and wearables, or the comparison of social behaviour/activities can benefit from the introduced method. Such data sets are e.g. temperature or energy measurements used in home automation systems or heart rates and speeds for activity trackers and many more applications in field of internet of things or for analysing human behaviour.

For this paper, we extended the well-known cross-correlation to an algorithm to calculate a similarity between measurement series. It allows to compare them in an meaningful way and with an adequate detection sensitivity for the complete series. We also showed how to avoid the compensation of measurement deviations between series and offered a normalisation concept to allow an easy interpretation of the calculated similarity values. This holds up even for users unconversant with our concept or the analysis process at all.

First tests on real data show promising results. For the tests a single computing cluster with uniform computing nodes and a user job (the CKM-Fitter, a program to calculate parameters of the standard model of particle physics) was used to minimise the not yet compensated time drift. An example for the automatically analysed jobs of one day is given in Fig. 11. One of the outliers in the diagram could be directly mapped to an additional reference which showed an overload situation for a computing node. Thus we had to analyse just one job which later on was identified as an additional error free execution pattern. Based on the design of the tests the result can not be generalised but it shows that the job centric monitoring data can be automatic analysed and the results can be easily presented to users, e.g. in a mail report.

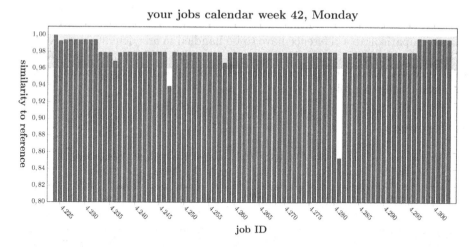

Fig. 11. Shown is the similarity of the jobs processed on one day in comparision to a reference which presents normal job execution. The acceptable similarity (shown in green for the similarity of $0,96$ to $1,00$) was computed based on historical information of the job execution observed for the last month.

With the work presented in this paper, we lay the foundation for an additional class of algorithms to process job centric monitoring data (our use case) and for measurement series in general.

Additional problems have to be solved to get good and reliable results in daily use. The most important challenge for job centric monitoring is to determine a time alignment of the measurement series. This is needed to compare the same section (same lines of code executed for distinct data) of program runs even when the section was executed at a different time in the job and the reference. We have to deal with more complex usage scenarios where job and reference are executed on different computing systems with different CPU speed, memory bandwidth, network connections, system noise or the influence of other user jobs. This leads to a complex and variable time drift between job and reference that has to be estimated because a direct measurement of this drift is not possible.

This clearly defines an important part of future work which was partly already done but not yet published. We plan to demonstrate an aligning strategy based on educated guessing which stepwise determines the time drift of job and reference. This is done by a maximisation process for the similarity values introduced here. One of the most challenging parts of the optimisation process is to realise a strategy with a low computing complexity to analyse many jobs with an adequate demand on computing time. For this article we reduced the number of different data sets to a minimum as we just wanted to demonstrate the basic principles of our method. Later we will use a larger pool of data for a more detailed analysis of the method.

In addition a comparison to other semi-automatic analysis strategies for measurement series has to be done in future work. Later on we have to show how to handle typical issues of job centric monitoring data like repeating segments based on loops or if-statements during program execution. Also missing segments of a job in comparison to a reference have to be handled in an adequate way. At least at this point we have to leave the generic description of the analysis method and show how we can improve its potential of analysing job centric monitoring data to find misbehaving jobs.

References

1. Andreeva, J., Calloni, M., Colling, D., Fanzago, F., D'Hondt, J., Klem, J., Maier, G., Letts, J., Maes, J., Padhi, S., Sarkar, S., Spiga, D., Mulders, P.V., Villella, I.: CMS analysis operations. J. Phys. Conf. Ser. **219**(7), 072007 (2010). http://stacks.iop.org/1742-6596/219/i=7/a=072007
2. Baur, T., Breu, R., Klmn, T., Lindinger, T., Milbert, A., Poghosyan, G., Reiser, H., Romberg, M.: An interoperable grid information system for integrated resource monitoring based on virtual organizations. J. Grid Comput. **7**, 319–333 (2009). doi:10.1007/s10723-009-9134-3. http://dx.doi.org/10.1007/s10723-009-9134-3
3. Beasley, D., Bull, D.R., Martin, R.R.: An overview of genetic algorithms: Part 1, fundamentals. Univ. Comput. **15**, 58–69 (1993)
4. Brunst, H., Hackenberg, D., Juckeland, G., Rohling, H.: Comprehensive performance tracking with vampir 7. In: Müller, M.S., Resch, M.M., Schulz, A., Nagel, W.E. (eds.) Tools for High Performance Computing 2009, pp. 17–29. Springer, Heidelberg (2010). doi:10.1007/978-3-642-11261-4_2
5. Chan, P., Stolfo, S.J.: Toward parallel and distributed learning by meta-learning. In: AAAI Workshop in Knowledge Discovery in Databases, pp. 227–240 (1993)

6. Denning, D.E.: An intrusion-detection model. IEEE Trans. Softw. Eng. **13**(2), 222–232 (1987)
7. Dickerson, J.E., Dickerson, J.A.: Fuzzy network profiling for intrusion detection. In: Proceedings of NAFIPS 19th International Conference of the North American Fuzzy Information Processing Society, Atlanta, pp. 301–306 (2000)
8. Eichenhardt, H., Müller-Pfefferkorn, R., Neumann, R., William, T.: User- and job-centric monitoring: analysing and presenting large amounts of monitoring data. In: Proceedings of the 2008 9th IEEE/ACM International Conference on Grid Computing. GRID 2008, pp. 225–232. IEEE Computer Society, Washington (2008). http://dx.doi.org/10.1109/GRID.2008.4662803
9. Elmroth, E., Gardfjll, P., Mulmo, O., ke Sandgren, Sandholm, T. : A Coordinated Accounting Solution for SweGrid Version: Draft 0.1.3, 7 October 2003
10. Ester, M., Kriegel, H.P., Sander, J., Xu, X.: A density-based algorithm for discovering clusters in large spatial databases with noise. In: KDD 1996, pp. 226–231 (1996)
11. Faloutsos, C., Ranganathan, M., Manolopoulos, Y.: Fast subsequence matching in time-series databases. In: Proceedings of the 1994 ACM SIGMOD International Conference on Management of Data, SIGMOD 1994, pp. 419–429. ACM, New York (1994). http://doi.acm.org/10.1145/191839.191925
12. Grefenstette, J.: Optimization of control parameters for genetic algorithms. IEEE Trans. Syst. Man Cybern. **16**(1), 122–128 (1986)
13. von Grünigen, D.: Digitale Signalverarbeitung: Mit einer Einführung in die kontinuierlichen Signale und Systeme. Fachbuchverlag Leipzig (2008)
14. Gusfield, D.: Algorithms on Stings, Trees, and Sequences. Computer Science and Computational Biology. Cambridge University Press, New York (1997)
15. Hartigan, J.A., Wong, M.A.: Algorithm AS 136: a K-means clustering algorithm. J. Roy. Stat. Soc. Ser. C (Appl. Stat.) **28**(1), 100–108 (1979)
16. Hilbrich, M., Müller-Pfefferkorn, R.: A scalable infrastructure for job-centric monitoring data from distributed systems. In: Bubak, M., Turala, M., Wiatr, K. (eds.) Proceedings Cracow Grid Workshop 2009, pp. 120–125. ACC CYFRONET AGH, ul. Nawojki 11, 30–950 Krakow 61, P.O. Box 386, Poland, February 2010
17. Hilbrich, M., Müller-Pfefferkorn, R.: Achieving scalability for job centric monitoring in a distributed infrastructure. In: Mühl, G., Richling, J., Herkersdorf, A. (eds.) ARCS Workshops. LNI, vol. 200, pp. 481–492. GI (2012)
18. Hilbrich, M., Weber, M., Tschüter, R.: Automatic analysis of large data sets: a walk-through on methods from different perspectives. In: 2013 International Conference on Cloud Computing and Big Data (CloudCom-Asia), pp. 373–380, December 2013
19. Hirschberg, D.S.: A linear space algorithm for computing maximal common subsequences. Commun. ACM **18**(6), 341–343 (1975)
20. Holland, J.: Genetic algorithms. Sci. Am. **267**(1), 44–50 (1992)
21. Huffmire, T., Sherwood, T.: Wavelet-based phase classification. In: Proceedings of the 15th International Conference on Parallel Architectures and Compilation Techniques, PACT 2006, pp. 95–104. ACM, New York (2006). http://doi.acm.org/10.1145/1152154.1152172
22. Jain, A.K.: Data clustering: 50 years beyond K-means. Pattern Recogn. Lett. **31**(8), 651–666 (2010)
23. Kaufman, L., Rousseeuw, P.J.: Finding Groups in Data: An Introduction to Cluster Analysis. Wiley Series in Probability and Statistics. Wiley, New York (2005). http://books.google.de/books?id=yS0nAQAAIAAJ

24. Keogh, E.J., Pazzani, M.J.: An enhanced representation of time series which allows fast and accurate classification, clustering and relevance feedback (1998)

25. Lee, W., Stolfo, S.J.: Data mining approaches for intrusion detection. In: Proceedings of the 7th Conference on USENIX Security Symposium, SSYM 1998, vol. 7, p. 6. USENIX Association, Berkeley (1998). http://dl.acm.org/citation.cfm? id=1267549.1267555

26. Lee, W., Stolfo, S., Mok, K.: A data mining framework for building intrusion detection models. In: Proceedings of the 1999 IEEE Symposium on Security and Privacy, pp. 120–132 (1999)

27. Lunt, T., Jagannathan, R., Lee, R., Whitehurst, A., Listgarten, S.: Knowledge-based intrusion detection. In: 1989 Proceedings of the Annual AI Systems in Government Conference, pp. 102–107 (1989)

28. MacQueen, J.B.: Some methods for classification and analysis of multivariate observations. In: Proceedings of the 5th Berkeley Symposium on Mathematical Statistics and Probability, vol. 1, pp. 281–297. University of California Press (1967)

29. Müller-Pfefferkorn, R., Neumann, R., Borovac, S., Hammad, A., Harenberg, T., Hüsken, M., Mättig, P., Mechtel, M., Meder-Marouelli, D., Ueberholz, P.: Monitoring of jobs and their execution for the LHC computing grid. In: Bubak, M., Turala, M., Wiatr, K. (eds.) Proceedings Cracow Grid Workshop 2006, pp. 224–231. ACC CYFRONET AGH, ul. Nawojki 11, 30–950 Krakow 61, P.O. Box 386, Poland (October 2006)

30. Myers, E.W.: An O(ND) difference algorithm and its variations. Algorithmica **1**, 251–266 (1986)

31. Needleman, S.B., Wunsch, C.D.: A general method applicable to the search for similarities in the amino acid sequence of two proteins. J. Mol. Biol. **48**(3), 443–453 (1970)

32. Paxson, V.: Bro: a system for detecting network intruders in real-time. Comput. Netw. **31**(2324), 2435–2463 (1999). http://www.sciencedirect.com/science/article/pii/S1389128699001127

33. Piro Rosario, M., Andrea, G., Giuseppe, P., Albert, W.: Using historical accounting information to predict the resource usage of grid jobs. Future Gener. Comput. Syst. **25**(5), 499–510 (2009)

34. Roesch, M., Telecommunications, S.: Snort - lightweight intrusion detection for networks, pp. 229–238 (1999)

35. Rousseeuw, P.J.: Silhouettes: a graphical aid to the interpretation and validation of cluster analysis. J. Comput. Appl. Math. **20**, 53–65 (1987)

36. Tan, P.N., Steinbach, M., Kumar, V.: Introduction to Data Mining. Addison-Wesley, Boston (2005)

37. Tang, K., Man, K., Kwong, S., He, Q.: Genetic algorithms and their applications. IEEE Signal Process. Mag. **13**(6), 22–37 (1996)

38. Vanderster, D.C., Brochu, F., Cowan, G., Egede, U., Elmsheuser, J., Gaidoz, B., Harrison, K., Lee, H.C., Liko, D., Maier, A., Mocicki, J.T., Muraru, A., Pajchel, K., Reece, W., Samset, B., Slater, M., Soroko, A., Tan, C.L., Williams, M.: Ganga: user-friendly grid job submission and management tool for LHC and beyond. J. Phys. Conf. Ser. **219**(7), 072022 (2010). http://stacks.iop.org/1742-6596/219/i=7/a=072022

39. Vlachos, M., Lin, J., Keogh, E., Gunopulos, D.: A wavelet-based anytime algorithm for K-means clustering of time series. In: Proceedings of the Workshop on Clustering High Dimensionality Data and Its Applications (2003)

Structural Properties of Linked RDF Documents

Weiyi Ge[(✉)], Chenglong He, Shiqiang Zong, and Wenke Yin

Science and Technology on Information Systems Engineering Laboratory,
Nanjing 210007, People's Republic of China
geweiyi@gmail.com, hechenglong@nuaa.edu.cn,
sqzong@yeah.net, iamywk@126.com

Abstract. There is an ever-growing number of diverse RDF documents available on the Web, and most of them are published following Linked Data principles. To indicate the current status of data interconnection, we analyze structural properties of these linked RDF documents. We propose a document link graph (DocGraph) to model links between documents, and analyze its structure from three aspects: degree distribution, morphological structure, and reachability. We report our experiments on structural properties of the graph using two crawls, each with about 10 million documents. We find that the DocGraph is scale-free, and with small average distance. Its structure in 2012 is close to that of the Hypertext Web in the years around 2001–2002, and is not as good as the structure of the Hypertext Web in later years. Therefore, we conclude that data interlinking is very necessary for the Web of Data.

Keywords: Semantic web · Document link graph · Structural properties · RDF data interlinking

1 Introduction

The Semantic Web vision and its related technologies have brought out the development of a Web of Data. In the past decade, we have witnessed the explosion of the Web of Data from an information repository of a few millions of RDF triples [1] into a Giant Global Graph [2]. Especially, the Linking Open Data (LOD) project[1] is populating the Web of Data with massive amounts of distributed yet interlinked RDF data. As of Sept. 2011, the LOD is estimated to contain 31 billion RDF triples and about 503 million cross dataset links [3].

As the Web of Data is filled with huge data, making them accessible over Web is an important issue. Linked Data Proposal [4] proposes that every entity, i.e. a person, company, should be identified with its own HTTP URI. And every URI should be made dereferencable into an RDF description of the entity, i.e. a document. Then, when interacting with the Web of Data, agents dereference URIs in the current document into other documents. Thus, directed links between documents are formed.

---

[1] http://www.w3.org/wiki/SweoIG/TaskForces/CommunityProjects/
LinkingOpenData.

© Springer International Publishing Switzerland 2015
W. Qiang et al. (Eds.): CloudCom-Asia 2015, LNCS 9106, pp. 283–294, 2015.
DOI: 10.1007/978-3-319-28430-9_21

Although more and more data are published on the Web, and we can enjoy the convenience on navigation that comes from Linked Data Proposal, we know nothing about the status of document link structure. For example, we may concern that how many directed links do exist, and whether the majority of documents can be reached from a given starting point by following directed links. It should be stressed that the status of document link structure indicates the navigability and searchability of the Web of Data. The traditional Hypertext Web can be considered to be a graph, where the directed edges are the hyperlinks between Web pages. And the source of its strength lies in its network structure. Many research studies have been carried out on analyzing the structure of linked Web pages, e.g. [5–7]. Similarly, we argue that the power of the Web of Data also lies in its structure, especially, the link structure of RDF documents.

In this paper, we use three structural properties to analyze the link structure:

- Degree Distribution: Given a graph, its average in/out-degree shows that it is sparse or dense. Its degree distribution indicates whether it is a scale-free network. Generally, a higher average degree indicates a denser graph, which is the necessary condition for a good connected structure.
- Morphological Structure: What is the size of its largest weakly/strongly connected component (LWCC/LSCC). Especially, the morphological structure [5] is an important and intuitive feature for understanding its link structure. A larger LWCC/LSCC usually indicates a more connected structure.
- Reachability: In a graph, from a random selected node as the start, how many nodes it can reach by following directed edges in average. And how many steps it should take in average or in most situations. If it can reach more nodes in fewer hops, a graph usually has a more connected structure.

We propose a document link graph (DocGraph) to model directed links between documents. In order to reduce the bias caused by different samples, we perform our experiments on two crawls both of which cover significant portions of the Web of Data. Our study indicates that the structure of the DocGraph in around 2012 is close to that of the Hypertext Web in the years around 2001–2002, but is not as good as the structure of the Hypertext Web in later years.

The rest of this paper is structured as follows. Section 2 gives some preliminary definitions. Section 3 introduces the structural properties that used to measure. Section 4 introduces experiment settings, especially the used datasets. Section 5 analyzes the structural properties of the DocGraph. Section 6 introduces related work. Section 7 concludes our work with future directions.

2   Definitions and Notations

In RDF, identifiers consist of three disjoint sets: URIs (U), blank nodes (B) and literals (L). An RDF triple t has three parts: a subject ($subj(t)$), a predicate ($pred(t)$), and an object ($obj(t)$), which is with the form $t \in (U \cup B) \times U \times (U \cup B \cup L)$.

An RDF document d is a non-empty set of RDF triples, and let D denotes all RDF documents. For the sake of simplicity, we use "document" to denote "RDF document" in this paper unless otherwise stated.

According to the Linked Data principles, when exploring the Web of Data to find other related data, an agent obtains all URI references contained in the current document, and dereferences them to obtain representations (i.e. RDF documents) of the identified resources. Thus, links between RDF documents are the composition of two relations: "contains" and "dereferences".

Definition 1 (Contains). *The "contains" relation $R_c \subseteq D \times U$ is a binary relation from D to U, where $(d, u) \in R_c$ means there is an RDF triple $t \in d$ such that $u \in \{subj(t), pred(t), obj(t)\}$.*

Definition 2 (Dereferences). *The "dereferences" relation $R_d \subseteq U \times D$ is a functional relation, where $(u, d) \in R_d$ means d is a representation of a resource identified by u.*

For example, as there is an RDF triple ⟨`owl:sameAs`, `rdf:type`, `rdf:Property`⟩ in OWL[2], OWL contains owl:sameAs, `rdf:type` and `rdf:Property`. And as RDF[3] is a representation of a resource identified by `rdf:type`, (`rdf:type`, *RDF*) $\in R_d$.

Definition 3 (Document Link Graph). *A document link graph (DocGraph), represented as (V, E), is a directed graph*

– $V \triangleq D$, *the set of documents,*
– $E \subseteq V \times V$, *the set of directed edges, where $E = R_c \circ R_d$.*

3   Structural Properties

We measure the structure of a given graph from three aspects.

3.1   Degree Distribution

In a graph, the average in/out-degree indicates its density, which is the necessary condition for a graph to have a good connected structure. Many networks are thought to be scale-free, e.g. Hypertext Web [8], biological networks [9], and social networks [10]. Therefore, whether a graph has a scale-free nature, i.e. the degree distribution follows a power law, is a basic question with regard to the link structure. The scale-free network is likely to have a small diameter [11], which results in a easy-to-traverse structure.

---

[2] http://www.w3.org/2002/07/owl.
[3] http://www.w3.org/1999/02/22-rdf-syntax-ns.

3.2   Morphological Structure

The distribution of connected components in the graph is another important measurement for the structure. A large strongly connected component (SCC) means that huge amounts of nodes can be reached following directed links from any node in the component. On the contrary, if there is no giant weakly connected component (WCC), the graph is very fragmented. Last but not least, the morphological structure [5] is an intuitive feature. If the largest weakly connected component (LWCC) has a dominant size, and the largest strongly connected component (LSCC) is far larger than the second one, we can use a method to get the morphological structure of the graph (cf. Fig. 2). In detail, we take the LSCC as the central of the structure. Then, there is an IN component: each node in IN has a directed path to all nodes in SCC, but nothing in SCC can reach nodes in IN. Similarly, each node in the OUT component can be reached by directed paths from SCC, but not vice versa. Apart from these, there are some nodes weakly connected to all of the above, but cannot reach SCC or be reached from SCC in directed fashion; they are in the TENDRILS. Left groups of nodes are in DISC (Disconnected Components).

Here we introduce a method to get its morphological structure. We use a breath first search (BFS) to compute the LWCC of the graph. And its time complexity is $O(|E|+|V|)$, where $|E|$ and $|V|$ are the number of edges and vertices in the graph, respectively. Moreover, the Tarjan's algorithm[4] is used to compute all SCCs of the graph, whose time complexity is also $O(|E| + |V|)$. Finally, from any node in the SCC component, we use the BFS to follow forward/backward links to get nodes in OUT or IN.

3.3   Reachability

The reachability at h is the number of pairs of nodes having a directed path within distance h. If we plot it on a coordinate system, where the x axis is h and the y axis is the reachability at h, the plot is called hop plot [12]. With the reachability, several terms related to distance can be computed: the average connected distance is the expected length of the shortest paths between connected pairs of nodes in the network as defined in [5]; the effective diameter is the minimum number of hops in which some fraction (e.g. 90 %) of all connected pairs of nodes can reach each other [12]; and the diameter is the maximal length of all the shortest paths between connected pairs of nodes. Especially, the reachability at ∞ divided by the number of all pairs of nodes is called the reach probability.

To compute the reachability, we have to find the shortest paths for every pair of nodes in the graph. And this is the all-pairs shortest path problem. For an unweighted digraph, no (quasi-)linear time algorithm is known. Therefore, given a large scale graph like ours, we resort to an approximate algorithm. In this paper, we use the HyperANF [13], as it is both accurate and efficient. Let $r(h)$ denote the reachability at h, the average connected distance

---

[4] http://en.wikipedia.org/wiki/Tarjan's_strongly_connected_components_algorithm.

Table 1. Top-10 sites with most documents in FC10 and CC12

(a) FC10		(b) CC12	
site	#documents	site	#documents
l3s.de	1,071,954	data.gov.uk	2,000,000
geonames.org	1,027,864	bio2rdf.org	1,356,642
hi5.com	1,023,098	bbc.co.uk	962,299
dbpedia.org	1,022,596	semantictweet.com	853,100
freebase.com	832,563	opencyc.org	500,912
linkedmdb.org	678,834	cornell.edu	366,539
bibsonomy.org	660,620	legislation.gov.uk	236,703
rdfabout.com	648,467	ufl.edu	144,846
dbtune.org	646,768	cyc.com	140,026
livejournal.com	633,732	freebase.com	135,537

is $\sum_{h=1}^{\infty} \frac{h*(r(h)-r(h-1))}{r(\infty)}$. The effective diameter is h such that $\frac{r(h)}{r(\infty)} = 0.9$. The diameter is $\arg\max\limits_{h} r(h)$. And the reach probability is $\frac{r(\infty)}{|V|^2}$.

4   Experiment Settings

RDF documents on the Web cannot be entirely fetched because of the dynamic and distributed features. In order to reduce the bias caused by different samples, we use two crawls both of which cover significant portions of the Web of Data. One crawl is collected by the Falcons search engine[5] till December 2010, named FC10. Another crawl is collected by a custom spider from January to May 2012, named CC12. This spider is fed by seeds sampled from Google Search Engine using a random walk based sampler [14]. This sampler ensures that all sampled seeds are uniform distributed on the Web. And this spider proceeds in a polite BFS manner. Furthermore, to avoid a dataset starts dominating the crawl, the crawler remove any document in this dataset from consideration when it has more than 2 million documents.

Statistics indicate that there are 12,587,771 RDF documents and 3,260,796,023 statements in FC10. The number of quadruples in FC10 approximates to 10.31 % statements on the LOD (31,634,213,770 statements by Sept. 2011 [3]). And it has a similar scale to BTC14[6] (4.1 billion statements). Documents in FC10 are distributed in 5,430 sites. In CC12, there are 7,660,773 RDF documents from 63,109 sites. Table 1 lists top-10 sites having most documents for these two crawls. From the table, we find that most data in them are also in the LOD cloud.

---

[5] http://ws.nju.edu.cn/falcons.

[6] http://km.aifb.kit.edu/projects/btc-2014/.

Table 2. Top 10 documents with the largest in-degrees for the DocGraph in FC10 and CC12

(a) FC10

Document	In-degree
http://xmlns.com/foaf/spec/	6,869,539
http://dublincore.org/2008/ 01/14/dcelements.rdf	3,634,871
http://dublincore.org/2008/ 01/14/dcterms.rdf	2,006,788
http://ontoware.org/swrc/ swrc_v0.3.owl	1,432,811
http://www.w3.org/2003/01/ geo/wgs84_pos	1,054,563
http://www.geonames.org/ ontology/	1,035,608
http://xmlns.com/foaf/0.1/ homePage	1,023,102
http://xmlns.com/foaf/0.1/ surName	969,547
http://dublincore.org/2008/ 01/14/dctype.rdf	837,097
http://rdf.freebase.com/rdf/ common/topic	832,561

(b) CC12

Document	In-degree
http://xmlns.com/foaf/spec/	4,575,813
http://dublincore.org/2010/ 10/11/dcterms.rdf	3,697,605
http://www.w3.org/2009/08/ skos-reference/skos.rdf	2,213,763
http://vocab.org/frbr/core.rdf	2,159,147
http://www.w3.org/2006/time	1,862,035
http://reference.data.gov.uk/ def/reference/uriSet	1,856,817
http://www.w3.org/2007/08/ pyRdfa/extract?uri=http:// trdf.sourceforge.net/provenance /ns.html	1,855,961
http://vocab.deri.ie/void	1,855,539
http://reference.data.gov.uk/ def/reference/URIset	1,855,353
http://vocab.deri.ie/scovo	1,496,274

5   Structure of Document Link Graph

We analysis the structure of the document link graph (DocGraph), whose nodes contain both ontologies and instance-level documents. Furthermore, we argue that agents in the community are familiar with all terms in the W3C base-ontologies RDF, RDF Schema[7], OWL and DAML[8]. Thus, they seldom fetch these ontologies. Therefore, we remove four ontologies from the DocGraph.

5.1   Degree Distribution

According to the definition in Sect. 3, we generate a DocGraph for each crawl. The DocGraph for FC10 has 12,587,767 nodes and 250,659,182 edges. And the DocGraph for CC12 has 7,660,770 nodes and 119,137,367 edges. Therefore, the average in/out-degree is 19.91 and 15.55 respectively. Both degrees are larger than the in/out-degree of the Hypertext Web in 1999–2001 (7.22–9.3 in Table 3). But after that, the Hypertext Web is becoming denser. For example, its average in/out-degree is 15.8 in 2002, 24.1 in 2003 (cf. Table 3). Therefore, we conclude that the density of the DocGraph in 2012 is close to the density of the Hypertext Web in 2001–2002, and is not as good as the Hypertext Web's in later years.

Nodes with the largest in-degrees for the DocGraph in FC10 and CC12 are listed in Table 2. As most of them are well-known ontologies, it's normal that there are many links to them. As there is no outstanding nodes with large out-degree, we do not list them here.

---

[7] http://www.w3.org/2000/01/rdf-schema.

[8] http://www.daml.org/2001/03/daml+oil.

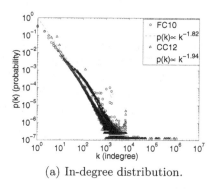

(a) In-degree distribution.

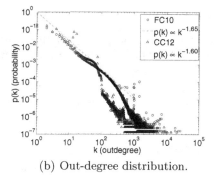

(b) Out-degree distribution.

Fig. 1. In-degree and out-degree distributions. In-degree follows a power law. Out-degree approximately follows a power law.

Figure 1(a, b) show the in-degree and out-degree distribution of the Doc-Graph, respectively. In-degree follows a power law with exponent 1.82 (FC10) and 1.94 (CC12). Out-degree follows approximately a power law with exponent 1.65 (FC10) and 1.60 (CC12). The same power law distributions are also found in the Hypertext Web, where in-degree exponent is 1.6–2.2 and out-degree exponent is 2.72 (cf. Table 3). An interesting observation is that nodes with the highest in-degree and nodes with the highest out-degree are quite distinct. Actually, the Kendall-tau correlation between the in-degree and outdegree is about zero. It implies that hubs (authorities) are not necessary to be authorities (hubs). As in/out-degree distributions follow power laws, the DocGraph is scale-free.

Table 3. Degree distribution

	#nodes	#edges	Average in/out-degree	γ_{IN}	γ_{OUT}
DocGraph(FC10)	12,587,767	250,659,182	19.91	1.82	1.65
DocGraph(CC12)	7,660,770	119,137,367	15.55	1.94	1.60
Web99[a]	203 m	1466 m	7.22	2.1	2.72
Web01[b]	80,571,247	752,527,660	9.3	1.9	N/A
UKWeb02[c] [6,7]	18,520,486	292,243,663	15.8	1.7	N/A
Web03[d] [7]	49,296,313	1,185,396,953	24.1	2.2	N/A
ITWeb04[e] [6,7]	41,291,594	1,135,718,909	27.5	1.6	N/A
IndochinaWeb04[f] [6]	7,414,866	194,109,311	26.2	N/A	N/A

[a]Crawled by AltaVista in 1999
[b]Crawled by WebBase in 2001
[c]Web of United Kingdom, crawled by UbiCrawler in 2002
[d]Crawled by WebBase in 2003
[e]Web of Italy, crawled by UbiCrawler in 2004
[f]Web of Indochina, crawled by UbiCrawler in 2004

5.2 Morphological Structure

To get the distribution of weakly connected components (WCCs), we treat the DocGraph as an undirected graph and perform a WCC algorithm to find all undirected connected components. In FC10, a giant WCC composed of 12,241,551 nodes is identified, which means that 97.25 % of the nodes are reachable from one another by following either forward or backward links. In CC12, the size of the largest WCC is 99.99 %. For a query engine or a crawler, it is not likely to follow backward links. Hence we should compute its strongly connected components (SCCs). We find that there is a large SCC in each crawl, which is much larger than the second one. For example, the largest SCC in FC10 consists of 49.29 % documents, and the largest one in CC12 consists of 60.94 % documents. While, the second largest SCCs have 986,405 and 1,121,828 nodes, respectively. A manual inspection shows that documents in SCC for both crawls come from many datasets and cover diverse topics.

Table 4. Morphological Structure

	SCC	IN	OUT	TENDRILS	DISC
DocGraph(FC10)	49.29 %	31.77 %	8.39 %	7.80 %	2.75 %
DocGraph(CC12)	60.94 %	36.40 %	2.12 %	0.53 %	0.01 %
Web99	27.74 %	21.29 %	21.20 %	21.52 %	8.24 %
Web01	56.46 %	17.24 %	17.94 %	N/A	N/A
UKWeb02	65.3 %	1.7 %	31.8 %	0.8 %	0.4 %
Web03	85.87 %	2.28 %	11.26 %	N/A	N/A
ITWeb04	72.3 %	0.03 %	27.6 %	0.01 %	0 %
IndochinaWeb04	51.4 %	0.66 %	45.9 %	0.66 %	1.4 %

As the largest SCC is much larger than the second one, we use a method introduced in Sect. 3 to analyze the morphological structure of the DocGraph. The sizes of various components in the morphological structure are listed in Table 4. From the table, we find that the DocGraph is mainly composed of two components: SCC and IN. The OUT component is almost non-existent, which is different from the Hypertext Web.

5.3 Reachability

With the HyperANF algorithm [13], we estimate the reachability of nodes in the DocGraph. Table 5 lists four measures related to the reachability of the DocGraph. For comparison purposes, we also list the reachability of the Hypertext Web in 1999. We cannot find values about the reachability of other Web graphs, so we do not list them in the table. From the table, we notice that the DocGraph in FC10 has a larger distance than that in CC12. This is because that CC12 is

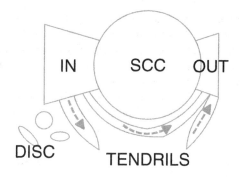

Fig. 2. Morphological structure of the DocGraph.

Table 5. Reachability

	Reach probability	Average connected distance	Effective diameter	$\gamma_{diameter}$
DocGraph(FC10)	42.73 %	13.32	16.02	258
DocGraph(CC12)	51.58 %	11.46	15.96	34
Web99	24 %	16	N/A	503

crawled by a limited-depth BFS crawler, while FC10 has a more complex crawling strategy. Though there are some differences in the distance between two crawls, their reach probabilities are almost the same, which in turn are much higher than the probability of the Hypertext Web in 1999.

Though there are no exact values on the reachability for other Web graphs, the relative sizes of the SCC and the IN and OUT components can give us some indications. Section 3.3 of [5] gives an equation to estimate the reach probability by sizes of three components, i.e. $\frac{(SCC+IN)*(SCC+OUT)}{V*(V-1)}$. Generally, the size of the SCC is of particular importance, since it constitutes the subset of reversible and complete access navigability. When an agent starts to surf the Web from the IN component, it maybe end up in the SCC, and maybe eventually in the OUT component, but can never go back to the IN component. If in the OUT component, it can never go back to the other components. But in the SCC, all nodes are reachable and can be revisited. Therefore, the size of SCC reflects the reach probability in some sense. Based on this, we find that the reach probability of the DocGraph is smaller than that of the Hypertext Web after 2002. In summary, its reachability is close to that of the Hypertext Web in around 2002, but is not as good as that in later years.

6   Related Work

Some work focus on analyzing the structural properties of entity link graph in the Web of Data. Gil et al. [15] find the RDF graph composed of ontologies from the DAML Ontology Library is a small-world and scale-free network. Cheng and Qu [16] study the graph structures of dependence between concepts and between vocabularies. The results characterize the current status of schemas in the Semantic Web in many aspects, including degree distributions, reachability, and connectivity. Guéret et al. [17] analyze the robustness of entity link structure in the Web of data. Ge et al. [18] find that the entity link graph in the instance-level is scale-free and has a small diameter. Ding et al. [19] give a structural analysis of the entity owl:sameAs network and focus discussion around its usage in the Web of Data.

Graph analysis techniques have also been applied to single ontologies. Hoser, et al. [20] illustrate the benefits of applying social network analysis to ontologies by measuring SWRC and SUMO ontologies. They discuss how different notions of centrality (degree, betweenness, eigenvector, etc.) describe the core content and structure of an ontology. Zhang [21] study NCI-Ontology, Full-Galen, and other five ontologies, and discover that the degree distributions of these entity networks fit power laws well. Theoharis, et al. [22] analyze graph features of 250 ontologies. For each ontology, they constructed a property graph and a class subsumption graph. The property graph is a directed graph whose nodes correspond to classes and literal types, and whose arcs point from the domain of a property to its range. They find that the majority of ontologies with a significant number of properties approximate a power law for total-degree distribution, and each ontology has a few focal classes that have numerous properties and subclasses.

To the best of our knowledge, there is no work analyzing the morphological structure of RDF document link graph, and thus there is no work assessing the reachability of RDF documents on the Web. Therefore, we want to fill this gap. We think that our analysis results can be used to engineer a better Web of Data in the future, just like what has happened in the Hypertext Web.

7   Conclusions and Future Work

To ensure the navigability and searchability of the Web of Data, we assess the link structure of RDF documents. All experiments are performed in two crawls each with about 10 million documents.

Degree distribution shows that RDF document link graph (DocGraph) is scale-free. Morphological structure shows that it is mainly composed of two components: SCC and IN, and its OUT component is very small. Reachabilty shows that it is with small average distance. Besides, results show that its structure properties (i.e. density, SCC size, reach probability and avg. distance) are close to those of the Hypertext Web in the years around 2001–2002, but are not as good as those of the Hypertext Web in later years. Therefore, to develop, it is recommended to add more links to connect documents. And we believe that sufficient data interconnections can improve the structure of the current Web of Data tremendously.

It is known that sampling cannot maintain all structural properties of the original graph. Thus, we need a sample of the Web of Data as large as possible.

In our experiment, both the Falcons 2010 crawl and our custom crawl are used. In the future, we will use other available large crawls to test the universality of our results. Besides, the relations between structural properties and crawling policies should be further investigated. In this paper, three structural properties are used to measure the structure of graphs. But there does not exist an exact and definitive description of structural properties, which could affect our judgment on the navigability and searchability of the Web of Data. Furthermore, we are interested in tracking the evolutionary of document link structure on the Web of Data.

Acknowledgments. This work is supported by the National Natural Science Foundation of China (NSFC) under Grants 61402426 and partially supported by Collaborative Innovation Center of Novel Software Technology and Industrialization.

References

1. Eberhart, A.: Survey of rdf data on the web. In: Proceedings of the 6th World Multiconference on Systemics, Cybernetics and Informatics (SCI 2002)
2. Berners-Lee, T.: Giant global graph. Decentralized Information Group Breadcrumbs, pp. 6–11 (2007)
3. Bizer, C., Jentzsch, A., Cyganiak, R.: State of the lod cloud. Version 0.3, 1803, September 2011
4. Berners-Lee, T.: Linked data-design issues (2006) (2011). http://www.w3.org/DesignIssues/LinkedData.html
5. Broder, A., Kumar, R., Maghoul, F., Raghavan, P., Rajagopalan, S., Stata, R., Tomkins, A., Wiener, J.: Graph structure in the web. Computer Netw. **33**(1), 309–320 (2000)
6. Donato, D., Leonardi, S., Millozzi, S., Tsaparas, P.: Mining the inner structure of the web graph. In: WebDB, pp. 145–150 (2005)
7. Serrano, M., Maguitman, A., Boguñá, M., Fortunato, S., Vespignani, A.: Decoding the structure of the www: a comparative analysis of web crawls. ACM Trans. Web (TWEB) **1**(2), 10 (2007)
8. Barabási, A.L., Albert, R., Jeong, H.: Scale-free characteristics of random networks: the topology of the world-wide web. Phys. A: Stat. Mech. Appl. **281**(1), 69–77 (2000)
9. Barabási, A.L., Oltvai, Z.N.: Network biology: understanding the cell's functional organization. Nat. Rev. Genet. **5**(2), 101–113 (2004)
10. Barabâsi, A.L., Jeong, H., Néda, Z., Ravasz, E., Schubert, A., Vicsek, T.: Evolution of the social network of scientific collaborations. Phys. A: Stat. Mech. Appl. **311**(3), 590–614 (2002)
11. Cohen, R., Havlin, S.: Scale-free networks are ultrasmall. Phys. Rev. Lett. **90**(5), 058701 (2003)
12. Chakrabarti, D., Faloutsos, C.: Graph mining: laws, generators, and algorithms. ACM Comput. Surv. (CSUR) **38**(1), 2 (2006)
13. Boldi, P., Rosa, M., Vigna, S.: Hyperanf: approximating the neighbourhood function of very large graphs on a budget. In: Proceedings of the 20th International Conference on World Wide Web, pp. 625–634. ACM (2011)

14. Bar-Yossef, Z., Gurevich, M.: Random sampling from a search engine's index. J. ACM (JACM) **55**(5), 24 (2008)
15. Gil, R., García, R., Delgado, J.: Measuring the semantic web. AIS SIGSEMIS Bull. **1**(2), 69–72 (2004)
16. Cheng, G., Qu, Y.: Term dependence on the semantic web. In: Sheth, A.P., Staab, S., Dean, M., Paolucci, M., Maynard, D., Finin, T., Thirunarayan, K. (eds.) ISWC 2008. LNCS, vol. 5318, pp. 665–680. Springer, Heidelberg (2008)
17. Guéret, C., Groth, P., van Harmelen, F., Schlobach, S.: Finding the achilles heel of the web of data: using network analysis for link-recommendation. In: Patel-Schneider, P.F., Pan, Y., Hitzler, P., Mika, P., Zhang, L., Pan, J.Z., Horrocks, I., Glimm, B. (eds.) ISWC 2010, Part I. LNCS, vol. 6496, pp. 289–304. Springer, Heidelberg (2010)
18. Ge, W., Chen, J., Hu, W., Qu, Y.: Object link structure in the semantic web. In: Aroyo, L., Antoniou, G., Hyvönen, E., Teije, A., Stuckenschmidt, H., Cabral, L., Tudorache, T. (eds.) ESWC 2010, Part II. LNCS, vol. 6089, pp. 257–271. Springer, Heidelberg (2010)
19. Ding, L., Shinavier, J., Shangguan, Z., McGuinness, D.L.: Sameas networks and beyond: analyzing deployment status and implications of owl:sameas in linked data. In: Patel-Schneider, P.F., Pan, Y., Hitzler, P., Mika, P., Zhang, L., Pan, J.Z., Horrocks, I., Glimm, B. (eds.) ISWC 2010, Part I. LNCS, vol. 6496, pp. 145–160. Springer, Heidelberg (2010)
20. Hoser, B., Hotho, A., Jäschke, R., Schmitz, C., Stumme, G.: Semantic network analysis of ontologies. In: Sure, Y., Domingue, J. (eds.) ESWC 2006. LNCS, vol. 4011, pp. 514–529. Springer, Heidelberg (2006)
21. Zhang, H.: The scale-free nature of semantic web ontology. In: Proceedings of the 17th International Conference on World Wide Web, pp. 1047–1048. ACM (2008)
22. Theoharis, Y., Tzitzikas, Y., Kotzinos, D., Christophides, V.: On graph features of semantic web schemas. IEEE Trans. Knowl. Data Eng. **20**(5), 692–702 (2008)

The Key Technologies of Real-Time Processing Large Scale Microblog Data Stream

Yunpeng Cao[1,2,3](✉) and Haifeng Wang[1,2,3]

[1] Communication Research Center, Linyi University, Linyi 276005, Shandong, China
[2] College of Information, Linyi University, Linyi 276005, Shandong, China
[3] Linda Institute, Shandong Provincial Key Laboratory of Network Based Intelligent Computing, Linyi 276005, Shandong, China
lyucyp@163.com, gadfly7@126.com

Abstract. Real-time monitoring microblog data can find sensitive information in time and provide help for public sentiment management and control. However, it needs processing large-scale data stream. MapReduce is a framework of processing large-scale data in batch mode, its purpose is to increase throughput, but its real-time performance is limited. Aiming at the real-time performance limitation of MapReduce, RT-SSP (Real-Time Staged Stream Processing), a hybrid staged real-time stream processing scheme both for batch and real-time processing was proposed. By this method large-scale high-speed data stream is locally processed in stages, the communication cost is reduced by storing intermediate results to local node, and key technologies such as cache optimization are used to realize high concurrent read and write. Experiments show that RT-SSP scheme can improve the real-time performance of processing large-scale microblog data stream and achieve speed-up ratio of about 2.3.

Keywords: Microblog monitoring · High-speed data stream · MapReduce · Real-time processing · RT-SSP

1 Introduction

Micro-blog is a convenient and quick social network of short messages and has vast users. Using computers, mobile phones and other devices, users can release, comment and forward micro-blog very quickly. So micro-blog information is large-scale, high-speed and continuous data stream. Processing micro-blog to discover sensitive information in time is very important for strengthening public opinion guidance and management, inhibiting harmful information spreading

Foundation items: National Natural Science Foundation of China (No. 60970012); Natural Science Foundation of Shandong Province (No. ZR2013FL005). Author introduction: Yunpeng Cao (1967-), male, master, associate professor, main research directions include large-scale data processing and parallel computing. Haifeng Wang (1976-), male, associate professor, doctor, main research directions include network computing, large-scale data processing and cloud computing.

© Springer International Publishing Switzerland 2015
W. Qiang et al. (Eds.): CloudCom-Asia 2015, LNCS 9106, pp. 295–306, 2015.
DOI: 10.1007/978-3-319-28430-9_22

and promoting beneficial information communication [1,2]. Facing high speed and continuous micro-blog data stream, strong processing ability and large storage capacity are needed. Because of the continuity and infinity of data stream, window mechanism (time and data amount) is used to divide processing boundary, and the accumulated data within boundary is historical data. With the rapid development of data acquisition and transmission technology, accumulating large amount of historical data in short time is possible. In micro-blog monitoring analysis, comparison, merge and so on between real-time data and results of processing historical data are required to complete quickly. So the contradictions between the need for processing real-time data stream on massive historical data and the shortage of processing and storage capacity has become a new challenge. This is the extensibility problem in data stream processing.

Researches on extensibility can be divided into two classes: centralized and distributed. In centralized environment, the extensibility is guaranteed mainly by admittance control, QoS reduced order, summary data and other methods at the expense of losing service quality. In distributed environment, extensibility is ensured mainly by balancing operator distribution among nodes [3], but the overall processing capacity is limited by the storage and processing capacity of single node. In order to break the limitation of single node and support storing and processing large-scale data, multi-core CPU cluster and four-layer storage structure including cache, memory, external memory and distributed storage are usually used. In this architecture, multi-core CPUs and Caches form local processing resources. Compared with distributed storage, the memory and external memory form a high-speed local storage. In cluster architecture, MapReduce [4] is often used to process large scale data. It provides a simple programming interface for parallel processing massive data and shields details such as task scheduling, data storage and transmission from programmers, thus reduces the requirements for programmers, so it is suitable for solving the extensibility problem. However, its existing specific implementations such as Hadoop, Phoenix [5] and so on belong to batch processing mode of persistent data. Every time runtime environment is initialized, large-scale data is reloaded and processed, Map and Reduce stage are executed synchronously, and massive data is transmitted among nodes. When continuous data stream is processed in batch mode, if small scale batch of data is processed each time, the system overhead is too large, real time is limited; if waiting until the batch reaches a certain scale, the processing delay increases, the real-time demand cannot be met similarly. Therefore, according to the real-time processing demand of large-scale high-speed data stream, how to use MapReduce model is worth considering [6].

In order to strengthen the data stream processing ability of MapReduce, methods such as preprocess, distributed cache and reusing intermediate results can be used to avoid repetitive processing overhead of historical data each time data stream arrives, make data stream processing localize, reduce overhead of data transmission among nodes. This paper analyzed and proved the reasonability of using MapReduce to solve this kind of problem, and put forward RT-SSP (Real-Time Staged Stream Processing) method, an improvement of MapReduce to adapt to the real-time processing scene of high-speed data stream.

RT-SSP method includes an optimization method of local staged processing based on system parameters and a local storage method supporting high concurrent read-write.

2 Real-Time Processing Method for Large-Scale High-Speed Data Stream

2.1 Research Background

Processing continuous data stream in batch mode cannot meet real-time demand. HOP [7] and S4 [8] extended the real-time processing ability of MapReduce by pipeline and distributed processing unit technology respectively, but lacked support for historical data preprocessing and synchronization and did not optimize data stream processing ability by making full use of intermediate result caching, local staged processing technology and so on.

The initialization overhead of each process can be reduced by thread pool technology and synchronization can be eliminated by transferring data asynchronously between stages. It can also improve real-time processing ability by controlling batch size within stage and adjusting resources between stages. RT-SSP local staged pipeline was constructed, global control for each stage was carried on by using CPU usage information in thread pool control, so CPU was utilized sufficiently and effectively and data stream processing ability was improved.

In this paper, the storage file of intermediate results was established on SSTable structure. However, existing file read-write strategies based on SSTable [9] only optimize write operation. For improving the hit ratio of replacement between internal and external memory, existing replacement algorithms such as LRU and clock [10] do not fully utilize the table information in the buffer of staged pipeline, so the replacement hit ratio is restricted. In this paper, by using read-write overhead estimation and buffer information, file read-write strategies and replacement algorithm are transformed, high concurrent read-write performance of intermediate results was further optimized.

2.2 RT-SSP Method

MapReduce abstracts parallel processing to two functions–Map and Reduce. Its procedure is: a large data set is divided into some small data sets, each (or several) small data set is processed by a node (usually a computer) running Map function and intermediate results are generated, then intermediate results are merged by many nodes running Reduce function and the final results are generated. Map and Reduce functions are implemented by user.

MapReduce model is defined as following [4]:

Map: $k1, v1 \rightarrow list(k2, v2)$
Reduce: $k2, list(v2) \rightarrow list(k3, v3)$

Map converts $[k1, v1]$ key-value pairs to $[k2, v2]$ key-value pairs, Reduce performs list operation for value list $list(V2)$ of each $K2$ and get the results, $[k3, v3]$

key-value pairs. If the data to be processed is D, the intermediate results of Map stage is I, then MapReduce process can be described as: $MR(D) = R(M(D)) = list(I)$. M represents Map method, R represents Reduce method, and $list$ represents operations done by Reduce. The property of MapReduce model is analyzed as following.

Definition 1. For function $F : I \rightarrow O$, if there exists a function $P : O \times O \rightarrow O$ which makes $F(D + \Delta) = P(F(D), F(\Delta))$, then F can be merged.

Definition 2. For n data subsets $D1, D2, ..., Dn$ of data set D, If $D1 \cap D2 \cap ... \cap Dn = \emptyset$ and $D1 \cup D2 \cup ... \cup Dn = D$, then $D1, D2, ..., Dn$ is a division of D.

Definition 3. For D, a collection of key-value pairs and key set K, the set $\{d|d.key \in K, d \in D\}$ is a projection of D on K, denoted as $\sigma K(D)$.

According to MapReduce model and above definitions, MapReduce has the following properties:

(1) Map method meets distribution law, i.e. the Map on the union of two data sets equals to the Union of Maps on each of the two sets: $M(D + \Delta) = M(D) + M(\Delta)$.
(2) Reduce method can be combined: $list(D + \Delta) = list(list(D), list(\Delta))$.
(3) Reduce method meets distribution law, i.e. if $K1, K2, ..., Kn$ is a division of I, the set of intermediate results, then: $list(I) = list(\sigma K1(I)) + list(\sigma K2(I)) + ... + list(\sigma Kn(I))$.

Theorem 1. MapReduce can be combined.
Proof: According to MapReduce's property, for data D and increment Δ:

$$MR(D + \Delta)$$
$$= R(M(D + \Delta))$$
$$= R(M(D) + M(\Delta))$$
$$= list(ID + I\Delta)$$
$$= list(list(ID), list(I\Delta))$$
$$= R(MR(D), MR(\Delta)).$$

Therefore, according to definition 1, MapReduce can be combined. Q.E.D.

Theorem 1 indicates that MapReduce can reduce reprocessing overhead every time data stream arrives by preprocessing historical data and caching intermediate results and improve real-time processing ability. The above process can be expressed as: $MR(D + \Delta) = list(ID + I\Delta) = MR(\Delta|ID)$.

Theorem 2. $K1, K2, ..., Kn$ is a division of the key set of I, the MapReduce intermediate results. For MapReduce of data increment Δ on I:

$$MR(\Delta|I) = MR(\Delta|\sigma K1(I)) + MR(\Delta|\sigma K2(I)) + ... + MR(\Delta|\sigma Kn(I)).$$

Proof: According to the property of MapReduce, for intermediate results I and data increment Δ:

$$MR(\Delta|I)$$
$$= list(I + I\Delta)$$
$$= list(\sigma K1(I) + \sigma K2(I) + ... + \sigma Kn(I) + I\Delta)$$
$$= list(\sigma K1(I) + I\Delta + \sigma K2(I) + I\Delta + ... + \sigma Kn(I) + I\Delta)$$

$$= list(\sigma K1(I) + I\Delta) + list(\sigma K2(I) + I\Delta) + ... + list(\sigma Kn(I) + I\Delta)$$
$$= MR(\Delta|\sigma K1(I)) + MR(\Delta|\sigma K2(I)) + ... + MR(\Delta|\sigma Kn(I))$$

Q.E.D.

Theorem 2 shows that MapReduce can process data stream and intermediate results only on local node by distributed caching intermediate results to avoid data transmission among nodes.

From practical view, processing micro-blog data stream on a cluster with 3.2 GHz CPU, 16 GB memory and 1 Gbps Ethernet connection, with existing 100 MB historical data, for each data stream of about 140 B, the average time required from receiving to completing Hash grouping was $2(\pm0.2)\,\mu s$, the average time required to complete string match and comparison and merge with historical data is $7(\pm0.1)$ ms, the average delay of data transmission is $30(\pm2)\,\mu s$. Experimental data shows that a 3.2 GHz CPU can receive and group data stream at the speed of more than 70 MB/s, and in practical application, limited by collecting bandwidth, etc., the data stream cannot reach this speed. Therefore receiving and grouping data stream (Map stage) takes a small part of CPU resources, the main works are string match of data stream and processing such as statistics, comparison, combination, etc. (Reduce stage) and data transmission. In order to reduce the overhead of reprocessing historical data and data transmission, the intermediate result of preprocessing historical data is cached in each node by distribution; each node receives micro-blog data stream redundantly, Reduce is performed in local cache, so inter-node processing from Map to Reduce is not needed, communication overhead is reduced. When local processing and storage resources of existing nodes cannot meet real-time demand, they can be extended by re-dividing and moving cached data to newly added nodes.

RT-SSP method ensures that, after dividing data set, for each small data set, Map and Reduce are completed by the same node, pipeline is formed, intermediate results are cached locally (mainly in memory) to improve processing speed and reduce transmission cost. In RT-SSP , the working node is responsible for maintaining local intermediate result caching and staged pipeline. The procedure of RT-SSP includes:

(1) Cache intermediate results. Preprocess relevant historical data to generate intermediate results, divide sensitive information according to the Hash value of $K2$, cache them to the local storage of various nodes.
(2) Local staged processing. Map stage gets the sensitive information processed only on local node by Hash function acting on $K2$, generates micro-blog-sensitive information pairs with micro-blog data received on local node, transfers them to Reduce stage asynchronously. According to string matching algorithm, Reduce stage will produce matching results of this batch of micro-blog, the matching results are cached in memory or written to disk.
(3) Data synchronization. Local processing results are synchronized to distributed storage.

3   Key Technologies of High Concurrency Control

In order to reduce the overhead of reprocessing historical data every time data stream arrives, RT-SSP supports caching intermediate results. The working threads of Map and Reduce will frequently read and write intermediate results. The optimization of high concurrent read-write performance of intermediate results is the key to improve the ability of processing data stream. After introducing the memory data structure of intermediate results and file structure of external storage, this section proposed one kind of local storage optimization method supporting high concurrent read and write.

3.1   Key Storage Structure

In MapReduce, $[k2, list(V2)]$ and $list(V2)$ are intermediate results. Intermediate results are stored with Hash linked list in memory. $K2$ with the same Hash value is organized with linked list in the same item of Hash table. If $K2$ can be predicted and has unique Hash value, conflict and lookup may be avoided by allocating sufficient items for Hash table, both insertion and search have the complexity of $O(1)$. If $K2$ does not have unique Hash value or is unpredictable, a linked list is maintained in Hash table item, insert and search only have the complexity of $O(1) + O(\log n)$.

In order to extend the local storage capacity of intermediate results, SSTable file [9] is constructed in external storage. The SSTable file includes an index block and many data blocks of 64 KB, disk space is allocated in block units for Hash table entries. During processing, if the required Hash table item of intermediate results is not in memory but in external disk and there is no space in memory, the replacement between memory and external storage will occur.

Aiming at large-scale historical data, in order to ensure extensibility in cluster environment, RT-SSP divides the Hash partitions of intermediate results for working nodes. The Hash partitions of $K2$ distributes in n working nodes $(P1, P2, ..., Pn)$. It is known from RT-SSP procedure that Map stage includes one Hash grouping operation for $K2$, each node is only responsible for processing the data within its partition. If the partition division uses the method of taking the remainder from node numbers, the extension (adding or removing nodes) will move large amount of data. In order to reduce the scale of moving data, RT-SSP uses consistent Hash algorithm [11] to divide intermediate results on nodes. For example, there are three partitions $P1$, $P2$ and $P3$ originally, when adding a node, only a part of $P1$ and $P3$ is needed to be divided into $P4$.

In staged pipeline, data stream real-time processing ability is restricted by Map and Reduce threads' concurrent read-write synchronization for memory structure and concurrent read-write overhead on the external files of intermediate results. In order to improve the concurrent read-write performance of local intermediate results, two aspects should be considered: (1) establish memory buffers to reduce synchronization among concurrent threads; (2) optimize read-write strategies and replacement algorithms to reduce the concurrent read-write overhead of external memory.

3.2   Buffer Technologies

Working nodes maintain a Hash linked list in memory to store intermediate results. Within local staged framework, Map and Reduce threads frequently read and write intermediate results, Map threads write processed results, Reduce threads read Map results, write or update the intermediate results. Therefore, in order to improve memory concurrent read-write performance, the key is to reduce the concurrent read-write synchronization overhead on the same area of memory between Map and Reduce threads. The read-write synchronization on intermediate results can be avoided by establishing buffers between Map and Reduce stages [12].

In order to avoid write synchronization on buffers among Map threads, each Map thread monopolizes a buffer. In order to avoid read-write synchronization on buffers between Map and Reduce threads, buffer area is designed as a FIFO queue, Map thread writes to queue tail, Reduce thread reads from queue head. Each Reduce thread is responsible for a Hash partition; there is no synchronization among Reduce threads. Buffers are established corresponding to Hash partitions Between Map and Reduce threads, Map threads filter out the data they are responsible for from data stream and write them to corresponding buffers, Reduce threads process buffered data within partitions and write results to Hash linked list. According to the number of Map and Reduce threads, buffered data in memory will form a $M \times N$ buffer matrix.

3.3   High Concurrent Read-Write Strategies

Existing file read-write strategy based on SSTable is optimized for writing. For example, BigTable [9] adopts additional write mode that writes to a new file directly when it writes cached data in memory to disk, and merges cached data and many small files when reading, and the overhead is higher. For the local storage file of intermediate results, read and write operations are relatively frequent and balanced, it is unsuitable only to optimize write operation. In order to improve the performance of concurrent read and write, read and write mode should be selected according to overhead. The following are the methods to estimate read and write overhead:

If the seek time is a constant Ts, data read and write overhead functions are Tr and Tw, data merging overhead function is Tm.

Additional writing data d' includes seek and write overhead, i.e. $Time_{dmic} = Ts + Tw(d')$.

Combining existing data d and newly added data d' includes two times of seeking, two times of reading and merging overhead, i.e. $Time_{rmc} = 2Ts + Tr(d) + Tr(d') + Tm(d, d')$.

Combined writing data d includes seek and write overhead, i.e. $Time_{dmec} = Ts + Tw(d)$.

Random reading data d includes seek and read overhead, i.e. $Time_r = Ts + Tr(d)$.

When the replacement between internal and external memory occurs, for the Hash table item to be replaced, firstly the buffer between Map and Reduce stage should be used to view whether the table item is about to be accessed. If the table item will not be accessed soon, additional writing mode with smaller overhead is used; if the table item will be accessed soon, $Time_1 = Time_{dmic} + Time_{rmc}$ and $Time_2 = Time_{dmec} + Time_r$ are compared; if $Time_1$ is bigger, merge write and random read are chosen, if $Time_2$ is bigger, additional writing and merge reading are chosen.

In addition, for small files generated by additional write, combination is done by management threads to optimize read operation, especially when CPU is idle due to low system load.

3.4   Replacement Algorithms

Improving the hit ratio of replacement algorithms is also important to reduce the read-write overhead of external storage. The optimal algorithm [10] determines the replaced page according to the page information to be accessed, so hit ratio is the highest; but in practical system, due to unpredictable pages to be accessed, this method is less used. In the local staged pipeline of RT-SSP, the buffer between Map and Reduce stages contains the Hash table item information to be accessed, so the idea of optimal algorithm can be used. In addition, locality of data access and replacement cost should be considered. Therefore, RT-SSP internal and external memory replacement algorithm determines the replaced Hash table items according to the order of whether is about to be accessed, whether is accessed most recently and whether the replace cost is the minimum. Among them, buffer data are retrieved to determine whether table item will be accessed soon, the last access time of table item is recorded based on LRU algorithm, the least table item which can accommodate for entered table item is chosen according to the amount of data.

During micro-blog processing, the program will continuously write results to buffer. When buffer is full, the longest Hash table item in the list is written to disk, buffer space is released; when reading, the longest linked list and the linked list that has not been used for the longest time are replaced.

3.5   Staged Optimized Processing

In order to improve data stream processing capacity, RT-SSP constructs staged pipeline in each working node, uses thread pool to reduce initialization overhead of every processing, and eliminate the synchronization between Map and Reduce stages by asynchronous transferring data. In stage division, in order to reduce staged overhead, data receiving stage and Map stage are merged, each stage is composed of working thread pools, input buffers and controllers within stage, and resource allocation is adjusted between stages by controllers.

In RT-SSP local staged pipeline, Map and Reduce stages occupy part of working threads respectively, the key shared resource is CPU. How to use CPU (including CPU Cache) fully and effectively is the key to improve the ability of data stream processing. The staged pipeline can optimize CPU usage by batch adjustment within stage and thread pool adjustment between stages.

4 Experiment and Analysis

This section validates RT-SSP method according to real-time monitoring micro-blog data as the benchmark test.

4.1 Benchmark Test

According to the statistical data of previous work, micro-blog data stream speed will reach 1 MB/s (each data is assumed 140 B, about 7000 data/s). At the same time, when micro-blog data is saved for a month, the storage amount will reach 3 TB.

Micro-blog monitoring uses the following RT-SSP algorithm: the item is found in the Hash table of all micro-blog-sensitive information pairs in Map stage, the position of its linked list is found in Reduce stage, each micro-blog-sensitive information pair is processed in turn and the linked list is updated. In case that the amount of micro-blog-sensitive information pairs is Ns, the Hash function $Hash(k) = k\ mod\ Ns$ whose output is Ns numbers can be used to group. There are totally Ns items in Hash table to store intermediate results, the matching result of one micro-blog-sensitive information pair is stored in one Hash table item averagely.

Data stream processing is built on cluster of 10 3.2 GHz Intel core i3930K CPU, 16 GB memory, 2 TB hard disk computers, Ubuntu13.04 and Hadoop2.2.0 are installed in each node; ordinary switches are used in network connection; data stream is simulated by using LoadRunner11.0 on a computer. In order to test data stream processing extensibility, historical data partitions are divided evenly in cluster nodes, monitoring sites are set randomly to record system time, add timestamps and control data stream speed.

4.2 Storage Performance Analysis

The optimization effect of RT-SSP on local storage is validated. a single node is used, the data stream speed is 1 MB/s (each data is 140 B, about 7000 data/s), the scale of intermediate results is 50 GB size, 10 tests are performed in every item, 10 min each time, average value is adopted to calculate experiment results. Table 1 shows the performance comparison before and after optimization on local storage. It can be seen that RT-SSP eliminates read-write synchronization by establishing buffers, so memory read-write performance is improved by 12.3 %; RT-SSP guides reading and writing strategies and replacement algorithms by using buffer information, so the read-write performance and the hit ratio of internal and external memory are improved by 15.8 % and 10.6 % respectively. Local read-write performance (read and write times per time unit) of intermediate results is increased by 1/4 according to combination of 3 methods.

Table 1. The storage performance optimization of intermediate results

Performance index	Test method	Experiment results	Effect /%
Memory read-write	read-write synchronization	143231 times/s	12.3
performance	synchronization elimination	160848 times/s	
external memory	BigTable	8408 times /s	15.8
read-write performance	RT-SSP	9736 times/s	
Memory hit ratio	LRU	67.7 %	10.6
	RT-SSP	74.9 %	
overall read-write	before improvement	140413 times/s	25.1
performance	RT-SSP	175657 times/s	

On the basis of above experiment, data scale is kept invariant and data flow speed is increased gradually to test the external memory read-write performance of RT-SSP method and the changing situation of hit ratio. Experiments show that, as data stream speed increases, because buffer queue becomes larger and larger, more adequate information of table items to be accessed is provided for external memory read-write strategies and replacement algorithms, so read-write performance and hit ratio increases continuously. But when data stream speed increases to a certain extent, the extension of buffer data scale will not improve read-write performance and hit ratio, but because retrieval cost increases, and additional internal and external replacements and read-write operations are caused due to occupying too much memory resources, so read-write performance and hit ratio decreases. Further analysis of the relationship among data stream speed, buffer queue size, read-write performance and hit ratio will be done in the next step to guide the dynamic changes of buffer size and load discarding strategies.

4.3   Real-Time Analysis

Three kinds of architectures–S4, HOP and RT-SSP are compared in real time analysis experiments. Because S4 and HOP do not support preprocess, historical data will be loaded and processed newly each time stream data is to be processed, a large number of unnecessary overhead restricts throughput. Therefore, in order to compare data stream processing ability, preprocessing logic is added in S4 and HOP implementation. Various data stream processing architectures are built, the data stream speed is 1 MB/s, each data scale is tested 10 times, 10 min each time, average value is adopted to calculate experiment results.

Experiments show that, when the size of intermediate results is less than 32 GB, because a node's memory can hold all the intermediate results, HOP and S4 also have very high throughput (greater than 12000 items/s), but because RT-SSP uses local staged pipeline and memory read-write optimization, throughput is higher; when the scale of intermediate results is more than 32 GB, intermediate results are distributed in the memory of two nodes, in HOP and S4 throughput decreases more rapidly as data transmission between nodes and synchronization overhead increases, while in RT-SSP data transmission is avoided due to using localization technique, throughput is still high; when the size of intermediate results is over 64 GB, as this is beyond the cache capacity for intermediate results of S4 and HOP, its throughput tends to be stable, but error rate increases as data size increases, while RT-SSP performs extension and optimization for the ability of caching intermediate results by using local disk, higher throughput (greater than 11500 items/s) is able to be kept while reducing error rate (less than 5 %).

4.4 Extendibility Analysis

Two experiments are conducted to test extensibility of RT-SSP on historical data size and data stream speed respectively.

In the first experiment data stream speed is 1 MB/s to test the historical data scale RT-SSP can process in the case that the number of nodes increases. Experiments show that, the improvement of RT-SSP processing capacity is approximately linear when the number of nodes increases; this is because RT-SSP divides intermediate results of history data and use localized processing to avoid data transmission and synchronization overhead. The reason that linear scaling is not achieved is that the read-write overhead of local files increases when historical data scale increases.

In the second experiment the size of intermediate results in each node is 50 GB to test the data stream RT-SSP can process in the case that the number of nodes increases. Experiments show that, in case that data stream speed is lower than 15 MB/s, the improvement of RT-SSP processing capacity is approximately linear when the number of nodes increases; in case that data stream speed is faster than 15 MB/s, the improvement of RT-SSP processing ability becomes slower when the number of nodes increases. This is because with the improvement of data stream speed, CPU overhead for each RT-SSP node to receive redundant data stream and perform Map stage increases, and the read-write performance of intermediate results and replacement hit ratio of RT-SSP begin to fall, so throughput is affected.

From the experience of monitoring micro-blog by RT-SSP, the stream processing application occupies only a very small portion of node multi-core CPU to complete receiving and grouping operations, and RT-SSP can effectively improve concurrent read-write performance of intermediate results and has a good extensibility in case that historical data scale is continuously extending.

5 Conclusion

The difficulties of processing high-speed large-scale data stream lie in the demands for extensibility and real-time. This paper proposes RT-SSP method to support such data processing. By distributed caching intermediate results and local staged pipeline technology, the data stream real-time processing ability of MapReudce is improved, the local pipeline is optimized in stages according to system parameters, CPU is utilized sufficiently and effectively, the real-time processing ability is improved; By transforming internal and external memory data structure, read-write strategies and replacement algorithms, high concurrent read-write performance of intermediate results is optimized. Experimental results show that RT-SSP method can guarantee real-time performance and extensibility of processing data stream on massive historical data. Compared with conventional MapReduce, the speedup ratio is about 2.3. In RT-SSP method, improving single node processing ability is the foundation of cluster extensibility, so this paper mainly solves the problem of local optimization. The key issue of guaranteeing RT-SSP extensibility–load balancing (including data distribution of intermediate results on heterogeneous working nodes and dynamic load scheduling) is the research focus of next step.

References

1. Cao, Y.: Monitoring large-scale microblog on GPUs. J. Comput. Inf. Syst. **10**(15), 6493–6500 (2014)
2. Cao, Y., Wang, H.: The key optimal parallel technologies of processing large-scale micro-blog data on GPUs. J. Comput. Inf. Syst. **10**(18), 7731–7738 (2014)
3. Abadi, D.J., Ahmad, Y., Balazinska, M., et al.: The design of the Borealis stream processing engine. In: Proceedings of the 2nd Biennial Conference on Innovative Data Systems Research (CIDR2005), pp. 277–289. Asilomar, USA (2005)
4. Dean, J., Ghemawat, S.: MapReduce: simplified data processing on large clusters. ACM Commun. **51**(1), 107–113 (2008)
5. Ranger, C., Raghuraman, R., Penmetsa, A., Bradski, G., Kozyrakis, C.: Evaluating MapReduce for multi-core, multiprocessor systems. In: Proceedings of the 13th International Conference on High-Performance Computer Architecture (HPCA2007), Phoenix, USA, pp. 13–24 (2007)
6. Qi, K., Han, Y., Zhao, Z., Ma, Q.: Real-time data stream processing and key techniques oriented to large-scalr sensor data. Comput. Integr. Manuf. Syst. **19**(3), 641–653 (2013)
7. Condie, T., Conway, N., Alvaro, P., Helerstein, J.M., Elmeleegy, K., Sears, R.: MapReduce online. In: Proceedings of the 7th USENIX Symposium on Networked Systems Design, Implementation (NSDI2010), San Jose, USA, pp. 313–328 (2010)
8. Neumeyer, L., Robbins, L., Nair, A., Kesari, A.: S4: distributed stream computing platform. In: Proceedings of the 10th IEEE International Conference on Data Mining Workshops (ICDMW2010), Sydney, Australia, pp. 170–177 (2010)
9. Chang, F., Dean, J., Ghemawat, S., et al.: Bigtable: a distributed storage system for structured data. In: Proceedings of the 7th Symposium on Operating Systems Design and Implementation (OSDI2006). Seattle, USA, pp. 205–218 (2006)
10. Lubomir, F.B., Show, A.C.: Operation System Principles. Prentice Hall, New Jersey (2003)
11. DeCandia, G., Hastorun, D., Jampani, M., et al.: Dynamo: amazon's highly available key-value store. In: Proceedings of the 21st ACM Symposium on Operating Systems Principles (SOSP2007). Stevenson, USA, pp. 205–220 (2007)
12. Qi, K., Han, Y., Zhao, Z., Fang, J.: MapReduce intermediate result cache for concurrent data stream processing. J. Comput. Res. Dev. **50**(1), 111–121 (2013)

An Efficient Index Method
for Multi-Dimensional Query in Cloud
Environment

Youzhong Ma[1,2]($\boxtimes$), Xiaofeng Meng[2], Shaoya Wang[3], Weisong Hu[3], Xu Han[2],
and Yu Zhang[2]

[1] School of Information and Technology, Luoyang Normal University, Luoyang, China
{mayouzhong,xfmeng,hanxumelody,zhangyu1990}@ruc.edu.cn
[2] School of Information, Renmin University of China, Beijing, China
[3] NEC Labs China, Beijing, China
{wang_shaoya,hu_weisong}@nec.cn

Abstract. The explosion of the data in many applications brings big
challenges to the traditional relational database management systems,
they are in trouble with the scalability when deal with very large volume
data. The cloud-based databases provide a promising approach to manage
massive data because of their native good scalability, fault tolerance and
high availability, while they can not provide efficient multi-dimensional
queries processing on the non-rowkeys. In real applications, many queries
are focused on many attributes, at the same time we can not predicate
all the query requirements and the query requirements always changes. In
this paper we proposed an efficient index solution layered on the key-value
store that can deal with multi-dimensional queries efficiently on large scale
data in cloud environment, and the solution can support adding new index
dynamically for the new query requirements. We implemented a prototype based on HBase and performed comprehensive experiments to test
the scalability and efficiency of our proposed solution.

Keywords: Multi-dimensional index · Cloud computing · HBase

1 Introduction

The explosion of the data in many applications brings big challenges to the
traditional relational database management system (RDBMS), the RDBMS can
provide efficient multi-dimensional query processing because of its lots of index
structures, such as KD trees [1], R-trees [2] et al., but RDBMS has big trouble
with the scalability when deal with massive data. The cloud-based databases
provide an effective way to deal with the big data because of its native good
scalability, fault tolerance and high availability, but the cloud-based databases,
such as Bigtable [3] and HBase, can only support efficient query on the rowkey,
they can not provide efficient multi-dimensional queries on the non-rowkeys.
This shortcoming limits the widespread usage of the cloud-based databases in
many applications.

© Springer International Publishing Switzerland 2015
W. Qiang et al. (Eds.): CloudCom-Asia 2015, LNCS 9106, pp. 307–318, 2015.
DOI: 10.1007/978-3-319-28430-9_23

With the development of GPS technology and widely spread of smart phones, location based service(LBS) and location based social network(LBSN) have been used by many users. LBS providers can collect users' location information via their smart phones, based on these information, LBS providers can offer many interesting service for the users, e.g. recommending nearest Peking Duck restaurant. The users can also find the nearby friends through LBSN. In order to conduct the above services, the system needs to support multi-dimensional query, such as on time, location and other attributes. In addition, in many other applications, the queries are focused on multiple attributes, such as e-commerce, the internet of things applications. At the same time we can not predicate all the query requirements and the query requirements always changes. In this paper we proposed an efficient index solution layered on the key-value store that can deal with multi-dimensional queries efficiently on large scale data in cloud environment, and the solution can support adding new index dynamically for the new query requirements. We implemented a prototype based on HBase and performed comprehensive experiments to test the scalability and efficiency of our proposed solution. The main contributions of the paper are as follows:

1. we propose an efficient index solution layered on the key-value store that can deal with multi-dimensional queries efficiently on large scale data in cloud environment, and the solution can support adding new index dynamically for the new query requirements.
2. We propose a Region Split Tree as the global index to reorganize the regions based on the selected key attributes, so we can locate the related regions through the Region Split Tree for a given query.
3. We develop a prototype system based on HBase and perform comprehensive experiments to test the scalability and efficiency of our proposed index solution.

The organization of the paper is as the follows: Sect. 2 describes the related research works about the index techniques for cloud data management; Sect. 3 gives the system overview of our proposed index solution; In Sects. 4 and 5, we introduce the details of global index and local index respectively; Sect. 6 mainly gives the detailed procedure of the range query algorithm; In Sect. 7, we perform comprehensive experiments to test our proposed index solution; We conclude this paper in Sect. 8.

2   Related Works

Many research works have been done to study the index techniques on the cloud data management, they can be divided into many different categories according to their index techniques:double-level index framework, distributed index, bitmap-based index, Hadoop framework based index, index solution based on key-value stores and other index techniques specialized in processing some kind of data type.

Wu et al. [4] are the first to explore the index techniques for the cloud data management, they firstly proposed a double-level index framework in the cloud environment. The double-level index framework includes two parts: global index and local index. In cloud environment, the data is always stored at different storage nodes in a distributed way, each local index is built for the data at every storage node, and the global index is built based on the local index. In order to improve the query efficiency and eliminate the bottleneck of the centralized index paradigm, the computer nodes are organized into overlay networks such as CAN and Chord.

Several index approaches have been proposed based on the double-level index framework. [5] is one kind of the double-level index solutions. [5] builds up one B$^+$-tree at each storage node, then some index nodes of each B$^+$-tree are selected based on a cost model, these index nodes are reorganized into the global index using BATON overlay network. But [5] can only support point query or range query on single attribute, can not deal with multi-dimensional queries. Wang et al. [6] and Ding et al. [7] proposed other different index solutions respectively to support multi-dimensional query for the cloud data management. Wang et al. [6] built one R-tree to index the local data on each compute node, and organized the compute nodes into a CAN overlay network, the global index was constructed by selecting portion of the local R-tree index nodes to publish into the CAN overlay network. Ding et al. [7] used MX-CIF quad tree as the local index and Chord overlay network as the global index. Efficient B-tree, Wang et al. [6] and Ding et al. [7] all use P2P overlay network to organize the global index, this scheme has good scalability, but it needs additional network cost when deal with a query. Zhang et al. [8] and A-tree [9] both adopt the centralized index scheme at the global index. Zhang et al. [8] also adopted the local index plus global index structure, it used the K-d tree for local data, and in the global index level he adopted the centralized index scheme by using R-tree to organize the portion of the local K-d tree nodes.

IHBase, THBase, CCIndex [10] and MD-HBase [11] proposed some index solutions based on the key-value store. IHBase [12] and ITHBase [13] are two open source projects that provide transactional and indexing extension for hbase. CCIndex [10] is a kind of secondary index solutions based on Key-value store proposed by Zhou et al., in [10], one secondary index table was built for each indexed column. And in order to reduce the random read, the detailed information of each record was pushed into the secondary index table, so that the random read can be changed into sequential read. The author also proposed some optimization methods to support multi-dimensional query based on the several secondary indexes. CCIndex is easily to be implemented, but it has several drawbacks. Firstly it needs much more additional storage space when there are many indexed columns; secondly CCIndex does not support adding or removing index after the table was created. MD-HBase, as a scalable multi-dimensional data infrastructure, was proposed in Shohi et al. [11]. In [11], the author transformed the data from multi-dimensional space into one dimension by using linearization techniques such as Z-ordering and used the z-order value as the rowkey. In order to reduce the false positive scan during the query, the author divided the space

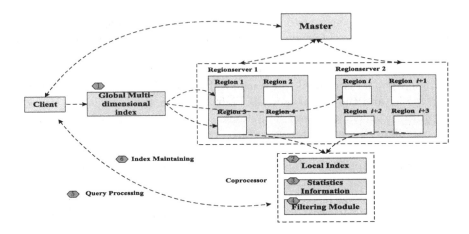

Fig. 1. System overview

into subspaces by using K-d tree and Quad-tree, and then constructed the index layer using the longest common prefix naming scheme.

3   System Overview

In order to provide efficient multi-dimensional query on large scale data and deal with the upcoming new query requirements on other attributes, we propose a hybrid index solution layered on the key-value store(HBase). The overview of the system is depicted in Fig. 1. There are mainly five components in our solution: global index, local index, statistical information, query processing and index maintaining. The global index is always a multi-dimensional index and mainly responsible for the selected key attributes; the local index is built on the non-key attributes or the new attributes for each region; statistical information module is used to collect the statistical information of the data in each region; query processing module is responsible for executing the queries based on the above index and statistical information; lastly, the index maintaining module is used to add new index on other attributes.

4   Global Index

It is well known that the query performance of the multi-dimensional index (R-tree) decreases dramatically as the dimensionality increases. Especially when the dimensionality is very high, the performance of the multi-dimensional query will be almost the same with that of full scanning the whole data. So in our solution we just create multi-dimensional index for the selected key attributes, not for all the attributes. The aim of the multi-dimensional index is to divide the data into several disjoint partitions on the selected attributes and each partition will be stored in a region in HBase, so we propose a Region Split Tree as the

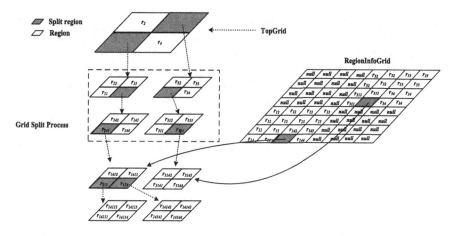

Fig. 2. Global index: region split tree

global multi-dimensional index. Figure 2 displays the overview of the Region Split Tree.

Actually Region Split Tree is a hybrid index by combining gird index and tree index. The basic idea of the grid index is to divide the k-dimensional space based on a orthogonal grid, the whole space is divided into many k-dimensional rectangle subspaces, and these rectangle subspaces are called grid directory. The strength of the grid index is that it can find the final results through limited times of access to the external storage. The main problem of the grid index is the storage of the grid directory, when the dimensionality is very high, the grid directory will be very large and the split will add many new grid directory items. So in our solution we make some modifications based on the original grid index.

TopGrid. Region Split Tree has two entrances: TopGrid and RegionInfoGrid. TopGrid is a coarse grained grid that is used to pre-split the space into partitions, each partition corresponds to a region of the HBase and the data of each partition is stored in the region. At the beginning the information that is stored in the TopGrid is the region names, as the data increases, if the number of the records in a specific partition exceeds a predefined threshold, we need to split the partition into several new partitions, at the same time we update the information that is stored in the TopGrid using the address of the new partitions. Finally all the regions can be indexed using the Region Split Tree. The granularity of the TopGird depends on the distribution of the original data, the data volume and other factors.

RegionInfoGird. When the data is skewed, the depth will be very high, it will cost too much time to locate the related regions by traversing the whole Region Split Tree from the TopGird to the bottom. In order to solve this problem, we proposed RegionInfoGrid that is a fine grained grid compared with TopGrid. RegionInfoGird is mainly used to store the information of the regions that lie at a given level (L) of the Region Split Tree. The names of the regions from $Level_2$

to $Level_L$ can be stored in the RegionInfoGrid. At $Level_L$, if a sub-grid is split again, then the address of the new sub-grids will be stored in the RegionInfoGrid. The granularity of the RegionInfoGird is the same as that of the Grid ar $Level_L$.

The information of RegionInfoGrid can be stored in main memory, If too large, we can store all the grids information into HBase table. We create an index table to store the grid directory items, the rowkey of the index table is the combination of the y-coordinate and x-coordinate, the value corresponding to each rowkey is the region name that the subspace belongs to. When we execute a query, if the query contains the selected attributes, we can get the names of the related regions by accessing the index table. Because each region may correspond to several different sub grids, we need to filter the duplicated region names.

Region Split Procedure. Given the TopGrid, we need to record the number of the records in each sub-grid, if it is over a threshold, we need to divide the sub-gird again, so on and so forth. The number of the new partitions the sub-gird is divided into must be suitable, not too big or too small. Supposing the dimensionality of the key-attributes is n, the split number is N, then:

$$N = 2^k; \quad k = \begin{cases} 2n, \ n = 2 \\ n, \ n > 2 \end{cases}$$

During the procedure of the gird splitting, the information stored in RegionInfoGrid needs to be updated accordingly.

Rowkey Formulation Scheme. Because the data in HBase is organized into different regions based on their rowkeys, and the rowkey is in each region is sequential, we need to design a suitable key formulation scheme. When we divide the space using Region Split Tree, we have to make sure that the data that is close in the original space is likely to be stored in the same region. So we use the z-order value as the rowkey of each record, such coding scheme can make sure that the rowkey is continuous in each region.

5   Local Index

In addition to the selected key attributes on which we build global multi-dimensional index, we have to consider other non-key attributes for satisfying the query requirements. We plan to create local index for the non-key attributes, and the local index is built for each region. The structure of the local index can be selected according to the characteristics of the attributes, we can select R-tree for those attributes which are always used together, if one attribute is always used dependently, we can use B-tree for such attribute. We have two kinds of solutions for the local index: real time processing and batch processing.

5.1   Real-Time Processing

In real time processing mode, we have to index each record when it is inserted into the region. In order to keep the consistency between the local index and the actual data inserted into the region, we can make use of the Coprocessor technique proposed in HBase recently. When a record is inserted into a region, it will trigger a new task that is used to insert the record into the local index through Coprocessor mechanism, the record will be inserted into the region only after the record has been inserted into the local index successfully.

5.2 Batch Processing

Although the real time processing mode can make sure that the data can be indexed timely, it will bring additional burden to the system and affect the performance of the insertion. In order to maintain the high inset throughput, we can adopt the batch processing mode. According to the batch processing mode, we don't create local index for the data when they are inserted into the region. After the data of one region becomes stable, that is to say no more data will be inserted in to this region again, we can create the local index for each region by scanning the whole data in the specific region. MapReduce paradigm can be used to speed up the batch processing procedure, map task is enough for the MapReduce job and each map task is responsible for one region.

5.3 Region and Local Index Localization

HBase always moves the regions among the region servers according to some predefined load balance strategy. While the local index is built for each region and it is stored as a file on the HDFS, at the beginning each region and its corresponding local index are at the same region server. In order to reduce the communication cost, we have to move the local index together with the region. We have two methods to choose: one is that we move the local index as long as the region has been moved; another one is that we can check the local index and regions occasionally and move the local index in batch.

6 Query Processing

In this section we mainly describe the detailed procedure of the range query processing based on our proposed index solution.

6.1 Range Query Processing

$Q(E_s, E_n)$ is a multi-dimensional query, E_s and E_n are the query conditions on the selected attributes and non-selected attributes respectively. Algorithm 1 display the detailed procedure of the multi-dimensional query. Firstly we need to decide whether the query contains the selected attributes or not, that is to say if E_s is empty, we have to send the query to all the regions, so we set the names of all the regions to the related region set S_Q (line 4), otherwise we can get the related regions through querying the global index (line 6). For each region R in the related regions set S_Q, we firstly decide whether R contains the desired results or not based on the statistical information, if not, region R can be skipped (line 9,10), otherwise we need to process R (line 11–21). When process region R, we need to decide whether it is necessary to use the local index or not, if yes, we can get the final results through the local index (line 12,13), if not, we need to scan the whole region to find the final results (line 14–19).

Algorithm 1. Range Query Processing

Input: $Q(E_s, E_n)$
$\quad\quad\quad$ E_s: conditions on selected attributes
$\quad\quad\quad$ E_n: conditions on non-selected attributes
Output: R_Q
1: $R_Q \leftarrow \emptyset$
2: $S_Q \leftarrow \emptyset$ /*Initialize the related region set to empty*/
3: **if** $(E_s == \emptyset)$ **then**
4: $\quad$ $S_Q \leftarrow$ the names of all the regions
5: **else**
6: $\quad$ $S_Q \leftarrow$ getRegions$(Q(E_s, E_n),\ RST)$
7: **end if**
8: **for** each region R in S_Q **do**
9: $\quad$ **if** (R doesn't contain the desired result) **then**
10: $\quad\quad$ continue
11: $\quad$ **else**
12: $\quad\quad$ **if** (need to query localindex) **then**
13: $\quad\quad\quad$ $R_Q \leftarrow R_Q \cup searchLocalIndex(R, E_s, E_n)$
14: $\quad\quad$ **else**
15: $\quad\quad\quad$ **for** each record r in R **do**
16: $\quad\quad\quad\quad$ **if** $r \in (E_s, E_n)$ **then**
17: $\quad\quad\quad\quad\quad$ $R_Q \leftarrow R_Q \cup r$
18: $\quad\quad\quad\quad$ **end if**
19: $\quad\quad\quad$ **end for**
20: $\quad\quad$ **end if**
21: $\quad$ **end if**
22: **end for**

6.2   Get Related Regions Through Global Index

Algorithm 2 displays how to get the related regions based on the global index: Region Split Tree, it mainly contains three steps. Firstly we get the sub-girds by querying the TopGird of the Region Split Tree, if the elements of the query result are all Regions (that is to say no region has been split), we return the result directly (line 3–6); if some regions have been split, we need to query the RegionInfoGird of Region Split Tree again (line 7–10). If the elements of the query result are all Regions, we combine the result and previous result S_Q, then retrun S_Q (line 11–13); otherwise we need to query the Region Split Tree continuously by using the entrances retrieved from the RegionInfoGird (line 14–18). Finally the related region set S_Q

is returned, the Range Query Processing Algorithm will continue to process the related regions.

Algorithm 2. Get Related Regions Through Global Index

Input: $Q(E_s, E_n)$

$\qquad$ E_s: conditions on selected attributes

$\qquad$ E_n: conditions on non-selected attributes

$\qquad$ RST: global index: Region Split Tree

Output: S_Q: the related regions set

1: $S_Q \leftarrow \emptyset$ /*Initialize the related region set to empty*/
2: $subGridSet \leftarrow \emptyset$ /*The temporal set to store the query result*/
3: $subGridSet \leftarrow getGirds(Q, RST.TopGird)$
4: **if** (*the regions in subGridSet are Region*) **then**
5: $\quad$ $S_Q \leftarrow subGridSet$
6: $\quad$ return S_Q
7: **else**
8: $\quad$ $S_Q \leftarrow Regions \ in \ subGridSet$
9: $\quad$ $subGridSet \leftarrow \emptyset$
10: $\quad$ $subGridSet \leftarrow getGirds(Q, RST.RegionInfoGird)$
11: $\quad$ **if** (*the regions in subGridSet are Region*) **then**
12: $\quad\quad$ $S_Q \leftarrow S_Q \cup subGridSet$
13: $\quad\quad$ return S_Q
14: $\quad$ **else**
15: $\quad\quad$ $S_Q \leftarrow S_Q \cup Regions \ in \ subGridSet$
16: $\quad\quad$ $S_Q \leftarrow S_Q \cup$
17: $\quad\quad$ $getRegions(Q, RST.RegionInfoGird.Entrances)$
18: $\quad$ **end if**
19: **end if**
20: return S_Q

---

7 Experiment Evaluation

In this section we will perform comprehensive experiments to test the performance of our prototype. We will compare our proposed index RegionSplitTree with other two index: UQE-Index [14] and EMINC [8], UQE-Index [14] is an Update and Query Efficient index for massive IOT data in cloud environment, EMINC [8] index refers to Efficient Multi-dimensional Index with Node Cube. The prototype is implemented based on HBase-0.94.5 and Hadoop-1.0.3 The experiments were performed on an in-house cloud platform, the cloud platform size varies from 4 to 16 nodes that are connected with 1 Gbit Ethernet switch, the configuration is: CPU: Q9650 3.00 GHz, memory: 4 GB, disk:1 TB, os: 64 bit Ubuntu 9.10 server. We mainly focus on the insert throughput, query processing performance such as rang query, point query.

The experiments are performed on two different data set: uniform distribution data set and skewed distribution data set, each data set contains 200 million records. The data sets are synthetic data, and they are generated in the following scheme: in the distribution uniform data set, each record has six attributes: time, latitude, longitude, att_1, att_2, contastantString. The values of longitude and latitude are uniformly distributed in the range [1, 10000], the time is the system time when the data is generated, ConstantString is used to tune the record size, att_1 and att_2 are two additional attributes on which we need to build local index. In the skewed distribution data set, each record has the same attributes as the uniform distribution data set, the difference is that longitude, latitude, att_1 and att_2 are skewed following *zipf* like distribution, and we set the skew factor as 0.5.

7.1   Performance of Insert Throughput

Figure 3 shows that the insert throughput of RegionSplitTree, UQE-Index and EMINC. We can see that the insert throughput of RegionSplitTree is up to 6000 rec/s, it is two times of EMINC. We also can see that UQE-Index is the best, it is two times of RegionSplitTree. The main reason is that UQE-Index can only create index on three attributes: time, latitude and longitute, while Region-SplitTree creates index on five attributes: time, latitude, longitude, att_1, att_2. So the cost of RegionSplitTree will be higher than that of UQE-Index.

7.2   Performance of Point Queries

Figures 4 and 5 show that the performance of point query for uniform data set and skewed data set. When the number of the computer nodes exceeds 8, the point query performance of RegionSplitTree is always better than that of UQE-Index for both uniform data set and skewed data set. But for the skewed data set, the performance of UQE-Index decreases as the number of the computer nodes increases, the main reason is that the data is skewed, although the computer nodes increase, the data is still inserted into some fixed computer nodes; on the other hand, communication cost will increase as the computer nodes increase. From the experiment result we can see that RegionSplitTree is more suitable for the skewed data set than UQE-Index.

7.3   Performance of Range Queries

Figure 6 shows the performance of range query for uniform data set. From the figure we can see that RegionSplitTree has the best performance when the selectivity is lower than 0.01 %, while the performance of RegionSplitTree becomes worst when the selectivity is more than 0.1 %. But Fig. 7 shows that the range query performance of RegionSplitTree is the always the best for all the selectivity on skewed data set.

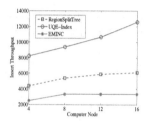 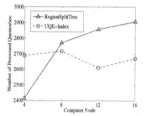 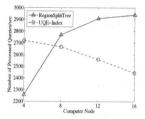

Fig. 3. Insert throughput   **Fig. 4.** Point query-uniform   **Fig. 5.** Point query-skewed

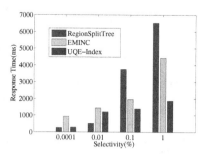 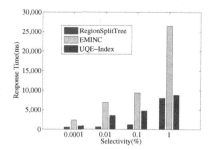

Fig. 6. Performance of range queries (uniform)

Fig. 7. Performance of range queries (skewed)

8   Conclusions and Future Work

In this paper we proposed an efficient index solution layered on the key-value store that can deal with multi-dimensional queries efficiently on large scale data in cloud environment, and the solution can support adding new index dynamically for the new query requirements. We pick up some important attributes which are often used in the queries as the key attributes and create global multi-dimensional index for the key attributes, we proposed a Region Split Tree as the global index. We build up local index for non-key attributes if needed, the ocal index can be R-tree or B-tree. Finally, we implemented a prototype based on HBase, and comprehensive experiment evaluations have been done to analyze our solution's efficiency and scalability.

In this paper we mainly focus on multi-dimensional range query, in the future we plan to extend our works to support other more complex query, such as KNN query, aggregate query. In addition to query on the simple data, we plan to deal with queries on the more complicated data, such as strings, vector data, graph data, an so on.

Acknowledgment. This work was partially supported by the grants from NEC Labs China; the Natural Science Foundation of China (No. 61070055, 91024032, 91124001); the Fundamental Research Funds for the Central Universities, and the

Research Funds of Renmin University (No. 11XNL010); National 863 High-tech Program (2012AA011001, 2013AA013204); Science and technology project of Henan Province (No. 152102210332).

References

1. Bentley, J.L.: Multidimensional binary search trees used for associative searching. Commun. ACM **18**, 509–517 (1975)
2. Guttman, A.: R-trees: A dynamic index structure for spatial searching. In: Conference on SIGMOD 1984, pp. 47–57 (1984)
3. Chang, F., Dean, J., Ghemawat, S., Hsieh, W.C., Wallach, D.A., Burrows, M., Chandra, T., Fikes, A., Gruber, R.E.: Bigtable: a distributed storage system for structured data. ACM Trans. Comput. Syst. **26**, 4 (2008)
4. Wu, S., Wu, K.L.: An indexing framework for efficient retrieval on the cloud. IEEE Data Eng. Bull. **32**, 75–82 (2009)
5. Wu, S., Jiang, D., Ooi, B.C., Wu, K.L.: Efficient B-tree based indexing for cloud data processing. In: VLDB, pp. 1207–1218 (2010)
6. Wang, J., Wu, S., Gao, H., Li, J., Ooi, B.C.: Indexing multi-dimensional data in a cloud system. In: SIGMOD Conference 2010, pp. 591–602 (2010)
7. Qiao, B., Wang, G., Chen, C., Ding, L.: An efficient quad-tree based index structure for cloud data management. In: Wang, H., Li, S., Oyama, S., Hu, X., Qian, T. (eds.) WAIM 2011. LNCS, vol. 6897, pp. 238–250. Springer, Heidelberg (2011)
8. Zhang, X., Ai, J., Wang, Z., Lu, J., Meng, X.: An efficient multi-dimensional index for cloud data management. In: CloudDB, pp. 17–24 (2009)
9. Papadopoulos, A., Katsaros, D.: A-tree: Distributed indexing of multidimensional data for cloud computing environments. In: CloudCom, pp. 407–414 (2011)
10. Zha, L., Wang, S., Xu, Z., Zou, Y., Liu, J.: CCIndex: a complemental clustering index on distributed ordered tables for multi-dimensional range queries. In: Ding, C., Shao, Z., Zheng, R. (eds.) NPC 2010. LNCS, vol. 6289, pp. 247–261. Springer, Heidelberg (2010)
11. Nishimura, S., Das, S., Agrawal, D., Abbadi, A.E.: MD-HBase: A scalable multi-dimensional data infrastructure for location aware services. In: Mobile Data Management 2011, pp. 7–16 (2011)
12. Kulbak, Y., Washusen, D.: IHBase (2010). http://github.com/ykulbak/ihbase
13. Kennedy, J., et al.: ITHBase:transactional and indexing extensions for HBase (2010). https://github.com/hbase-trx/hbase-transactional-tableindexed
14. Ma, Y., Rao, J., Hu, W., Meng, X., Han, X., Zhang, Y., Chai, Y., Liu, C.: An efficient index for massive IOT data in cloud environment. In: CIKM 2012, pp. 2129–2133 (2012)

Time Series Similarity Evaluation Based on Spearman's Correlation Coefficients and Distance Measures

Jiaqi Ye[1(✉)], Chengwei Xiao[1], Rui Máximo Esteves[2], and Chunming Rong[1]

[1] Department of Electrical Engineering and Computer Science, University of Stavanger, Stavanger, Norway
{j.ye, c.xiao}@stud.uis.no, chunming.rong@uis.no
[2] Department of Condition Monitoring, National Oilwell Varco, Stavanger, Norway
rui.esteves@nov.com

Abstract. This paper evaluates the similarity between two time series generated by two sensors manufactured by different companies, trying to provide some valuable information upon choosing sensors of different brands. Spearman correlation coefficient analysis and Euclidean distance measurement have been applied. Experiment is carried out on R. Visualization of the studied time series and results of similarity measured over time series by Spearman correlation coefficient and Euclidean distance are presented. Besides, the consistency and inconsistency in the analysis results of two measurements have been discussed in this paper.

Keywords: Time series · Similarity · Spearman correlation coefficient · Euclidean distance · R · Visualization · Consistency · Inconsistency

1 Introduction

Today, time series are ubiquitous and the interest in querying and mining such data has increased dramatically in the last decade [1]. This project aims to measure the similarity of time series generated by two sensors manufactured by different companies. The sensors are both installed in the Pump and provide the vibration information, which can be used to analyze the existing condition of the machinery, and to provide some predictive information such as what parts may be on the way to failure, and when the failure is likely to occur [2]. Thus, the performance comparison of the sensors provides information which can be used as an important reference factor for the company when choosing the sensors.

A time series is a sequence of data points, usually collected at regular intervals over a period of time. Similarity measure over time series is indispensable for time series clustering and classification systems [3]. There are many algorithms proposed to solve this problem [3–7], including Spearman correlation coefficient and various kinds of distance measure. This paper applies Spearman correlation coefficient and the most

© Springer International Publishing Switzerland 2015
W. Qiang et al. (Eds.): CloudCom-Asia 2015, LNCS 9106, pp. 319–331, 2015.
DOI: 10.1007/978-3-319-28430-9_24

famous Euclidean distance measurement on real world datasets to evaluate the similarity between two time series of two kinds of sensors. Spearman correlation coefficient is a simple and efficient way to analyze the similarity of the shape of two time series. It operates on raw data and it is based on the ranks of the data, besides it is insensitive to outliers. However, it loses information when converting data to ranks [8]. In contrast, Euclidean distance calculates root of the sum of the distance between each observation points, it is sensitive to extreme large values while caring about details at the same time.

The conclusion is made by considering the analysis results obtained from two methods, and the consistency and inconsistency in the results have been discussed. Through comprehensive analysis, this paper provides reliable supporting data for the company to make decisions while choosing sensors.

The rest of this paper is structured as follows. Section 2 presents the description of the dataset, including the sample size and the number of studied variables. In Sect. 3, we present the standard definition of Spearman correlation coefficient and Euclidean distance, and describe the definition of similarity measured by them. Section 4 reports the experiment results including Spearman correlation coefficient and Euclidean distance of examined variables. Number and percentage of variables with high Spearman correlation coefficient and small Euclidean distance are reported. The visualization of time series and their correlation are presented. Besides, the consistency and inconsistency in the analysis results made by Spearman correlation coefficient and Euclidean distance measure are discussed in Sect. 4. Lastly, the conclusion is presented, considering the analysis results of both Spearman correlation coefficient and Euclidean distance.

2   Data Description

In this paper, sensors manufactured by different companies are named sensor A and sensor B. Besides, sensor A and sensor B both have two types of sensors, which are acceleration sensor and velocity sensor. Acceleration sensor can be used to detect both low frequency and high frequency signal, while velocity sensor is ideal for sensing low and mid-frequency signal. In our experiment, the acceleration sensor can detect signal whose frequency ranges from 2 Hz up to 50 kHz, and the velocity sensor covers frequencies from 2 Hz to 1 kHz. Sensors detect and record the runtime parameters and send them back to the data center for further analysis.

The time series generated by sensor A and sensor B are sampled at the same rate and have equivalent length, thus it is able to apply Spearman correlation analysis and Euclidean distance measurement on it. There are some faults exist in the dataset, the errors act like spikes, which are abnormal large values. The spikes might be large particles being injected or pressure bursts from the pressurized air. These abnormal large values affect data scaling, and then they would affect Euclidean measurement results, so they should be deleted before performing analysis. The summary of datasets of sensor A and sensor B after pre-processing is given in Table 1.

Table 1. Summary of datasets

Dataset type	File name	Type	Num. of variables	Sample size	Name of variables
Overall dataset	overall_acc_2Hz-10kHz	Acceleration sensor	2	4105	RMS; 0-P
	overall_vel_2Hz-1kHz	Velocity sensor	12800	63	Frequency variables
Spectrum dataset	spec_acc_2Hz-50kHz	Acceleration sensor	2	4105	RMS; 0-P
	spec_vel_2Hz-1kHz	Velocity sensor	6400	63	Frequency variables

3 Methodology

3.1 Spearman Correlation Coefficient

The Spearman correlation coefficient is a nonparametric (distribution free) rank statistic for evaluating the strength of monotone association between two independent variables [8]. Compared to Pearson correlation coefficient, Spearman correlation coefficient operates on the ranks of the data rather than the raw data, and it does not require the relationship between variables is linear. Since it is based on the ranks of the data, it can well represent the similarity of the trend of the time series. However, it loses information when the data is converted to ranks. For example, Spearman correlation coefficient is insensitive to extreme data. In our case, this property would become a disadvantage since the performance of sensors in extreme situation is concerned.

Spearman correlation analysis ranks each variable separately from lowest to highest and record the difference between ranks of each data pair [9]. The sum of the square of the difference between ranks denotes the strength of the correlation between variables. If the data is strong correlated, then the sum will be small, vice versa. Besides, the magnitude of the sum is related to the significance of the correlation.

The Spearman ranks correlation coefficient can be calculated using the following equations [10]:

$$r_s = 1 - \frac{6 \sum d_i^2}{N(N^2 - 1)} \tag{1}$$

Where d_i is the difference between ranks for each data pair, and N is the number of data pairs.

3.2 Euclidean Distance Measurement

Euclidean distance is a shape based distance measures that operates on raw representation of time series [11]. It is based on directly comparing the raw values and the shape of the series. Euclidean distance has the advantage of being easy to compute and the computation cost is linear in terms of sequence length [12]. However, Euclidean distance requires that the two time series be of the same length and it does not support local time shifting.

Euclidean distance derives from L_p norms where $p = 2$. L_p distances have been widely used in many tasks related to time series analysis and mining due to their simplicity [13].

Definition 3.1 $L_p - norm$ **[11]**: *Given two time series with one dimension* $X = \{x_0, x_1, \ldots, x_{N-1}\}$ *and* $Y = \{y_0, y_1, \ldots, y_{N-1}\}$, *the* L_p *distance between X and Y is*

$$L_p - norm(X, Y) = \sqrt[p]{\sum_{i=1}^{N} (x_i - y_i)^p} \tag{2}$$

It is not hard to find out that the larger p is, the smaller the L_p distance will be, and the L_p distance would be closer to the large difference between pairs in the observations. In other words, L_p distance measurement with large p value is more sensitive to extraordinary large observations. In our case, Euclidean distance is considered to be most suitable, for it is sensitive to extreme observations while caring other details at the same time.

Figure 1 shows an example of two time series with small and large Euclidean distance respectively. According to the figure, it is obvious that the two time series are similar if their Euclidean distance is small.

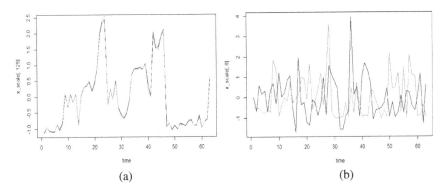

(a)   (b)

Fig. 1. Examples of time series with small (a) and large (b) Euclidean distance

Definition 3.2 Similarity Measured by Distance Function [1]: *Given two time series X and Y, along with a Euclidean distance function Euc, X is similar to Y if*

$$Euc(x, y) < \varepsilon \tag{3}$$

Where ε is a predefined threshold.

For most applications, the threshold definition is left to the user to choose a value to be applied to queries [14]. Were threshold chosen too high, there would be a risk of not retrieving enough results or even no results. On the other hand, were threshold chosen too low, many retrieved items would be irrelevant.

4   Experiment and Results

This paper applies both Spearman correlation coefficient and Euclidean distance measurement to evaluate the similarity between two time series generated by two sensors manufactured by different companies. The experiment is implemented in R, and the experiment results are presented in the order of dataset files.

4.1   Overall Dataset

- **overall_acc_2Hz-10kHz (Acceleration sensor)**

Summary of file "overall_acc_2Hz-10kHz" is listed in the Table 2, each variable (RMS and 0-P) has 4105 observations. The Spearman correlation coefficients of RMS and 0-P are 0.9335 and 0.7926, which are quite high in such a large number of observations. The average Euclidean distance of RMS and 0-P are small, which are 0.002 and 0.003 respectively. Figure 2 shows the visualization of the two time series, the red points and black points represent observations of sensor A and sensor B. From Fig. 2(a) and (b) we can see that the two sensors' records are similar during the sampling period. Figure 2(c) and (d) shows correlation between two time series of two sensors, the relationship between them is monotonous and approximately linear, which again shows the high similarity between sensor A and sensor B's time series of RMS and 0-P.

Table 2. Summary of file "overall_acc_2Hz-10kHz"

File name	Variables	Num. of observations	Data range (scaled)	Spearman correlation coefficient	Average Euclidean distance
overall_acc_2Hz-10kHz	RMS	4105	$-0.62 \sim 5.12$	0.9335	0.002
	0-P	4105	$-0.57 \sim 3.40$	0.7926	0.003

- **overall_vel_2Hz-1kHz (Velocity sensor)**

Similarly, Table 3 describes the summary of file "overall_acc_2Hz-10kHz". The corresponding Spearman correlation coefficients of RMS and 0-P are 0.9689 and 0.9347, which are higher than those of acceleration sensors. However, the average Euclidean distance of RMS and 0-P are 0.02 and 0.03 respectively, which are much higher than that of acceleration sensors.

The Spearman correlation analysis results and Euclidean distance seems inconsistent, this can be explained by Fig. 3. From Fig. 3(a) and (c), it is not hard to find that at some time domain, sensor A and sensor B behave quite different. For example, around time index 1000, sensor B is much more "active" than sensor A. This difference can also be seen from figure (c) and (d), there are some extreme large observations at which two sensors behave differently. The inconsistence between the Spearman correlation coefficient and Euclidean distance is mainly because the two sensors generally have similar

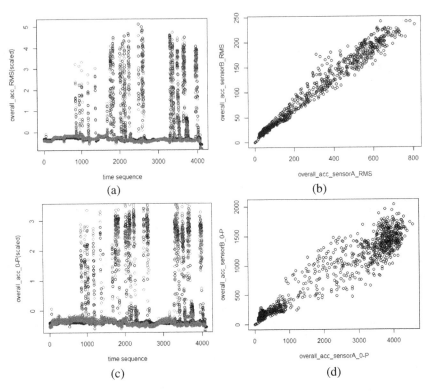

Fig. 2. Visualization of Time series generated by acceleration sensor A and sensor B and their correlation; (a) Time series of variable RMS, black point represents time series of sensor A, and red point stands for time series of sensor B; (b) Correlation of RMS time series generated by sensor A and sensor B; (c) Time series of variable 0-P, black point represents time series of sensor A, and red point stands for time series of sensor B; (d) Correlation of 0-P time series generated by sensor A and sensor B (Color figure online).

trend while reacting quite differently at some time domain. The similarity evaluation between them depends on how users care about the extreme large values.

4.2   Spectrum Dataset

The spectrum dataset contains frequency variables and provides valuable information which arouses much attention. The summary of two spectrum datasets is listed in Table 4. In this section, in order to examine the performance in an all-around way, this

Table 3. Summary of file "overall_vel_2Hz-1kHz"

File name	Variables	Num. of observations	Data range (scaled)	Spearman correlation coefficient	Average Euclidean distance
overall_vel_2Hz-1kHz	RMS	4105	$-0.13 \sim 38.72$	0.9689	0.02
	0-P	4105	$-0.13 \sim 38.37$	0.9347	0.02

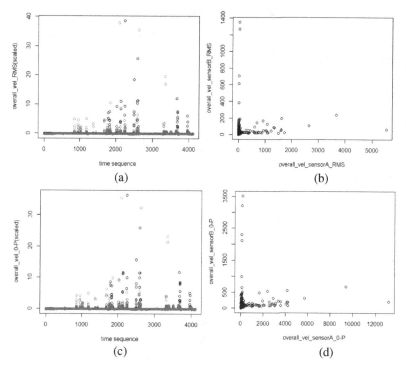

Fig. 3. Visualization of Time series generated by velocity sensor A and sensor B and their correlation; (a) Time series of variable RMS, black point represents time series of sensor A, and red point stands for time series of sensor B; (b) Correlation of RMS time series generated by sensor A and sensor B; (c) Time series of variable 0-P, black point represents time series of sensor A, and red point stands for time series of sensor B; (d) Correlation of 0-P time series generated by sensor A and sensor B; Spectrum dataset (Color figure online)

paper performs two kinds of experiment. The first experiment is to compare the time series of every frequency pairs. The other experiment compares frequency spectrum sampled at the same time, in this sense, the observations is the frequency variables. The first experiment is aimed to examine the performance of sensors at specified frequency, while the second experiment can reflect overall property of the spectrum.

- **spec_acc_2Hz-50kHz**

Table 5 shows the number of frequency variables and the corresponding percentage with high Spearman correlation coefficient in the spectrum dataset of acceleration sensors. Just about 5.4 % of frequency variables that has Spearman correlation coefficient are higher than 0.7, which shows weak Spearman correlation between the time series of sensor A and sensor B.

Table 6 describes the average Euclidean distance of two time series of frequency variables, the threshold for distance similarity measurement is defined as 0.09 (average Euclidean distance). The threshold is decided through many trials and observations, of course it can be adjusted by users to meet their requirement and application. As Table 6

Table 4. Summary of spectrum datasets

File name	Num. of frequency variables	Num. of observations	Data range (scaled)
spec_acc_2Hz-50kHz	12800	63	$-3 \sim 8$
spec_vel_2Hz-1kHz	6400	63	$-3 \sim 8$

Table 5. Spearman correlation coefficients summary

Spearman coefficient (p-value < 0.05)	>0.85	>0.8	>0.75	>0.7
Num. of variables	118	233	398	690
Percentage	0.92 %	1.82 %	3.11 %	5.4 %

Table 6. Euclidean distance summary

Average Euclidean distance	<0.015	<0.03	<0.045	<0.06	<0.075	<0.09
Num. of variables	6	21	63	140	300	600
Percentage	0.047 %	0.16 %	0.49 %	1.09 %	2.23 %	4.69 %

shows, there is just about 600 frequency variables whose average Euclidean distance is below 0.09 (threshold), accounts for only 4.69 % in the dataset. This result is consistent with the result get from Spearman correlation analysis.

Figure 4 shows an example of the spectrums of two acceleration sensors sampled at the same timestamp. According to Fig. 4, the two sensors behave quite differently at some frequency domain. Sensor A (black points) is more "active" around frequency 1000 (index), while sensor B (red points) is more "active" around $3500 \sim 8500$ frequency domain. Again, Fig. 4 shows the dissimilarity between acceleration sensor A and sensor B in spectrum dataset.

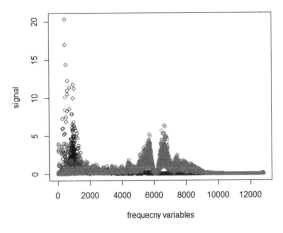

Fig. 4. Spectrum of two sensors sampled at a same time point (black points refer to observations of acceleration sensor A while red points refer to observations of sensor B) (Color figure online)

- **spec_vel_2Hz-1kHz.**

Compared with acceleration sensors, time series of velocity sensors are much more similar as Tables 7 and 8 shows. According to Table 7, the number of variables which has Spearman correlation coefficient larger than 0.7 is 6283, accounting for about 98.2 %. And the percentage of variables whose average Euclidean distance is small than 0.09 (threshold) is about 96.6 %. Both Spearman correlation analysis and Euclidean distance similarity measurement show that there is high similarity between the spectrum time series of sensor A and sensor B.

Table 7. Spearman correlation coefficient summary

Spearman coefficient (p-value < 0.05)	>0.85	>0.8	>0.75	>0.7
Num. of variables	5075	5795	6133	6283
Percentage	79.3 %	90.5 %	95.8 %	98.2 %

An example of the spectrums of two velocity sensors sampled at the same timestamp is illustrated in Fig. 5. The two velocity sensors behave almost the same through the frequency domain. Thus, it proves again that the performance of two velocity sensors of different brands is similar according to this spectrum dataset.

Table 8. Euclidean distance summary

Average Euclidean distance	<0.015	<0.03	<0.045	<0.06	<0.075	<0.09
Num. of variables	421	1858	3772	5155	5878	6180
Percentage	6.6 %	29.0 %	58.9 %	80.5 %	91.8 %	96.6 %

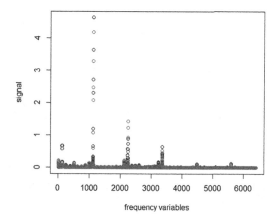

Fig. 5. Spectrum of two sensors sampled at a same time point (black points refers to velocity sensor A while red points refers to sensor B) (Color figure online)

4.3   Spearman Correlation VS Euclidean Distance

Spearman correlation performs analysis based on the ranks of data, thus it can represent the similarity of the shape of two distributions, while the Euclidean distance calculates the distance between two time series and represents the similarity according to a predefined threshold. These two methods represent the similarity of two time series from different aspects. They are both sensitive to the shape of time series. However, Spearman correlation cares only about the ranks of data while Euclidean distance cares more about the overall trend. Besides, Euclidean distance is more sensitive to the extreme data.

The general conclusion get from Spearman correlation analysis and Euclidean distance measurement is consistent, while there is some inconsistency exists if we look into the some specified variables. This paper divide the data into different types according to the results get from Spearman correlation analysis and Euclidean distance evaluation.

- Type 1: High Spearman correlation coefficient & small Euclidean distance
  This type of data has high similarity between two time series, and the similarity has been proved by both two methods.
- Type 2: High Spearman correlation coefficient & large Euclidean distance

As the issues described in the file "overall_vel_2Hz-1kHz", two sensors' different reaction to extreme data can cause this problem. Figure 6 shows an example of this kind of data. The general trend of two time series is similar while there is a big difference between two sensors observations around time index 49. This large difference causes the inconsistency between Spearman correlation coefficient and Euclidean distance analysis result.

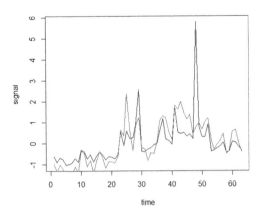

Fig. 6. An example of type 2 data (Spearman correlation coefficient = 0.95 and Average Euclidean distance = 0.10)

- Type 3: Low Spearman correlation coefficient & small Euclidean distance
  In contrast to type 2 data, type 3 data may react similarly to large "signal". However their "background noise" may fluctuate within a narrow range. Thus the ranks of data may be different while the Euclidean distance is small. Figure 7 shows an example of this kind of data.

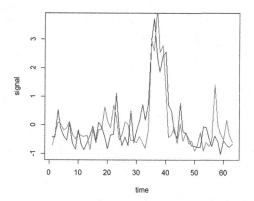

Fig. 7. An example of type 3 data (Spearman correlation coefficient = 0.35 and Average Euclidean distance = 0.08)

- Type 4: Low Spearman correlation coefficient & large Euclidean distance

This type of data reflects the variables at which two sensors have different time series, this dissimilarity is proved by both methods.

Table 9 describes the summary of four types of data in spectrum files of acceleration sensors and velocity sensors. For the acceleration sensors, the largest proportion of data (about 93.3 %) falls into type 4, which shows the dissimilarity between the performance of acceleration sensor A and sensor B. In contrast, for the velocity sensors, there is about 96.0 % data falls into type 1, which shows the high similarity between performance of velocity sensor A and sensor B.

Table 9. Summary of four data types in spectrum datasets

File name	Num. and percentage of Type 1 data	Num. and percentage of Type 2 data	Num. and percentage of Type 3 data	Num. and percentage of Type 4 data
spec_acc_2Hz-50kHz	437 (3.4 %)	253 (2.0 %)	163 (1.3 %)	11947 (93.3 %)
spec_vel_2Hz-1kHz	6143 (96.0 %)	138 (2.16 %)	37 (0.58 %)	80 (1.25 %)

5 Conclusion

This paper is aimed at comparing the performance of two sensors manufactured by different companies. This is done by evaluating the similarity of time series generated by these two sensors (sensor A and sensor B) using Spearman correlation coefficient analysis and Euclidean distance measurement. Both sensor A and sensor B have two kinds of sensors —Acceleration sensor and Velocity sensor respectively. Through experiment and analysis, it has been proven that the data generated by these acceleration sensors and velocity sensors have quite different statistical properties. For the overall variables (e.g. RMS, 0-P), time series of acceleration sensor A and sensor B has much higher similarity than those of velocity sensor A and sensor B. In contrast, for the frequency variables in spectrum dataset, it can be concluded that the performance of velocity sensor A and B is similar, while the performance of acceleration sensor A and B is dissimilar. This paper also discusses the consistency and inconsistency of analysis results of Spearman correlation analysis and Euclidean distance measurement, and divides the data into four types. These four types of data have different attributes and can be used in applications of different purposes. Further work such as applying other efficient time series similarity measurements to more complicated time series (e.g. multidimensional time series) can be done in the future.

References

1. de Jong, P.: Time series analysis. In: Predictive Modeling Applications in Actuarial Science. Predictive Modeling Techniques, vol. 1, p. 427 (2014)
2. Reimche, W., Südmersen, U., Pietsch, O., et al.: Basics of vibration monitoring for fault detection and process control. In: 3rd Pan-American Conference for Non-Destructive Testing. Rio de Janeiro, Brazil (2003)
3. Serrà, J., Arcos, J.L.: An empirical evaluation of similarity measures for time series classification. Knowl.-Based Syst. (2014)
4. Ding, H., Trajcevski, G., Scheuermann, P., et al.: Querying and mining of time series data: experimental comparison of representations and distance measures. Proc. VLDB Endowment 1(2), 1542–1552 (2008)
5. Liao, H., Xu, Z., Zeng, X.J.: Distance and similarity measures for hesitant fuzzy linguistic term sets and their application in multi-criteria decision making. Inf. Sci. 271, 125–142 (2014)
6. Gunopulos, D., Das, G.: Time series similarity measures (tutorial PM-2). In: Tutorial Notes of the Sixth ACM SIGKDD International Conference on Knowledge Discovery and Data Mining, pp. 243–307. ACM (2000)
7. Batista, G.E., Keogh, E.J., Tataw, O.M., et al.: CID: an efficient complexity-invariant distance for time series. Data Min. Knowl. Disc. 28(3), 634–669 (2014)
8. Gauthier, T.D.: Detecting trends using spearman's rank correlation coefficient. Environ. Forensics 2(4), 359–362 (2001)
9. Mudelsee, M.: Estimating Pearson's correlation coefficient with bootstrap confidence interval from serially dependent time series. Math. Geol. 35(6), 651–665 (2003)
10. Seneviratne, S.I., Donat, M.G., Mueller, B., et al.: No pause in the increase of hot temperature extremes. Nat. Clim. Change 4(3), 161–163 (2014)

11. Pearson's. Comparison of Values of Pearson's and Spearman's Correlation Coefficients... (2007)
12. Breu, H., Gil, J., Kirkpatrick, D., et al.: Linear time Euclidean distance transform algorithms. IEEE Trans. Pattern Anal. Mach. Intell. **17**(5), 529–533 (1995)
13. Mori, U., Mendiburu, A., Lozano, J.A.: Distance measures for time series in R: the TSdist package
14. Da Silva, R., Stasiu, R., Orengo, V.M., et al.: Measuring quality of similarity functions in approximate data matching. J. Informetrics **1**(1), 35–46 (2007)

Security and Privacy

Keyword Search Over Encrypted Data in Cloud Computing from Lattices in the Standard Model

Chunxiang Gu[✉], Yonghui Zheng, Fei Kang, and Dan Xin

State Key Laboratory of Mathematical Engineering and Advanced Computing,
Zhengzhou 450001, China
gcxiang5209@aliyun.com

Abstract. As cloud computing becomes popular, more and more sensitive information are being centralized into the cloud. Hence, we need some mechanism to support operations on encrypted data in many applications in cloud. Public key encryption with keyword search (PEKS) is a mechanism for searching on encrypted data. It enables user Alice to send a secret value T_w to a cloud server that will enable the server to selectively retrieve encrypted messages containing the keyword w, but learn nothing else. In this paper, we propose a PEKS scheme using lattices. Lattice-based cryptosystems are becoming an increasingly popular in the research community. Our scheme can be proven secure in the standard model with the hardness of the standard Learning With Errors problem.

Keywords: Lattice-based cryptography · Cloud computing · Keyword search · Provable secure

1 Introduction

With the fast development of cloud computing, more and more sensitive information are being centralized into the cloud, such as emails, personal health records, government documents, etc. By storing their data into the cloud, the data owners can be relieved from the burden of data storage and maintenance so as to enjoy the on-demand high quality data storage service. In many cloud computing scenarios, the individual users might want to retrieve certain specific data files that they are interested. Keyword-based search technique has been widely applied in search scenarios, such as Google search. However, due to the fact that the data owners and cloud server are not in the same trusted domain, sensitive data usually should be encrypted for data privacy and authorized accesses. What's more, for privacy, data encryption may also demands the protection of keyword since keywords usually contain important information related to the data files. This leads to the traditional plaintext search techniques useless in this scenario.

Foundation: Henan province outstanding youth science and technology innovation (134100510002); Henan province basis and research in cutting-edge technologies (142300410002); State Key Laboratory of Mathematical Engineering and Advanced Computing innovation

To securely search over encrypted data, some searchable encryption techniques have been developed [1–8]. In 2004, Boneh et al. [2] proposed the concept of public-key encryption with keyword search (PEKS). The mechanism enables one to give trapdoors for the keywords he wants the server to search for. The trapdoors can be used to test the existence of keywords in the encrypted data without compromising the security of the original data. PEKS mechanism can be widely used in many practical applications, such as encrypted emails extraction [2], encrypted and searchable audit logs [3], and encrypted files extraction in Cloud [6].

Recently, lattice-based cryptosystems are becoming an increasingly popular in the research community due to their attractive and distinguishing features: their resistance so far to cryptanalysis by quantum algorithms, their asymptotic efficiency and conceptual simplicity, and the guarantee that their random instances are as hard as the hardness of lattice problems in worst case. Many new constructions with lattices have been proposed, such as one-way functions and hash functions [9, 10], trapdoor functions [11], public-key encryption schemes [11], ID-based encryption schemes [11, 13, 14], fully homomorphic encryption schemes [15, 16], and so on. In this paper, we propose a public key encryption with keyword search scheme using lattices. The scheme can be proven secure with the hardness of the standard Learning With Errors (LWE) problem in the standard model.

The rest of this paper is organized as follows: In Sect. 2, we recall some preliminary works. In Sect. 3, we describe the details of our new schemes and their security proofs. Finally, we conclude in Sect. 4.

2 Preliminaries

2.1 Public-Key Encryption with Keyword Search

A PEKS scheme consists of four polynomial-time algorithms described as below [2]:

- KeyGen(k): Take a security parameters k, and generate a public/private key pair (pk, sk).
- PEKS(pk,w): Take a public key pk and a keyword w as input, and output a searchable encryption for w.
- Trapdoor(sk,w): Take the private key sk, and a keyword w as input, and output a trapdoor T_w for w.
- Test(pk, c,T_w). Given the public key pk, a searchable encryption c where c = PEKS(pk, w'), and the trapdoor T_w to a keyword w. Determine whether or not w = w'.

The general notion of security of PEKS scheme is indistinguishability against chosen keyword attack (IND-CKA) [2]. In this paper, we consider a weak notion of PEKS security called indistinguishability against selective keyword attack (IND-SKA). A PEKS scheme is IND-SKA security if no polynomial time adversary A has a non-negligible advantage against a challenger C in the following game.

Security Game:

Init. The adversary A selects and outputs target key words w^*.

KeyGen. C runs KeyGen algorithm to generates a key pair (pk, sk), and gives pk to A.

Phase 1. A can adaptively ask the challenger for the trapdoor T_w for any keyword w of his choice as long as $w \neq w^*$.

Challenge. The adversary decides when phase 1 ends. Then the challenger C picks a random bit $b \in \{0, 1\}$ and a random cipher-text c in the cipher-text space. If $b = 0$ it sets the challenge to $c^* = \text{PEKS}(pk, w^*)$, otherwise it sets $c^* = c$. It gives c^* as the challenge to the attacker.

Phase 2. A can adaptively ask the challenger for the trapdoor T_w for any keyword w of his choice as long as $w \neq w^*$.

Guess. Eventually, the attacker A outputs $b' \in \{0, 1\}$ and wins the game if $b = b'$.

The advantage of A in this game is defined as $|\Pr[b' = b] - \frac{1}{2}|$.

Recall that Waters [18] showed how to convert the selectively-secure identity-based encryption to an adaptively secure identity-based encryption. A similar technique can also use to convert a IND-SKA secure PEKS scheme to a IND-CKA secure one.

2.2 Lattices and Sampling Algorithms

Given n linearly independent vectors $b_1, b_2, \ldots, b_n \in R^m$, the lattice Λ generated by them is denoted $L(b_1, b_2, \ldots, b_n)$ and define as: $L(b_1, b_2, \ldots, b_n) = \{\sum_{i=1}^{n} x_i b_i | x_i \in Z\}$. The vectors $b_1, b_2, \ldots, b_n$ are called the basic of the lattice. Let $\tilde{B}$ denote its Gram-Schmidt orthogonalization of the vectors $b_1, b_2, \ldots, b_n$ taken in that order, and we let $||B||$ denote the length of the longest vector in B for the Euclidean norm.

For q prime and $A \in Z_q^{n \times m}$ and $u \in Z_q^n$, define $\Lambda_q^{\perp}(A) = \{e \in Z^m | A \cdot e = 0 \mod q\}$ and $\Lambda_q^u(A) = \{e \in Z^m | A \cdot e = u \mod q\}$. Ajtai [9] and later Alwen and Peikert [20] showed that for any prime $q \geq 2$ and $m \geq 5n\log q$, there exists a probabilistic polynomial-time algorithm TrapGen(q,n) that outputs a pair $A \in Z_q^{n \times m}$ and T_A such that A is statistically close to uniform and T_A is a basis for $\Lambda_q^{\perp}(A)$ with length $L = ||\tilde{T}_A|| \leq m\omega(\sqrt{\log m})$ with all but $n^{-\omega(1)}$ probability.

We further we review Gaussian functions used in lattice based cryptographic constructions. Let m be a positive integer and $\Lambda \in R^m$ be a m dimensional lattice. For any vector $c \in R^m$ and any positive parameter $\sigma \in R_{>0}$, we define:

$\rho_{\sigma,c}(x) = \exp(-\pi||x - c||^2/\sigma^2)$ and $\rho_{\sigma,c}(\Lambda) = \sum_{x \in \Lambda} \rho_{\sigma,c}(x)$

The discrete Gaussian distribution over Λ with center c and parameter σ is

$$\forall y \in \Lambda \quad D_{\Lambda,\sigma,c}(y) = \frac{\rho_{\sigma,c}(y)}{\rho_{\sigma,c}(\Lambda)} \tag{1}$$

For notational convenience, $\rho_{\Lambda,\sigma,0}$ and $D_{\Lambda,\sigma,0}$ are abbreviated as ρ_σ and $D_{\Lambda,\sigma}$. Gentry et al. [11] construct the following algorithm for sampling from the didcrete Gaussian $D_{\Lambda,\sigma,c}$, given a basis T_A for the m-dimensional lattice Λ with $\sigma \geq ||\tilde{T}_A||\omega(\sqrt{\log m})$. Specialized to the case of random lattices, they show an algorithm:

SamplePre (A, T_A, u, σ): On input a matrix $A \in Z_q^{n \times m}$ with 'short' trapdoor basis T_A for $\Lambda_q^\perp(A)$, a target image $u \in Z_q^n$ and a Gaussian parameter $\sigma \geq ||\tilde{T}_A||\omega(\sqrt{\log m})$, outputs a sample $e \in \Lambda_q^u(A)$ from a distribution that is within negligible statistical distance of $D_{\Lambda_q^u(A),\sigma}$.

Let A, B be matrices in $Z_q^{m \times m}$, R be a matrix in $\{-1, 1\}^{m \times m}$, $F = A|AR + B$. Agrawal et al. [13] proposed two polynomial time sample algorithms to sample short vectors in $\Lambda_q^u(F)$ for some $u \in Z_q^n$ using a trapdoor for $\Lambda_q^u(A)$ or $\Lambda_q^u(B)$:

SampleLeft (A, M, T_A, u, σ): On input a matrix $A \in Z_q^{n \times m}$ with 'short' trapdoor basis T_A for $\Lambda_q^\perp(A)$, $M \in Z_q^{n \times m}$, a target image $u \in Z_q^n$ and a Gaussian parameter $\sigma \geq ||\tilde{T}_A|| \cdot \omega(\sqrt{\log 2m})$, outputs a sample $e \in \Lambda_q^u(A|M)$ from a distribution that is within negligible statistical distance of $D_{\Lambda_q^u(A|M),\sigma}$.

SampleRight $(A, B, R, T_B, u, \sigma)$: On input matrices $A, B \in Z_q^{n \times m}$ with 'short' trapdoor basis T_B for $\Lambda_q^\perp(B)$, a matrix $R \in \{-1, 1\}^{m \times m}$, a target image $u \in Z_q^n$ and a Gaussian parameter $\sigma \geq ||\tilde{T}_B|| \cdot s_R \cdot \omega(\sqrt{\log m})$ where $s_R = ||R||$, outputs a sample $e \in \Lambda_q^u(A|AR + B)$ from a distribution that is within negligible statistical distance of $D_{\Lambda_q^u(A|AR+B),\sigma}$.

2.3 The LWE Hardness Assumption

Security of all our construction reduces to the LWE problem, a classic hard problem on lattices defined by Regev [12].

Consider a prime q, a positive integer n, some probability distribution χ over Z_q, and a vector $s \in Z_q^n$, define $A_{s,\chi}$ as the distribution on $Z_q^n \times Z_q$ of the variable of $(a, a^T \cdot s + x)$ where $a \leftarrow Z_q^n$ is uniform and $x \leftarrow \chi$ are independent.

The Decisional LWE Problem [12]. The goal of the decisional (Z_q, n, χ)-LWE problem is to distinguish between the distribution $A_{s,\chi}$ some uniform secret $s \leftarrow Z_q^n$ and uniform distribution on $Z_q^n \times Z_q$(via oracle access to the given distribution). In other words, if LWE problem is hard, then the collection of distribution is pseudorandom. Regev has showed that for certain noise distributions χ, the LWE problem is as hard as the worst-case SIVP and GapsSVP under a quantum reduction.

Propositon 2 [12]. For an $\alpha \in (0, 1)$ and a prime $q > 2\sqrt{n}/\alpha$, let $\bar{\psi}_\alpha$ denote the distribution over Z_q of the random variable $\left\lfloor qX + \frac{1}{2} \right\rfloor \bmod q$ where X is a normal random variable with mean 0 and standard deviation $\alpha/\sqrt{2\pi}$. Then, if there exists an efficient, possibly quantum, algorithm for deciding the (Z_q, n, χ)-LWE problem, there exists a quantum poly-time algorithm for approximating the SIVP and GapSVP problems, to within $\tilde{O}(n/\alpha)$ factors in the l_2 norm, in the worst case.

3   A PEKS Scheme from Lattice

Our construction uses an encoding function $H: Z_q^n \rightarrow Z_q^{n \times n}$ to map identities in Z_q^n to matrices in $Z_q^{n \times n}$. The function should be an encoding with full-rank differences (FRD). That is, for all distinct $u, v \in Z_q^n$, the matrix $H(u) - H(v) \in Z_q^{n \times n}$ is full rank, and H is computable in polynomial time (in $n \log q$). Agrawal et al. [13] has presented an efficient and explicit FRD construction.

3.1   The Construction

Let n, m, k be positive integers, q be prime, with $q \geq 2$ and $m \geq 5n \log q$, $H: Z_q^n \rightarrow Z_q^{n \times n}$ be a FRD function. The scheme is described as follows:

- KeyGen(1^n): On input a security parameter n, invoke TrapGen(q,n) to generate a uniformly random matrix $A \in Z_q^{n \times m}$ along with a short basis T_A for $\Lambda_q^\perp(A)$. Select uniformly random matrices $A_1, B \in Z_q^{n \times m}$, and $U = [u_1, \ldots u_k] \in Z_q^{n \times k}$. Output the public key $pk = (A, A_1, B, U)$ and secret key $sk = T_A$.
- PEKS(pk, w): On input public key pk and keyword $w \in Z_q^n$, do:
  1. Compute $F_w = A|A_1 + H(w) \cdot B$.
  2. Select a random $s \in Z_q^n$ and a random matrix $R \in \{-1, 1\}^{m \times m}$ uniformly.
  3.
     Compute $c_0 = U^T \cdot s + x \in Z_q^k$, $c_1 = F_w^T \cdot s + \begin{bmatrix} y \\ R^T \cdot y \end{bmatrix} \in Z_q^{2m}$,   where
     $x^T = (x_1, \ldots, x_k)$ with $x_i \leftarrow \chi$ $(i = 1, \ldots, k)$, and $y \leftarrow \chi^m$ are noise vectors. Output the cipher-text (c_0, c_1).
- Trapdoor(pk, sk, w): Compute $M_w = A_1 + H(w) \cdot B$, for $i = 1, \ldots, k$, choose a $e_i \in Z_q^{2m}$ satisfying $(A|M_w) \cdot e_i = u_i$ by $e_i \leftarrow$ SampleLeft$(A, M_w, T_A, u_i, \sigma)$, where σ is a Gaussian parameter. Output $T_w = [e_1, \ldots, e_k]$ as the trapdoor.
- Test($(c_0, c_1), pk, T_w$): Let $c_0^T = (c_{0,1}, \ldots, c_{0,k})$, for $i = 1, \ldots, k$, compute
- $b_i = c_{0,i} - e_i^T \cdot c_1 \in Z_q$. If for all $i = 1, \ldots, k, |b_i| < \lfloor q/4 \rfloor$, output 1 ("yes"); otherwise, output 0 ("no").

We should note that the construction can only be proven secure stands on a weak IND-SKA security notion. However, with a similar technique in [13, 19], the scheme can be convert to an adaptively secure PEKS scheme with the cost of larger size of keys and cipher-texts.

3.2 Consistency

The consistency of the scheme requires that for any keyword w w and w', Test(PEKS$(pk, w), pk, T_{w'}$), output 1 if $w = w'$, and 0 otherwise.

Suppose $(c_0, c_1) = $ PEKS(pk, w) and $T_{w'} = [e_1, \dots, e_k]$, then

$$b_i = c_{0,i} - e_i^T c_1 = u_i^T \cdot s + x_i - e_i^T (F_w^T \cdot s + \begin{bmatrix} y \\ R^T y \end{bmatrix}) \tag{2}$$

It is easy to see that if $w = w'$, then $|b_i| = \left| x_i - e_i^T \cdot \begin{bmatrix} y \\ R^T y \end{bmatrix} \right|$, just as discussion in

[13], with parameters $q = \Omega(\sigma \cdot m^{3/2})$, $\alpha < [\sigma \cdot m \cdot \omega(\sqrt{\log m})]^{-1}$, and $\chi = \bar{\Psi}_\alpha$, with overwhelming probability we have $b_i < q/4$.

However, if $w \neq w'$, with the randomness of u_i and e_i, $u_i^T - (F_w \cdot e)^T$ can be seen as a random vector in Z_q^n. That is $|b|$ is a random in $[0, q]$ (over the random choices of u_i and e_i). Hence, for each $i = 1, \dots, k$ the probability of $|b_i| < \lfloor q/4 \rfloor$ is at most $\frac{1}{4}$. That is, the Test algorithm output 0 with probability $1 - 2^{-2k}$. When k is large enough, we can ensure the consistency of the scheme.

3.3 Security Proof in the Standard Model

For the simpleness, we only consider the parameter $k = 1$ in the proof. The proof for $k > 1$ is almost the same.

Proposition 3. The PEKS scheme is IND-SKA secure provided that (Z_q, n, χ)-LWE assumption holds.

Proof. The proof shows that suppose there is a IND-SKA adversary F that have non-negligible advantage ε in attacking the scheme, then we can construct a polynomial-time algorithm S that can solve the LWE problem with the same probability.

With the adversary F, we construct the algorithm S as follows:

S requests from LWE oracle Ψ and receives, for each $i = 0, \dots, m$, a fresh pair $(u_i, v_i) \in Z_q^n \times Z_q$. S chooses a random matrix $R^* \in \{-1, 1\}^{m \times m}$ uniformly, and invokes TrapGen(q,n) to generate a uniformly random matrix $B \in Z_q^{n \times m}$ along with a short basis T_B for $\Lambda_q^\perp(B)$.

Targeting. The adversary F announces to S the keywords w^* that it intends to attack.

KeyGen. S assembles the random matrix $A \in Z_q^{n \times m}$ from m of given LWE samples by letting the i _th column of A be the vector u_i for all $i = 1, \dots, m$, and sets $A_1 = A \cdot R^* - H(w^*)B$, $U = [u_0]$, and gives the public key $pk = (A, A_1, B, U)$ to F.

Phase 1. S answers the queries of F as follows:

Trapdoor(.) query: For input $w \in Z_q^n$ with $w \neq w^*$, $H(w) - H(w^*) \in Z_q^{n \times n}$ is full rank $(H:Z_q^n \to Z_q^{n \times n}$ be a FRD function). Hence T_B is also a trapdoor for $\Lambda_q^{\perp}(B')$, where $B' = (H(w) - H(w^*)) \cdot B$. Therefore, S can respond the query by running $e \leftarrow$ SampleRight$(A, (H(w) - H(w^*))B, R^*, T_B, u_0, \sigma) \in Z_q^{2m}$, and sending $T_w = e$ as the trapdoor to the adversary.

Challenge. When Phase 1 ends, S retrieves $v_0, \ldots, v_m \in Z_q$ from the LWE instance, and sets $v^* = [v_1, \ldots, v_m]^T \in Z_q^m$, $c_0^* = v_0 \in Z_q$, $c_1^* = \begin{bmatrix} v^* \\ (R^*)^T v^* \end{bmatrix} \in Z_q^{2m}$. S chooses a random $b \in \{0, 1\}$. If $b = 0$ it responds with the challenge cipher-text (c_0^*, c_1^*) to the adversary, otherwise it chooses a random $(c_0, c_1) \in Z_q \times Z_q^{2m}$, responds with (c_0, c_1) to the adversary. (We should note that $F_{w^*} = A|A \cdot R^* - H(w^*)B + H(w^*) \cdot B = A|A \cdot R^*$. And when Ψ is a pseudo-random LWE oracle, then $v^* = A^T \cdot s + y$ for some noise vector $y \in \chi^m$. Therefore

$$c_1^* = \begin{bmatrix} v^* \\ (R^*)^T v^* \end{bmatrix} = \begin{bmatrix} A^T \cdot s + y \\ (A \cdot R^*)^T s + (R^*)^T y \end{bmatrix}$$
$$= (F_{w^*})^T \cdot s + \begin{bmatrix} y \\ (R^*)^T y \end{bmatrix}, \tag{3}$$

and $c_0^* = u_0^T \cdot s + x$ for some random noise values $x \in \chi$. In this case (c_0^*, c_1^*) is a valid encryption for w^*. When Ψ is a random oracle then (c_0^*, c_1^*) is uniform in $Z_q^{2m} \times Z_q$.)

Phase 2. S answers queries of F as follows: **Trapdoor(.)** query: For input $w \in Z_q^n$ with $w \neq w^*$, $H(w) - H(w^*) \in Z_q^{n \times n}$ is full rank $(H:Z_q^n \to Z_q^{n \times n}$ be a FRD function). Hence T_B is also a trapdoor for $\Lambda_q^{\perp}(B')$, where $B' = (H(w) - H(w^*)) \cdot B$. Therefore, S can respond the query by running

$$e \leftarrow \text{SampleRight}(A, (H(w) - H(w^*))B, R^*, T_B, u_0, \sigma) \in Z_q^{2m} \tag{4}$$

and sending $T_w = e$ as the trapdoor to the adversary.

Guess. Eventually F outputs $b' \in \{0, 1\}$.

Finally, S output F's guess as the answer to his own the LWE challenge. The distribution of the public parameters is identical to its distribution in the real system as are responses to private key queries. The challenge cipher-text is distributed either as in the real system or is independently random in $Z_q \times Z_q^{2m}$. Hence, the advantage of S in solving LWE is the same as F's advantage in attacking the scheme. This completes the security proof..

4   Conclusion

Lattice-based cryptosystems are becoming an increasingly popular in the research community. In this paper, we propose a PEKS scheme using lattices. To the best of our knowledge, this is the first lattices-based PEKS scheme. The scheme can be proven secure with the hardness of the standard LWE problem in the random oracle model. We should note that although our security proof stands on a weak IND-SKA security notion, our scheme can be convert to an adaptively secure PEKS scheme with a similar technique in [13, 19], with the cost of larger size of keys and cipher-texts.

References

1. O'Neill, A., Boldyreva, A., Bellare, M.: Deterministic and efficiently searchable encryption. In: Menezes, A. (ed.) CRYPTO 2007. LNCS, vol. 4622, pp. 535–552. Springer, Heidelberg (2007)
2. Boneh, D., Crescenzo, G.D., Ostrovsky, R., Persiano, G.: Public key encryption with keyword search. In: Proceedings of EUROCRYP 2004 (2004)
3. Waters, B., Balfanz, D., Durfee, G., Smetters, D.: Building an encrypted and searchable audit log. In: Proceedings of 11th Annual Network and Distributed System (2004)
4. Chang, Y.-C., Mitzenmacher, M.: Privacy preserving keyword searches on remote encrypted data. In: Ioannidis, J., Keromytis, A.D., Yung, M. (eds.) ACNS 2005. LNCS, vol. 3531, pp. 442–455. Springer, Heidelberg (2005)
5. Boneh, D., Waters, B.: Conjunctive, subset, and range queries on encrypted data. In: Vadhan, S.P. (ed.) TCC 2007. LNCS, vol. 4392, pp. 535–554. Springer, Heidelberg (2007)
6. Li, J., Wang, Q., Wang, C., Cao, N., Ren, K., Lou, W.: Fuzzy keyword search over encrypted data in cloud computing. In: IEEE INFOCOM (2010)
7. Gu, C., Zhu, Y.: New efficient searchable encryption schemes from bilinear pairings. Int. J. Netw. Secur. 10(1), 25–31 (2010)
8. Ding, X., Yang, Y., Deng, R.H., Bao, F.: Private query on encrypted data in multi-user settings. In: Chen, L., Mu, Y., Susilo, W. (eds.) ISPEC 2008. LNCS, vol. 4991, pp. 71–85. Springer, Heidelberg (2008)
9. Ajtai, M.: Generating hard instances of lattice problems (extended abstract). In: STOC 1996: Proceedings of the Twenty-Eighth Annual ACM Symposium on Theory of Computing, pp. 99–108. ACM, New York (1996)
10. Micciancio, D.: Generalized compact knapsacks, cyclic lattices, and efficient one-way functions from worst-case complexity assumptions. In: FOCS, pp. 356–365 (2002)
11. Gentry, C., Peikert, C., Vaikuntanathan, V.: Trapdoors for hard lattices and new cryptographic constructions. In: Ladner, R.E., Dwork, C. (eds.) STOC, pp. 197–206. ACM (2008)
12. Regev, O.: On lattices, learning with errors, random linear codes, and cryptography. In: STOC 2005: Proceedings of the Thirty-Seventh Annual ACM Symposium on Theory of Computing, pp. 84–93. ACM, New York (2005)
13. Boneh, D., Agrawal, S., Boyen, X.: Efficient lattice (H)IBE in the standard model. In: Gilbert, H. (ed.) EUROCRYPT 2010. LNCS, vol. 6110, pp. 553–572. Springer, Heidelberg (2010)
14. Boneh, D., Boyen, X., Agrawal, S.: Lattice basis delegation in fixed dimension and shorter-ciphertext hierarchical IBE. In: Rabin, T. (ed.) CRYPTO 2010. LNCS, vol. 6223, pp. 98–115. Springer, Heidelberg (2010)
15. Gentry, C.: Fully homomorphic encryption using ideal lattices. In: STOC, pp. 169–178 (2009)

16. Gentry, C.: Toward basing fully homomorphic encryption on worst-case hardness. In: Rabin, T. (ed.) CRYPTO 2010. LNCS, vol. 6223, pp. 116–137. Springer, Heidelberg (2010)

17. Boyen, X.: Expressive encryption systems from lattices. In: Lin, D., Tsudik, G., Wang, X. (eds.) CANS 2011. LNCS, vol. 7092, pp. 1–12. Springer, Heidelberg (2011)

18. Waters, B.: Efficient identity-based encryption without random oracles. In: Cramer, R. (ed.) EUROCRYPT 2005. LNCS, vol. 3494, pp. 114–127. Springer, Heidelberg (2005)

19. Boyen, X.: Lattice mixing and vanishing trapdoors: a framework for fully secure short signatures and more. In: Nguyen, P.Q. (ed.) PKC 2010. LNCS, vol. 6056, pp. 499–517. Springer, Heidelberg (2010)

20. Alwen, J., Peikert, C.: Generating shorter bases for hard random lattices. In: STACS, pp. 75–86 (2009)

Integrity Verification of Replicated File with Proxy in Cloud Computing

Enguang Zhou[1], Jiangxiao Zhang[2], Zhoujun Li[1(✉)], and Chang Xu[3]

[1] State Key Laboratory of Software Development Environment, Beihang University,
Beijing 100191, China
zhoujun.li@263.net

[2] College of Mathematics and Information Technology, Xingtai University,
Xingtai 054001, China

[3] School of Computer Science and Technology, Beijing Institute of Technology,
Beijing 100081, China

Abstract. In 2013, Wang designed an efficient proxy provable data possession scheme. However, Wang's scheme focuses on a single copy of the file and the client just delegates the integrity checking task to only one proxy, not multiple ones. In this paper, we extend Wang's scheme to apply to multiple replicas. We propose two proxy multiple-replica provable data possession schemes. In the first scheme, the client delegates the integrity checking task to only one proxy. In the second scheme, multiple proxies form a group, and the client delegates the integrity checking task to this group. In the two proposed schemes, only the client and the proxy (proxies) can guarantee that multiple replicas are correctness. Finally, we analyze the security and performance of the two schemes.

Keywords: Data integrity · Multiple-replica · Cloud security · Cloud computing

1 Introduction

In cloud computing, clients outsource the large data files to the untrusted cloud storage server. As clients no longer possess the local copy of their data, protecting the integrity of outsourced data is very important in cloud computing.

In recent years, researchers have proposed two novel approaches, called provable data possession (PDP) [1] and proofs of retrievability (POR) [2], for validating the integrity of data at untrusted remote servers. Ateniese et al. [1] firstly proposed a provable data possession model to guarantee the integrity of the files at remote servers. They also presented a scalable PDP scheme [3] to support dynamic operations, such as block modification, deletion and append. However, this scheme does not support block insertion. Juels et al. [2] presented a POR model to ensure not only data possession but also retrievability. Shacham and Waters [4] proposed two compact POR schemes. The first one is constructed based on BLS signatures [5] and is proved in the random oracle model. The second one is constructed based on

ⓒ Springer International Publishing Switzerland 2015
W. Qiang et al. (Eds.): CloudCom-Asia 2015, LNCS 9106, pp. 344–353, 2015.
DOI: 10.1007/978-3-319-28430-9_26

pseudo-random functions and is proved in the standard model. Wang [6] proposed the concept of proxy provable data possession. In Wang's scheme, the client does not allow anyone to act as a verifier, and delegates the remote data integrity checking task to some proxy. The proxy has the capacity to verify the integrity of the data stored in the cloud according to a warrant.

Erway et al. [7] proposed a dynamic PDP scheme based on the rank-based authenticated skip list (RASL). This scheme was the first PDP scheme that can support fully dynamic data operations, including operations of modification, deletion, and insertion. Wang et al. [8] proposed a POR scheme based on the Merkle Hash Tree (MHT) [9] that can support public auditing and fully dynamic data operations. Zhu et al. [10] presented a cooperative provable data possession (CPDP) scheme in multicloud Storage. They indicated that their CPDP scheme held completeness, knowledge soundness, and zero-knowledge properties. Unfortunately, Wang et al. [11] pointed out that the malicious cloud service provider or the malicious organizer can deceive the verifier in Zhu et al.'s CPDP scheme. Yang et al. [12] described an efficient and secure dynamic auditing protocol. Subsequently, Ni et al. [13] pointed out that the protocol in [12] is vulnerable to the attack from the active adversary. Liu et al. [14] proposed a public auditing scheme based on the ranked Merkle hash tree (RMHT), which can support fine-grained update requests. Wang et al. [21] adopted random masking technique to propose a privacy-preserving public auditing protocol.

However, these schemes mentioned above focus on a single copy of the file, which are unsuitable for the environment that the cloud storage servers (CSS) store multiple copies of the client's file. Curtmola et al. [15] firstly proposed a multiple-replica provable data possession (MR-PDP) scheme. In MR-PDP scheme, the client can check the integrity of all outsourced copies. However, Curtmola et al.'s scheme only achieves private verifiability, i.e., only the client itself can check the integrity of the multiple replicas. Hao et al. [16] proposed a multiple-replica remote data possession scheme, which achieved public verifiability and allowed anyone to check the possession of the multiple replicas. However, Hao et al.'s scheme suffers from a drawback: the key k_r that is used in the generation of multiple replicas is publicly known, so that the cloud storage servers can derive the encrypted file F' and store only one replica. Barsoum et al. [17] proposed a pairing-based provable multi-copy data possession (PB-PMDP) scheme, which supports public verifiability. Xiao et al. [18] proposed two efficient multiple-file remote data checking (MF-RDC) schemes. However, these schemes in [15–18] could not resist denial of service (DoS) attacks or distributed denial of service (DDoS) attacks, in which malicious auditors send large amounts of challenge requests to exhaust the server's resources. Hwang et al. [19] proposed a DoS-resistant ID-based password authentication scheme with client puzzles. We apply the strategy in [19, 23] to prevent DoS/DDoS attacks.

Wang [6] proposed a proxy provable data possession scheme. However, Wang's scheme is unsuitable for the multiple-replica environment, and the client just delegates the integrity checking task to only one proxy, not multiple ones. In this paper, we extend Wang's scheme and propose two proxy multiple-replica provable data possession schemes. The two schemes use the client puzzle protocol in [19, 23] to prevent DoS/DDoS attacks. In the first scheme, the client delegates the integrity checking task to only one proxy, and the proxy is able to verify the integrity of client's multiple replicas. In the first scheme, if the proxy fails, then only the client itself can check the data integrity, so in some situations the client needs multiple proxies to perform the data integrity checking. In the second scheme, multiple proxies form a group, and the client delegates the integrity checking task to the group. In the two schemes, only the client and the delegable proxy (proxies) can check the data integrity and guarantee that multiple replicas are correctness. Finally, we analyze the security and performance of the two schemes.

2   Preliminaries

We describe the bilinear map in the following. Let G and G_T be two cyclic multiplicative groups with the same prime order q. For any $g_1, g_2 \in G$ and all $a, b \in Z_q^*$, let $e : G \times G \rightarrow G_T$ be a bilinear map which satisfies the following properties:

(1) bilinearity: $e(g_1^a, g_2^b) = e(g_1, g_2)^{ab}$;
(2) non-degeneracy: $e(g_1, g_2) \neq 1$ unless g_1 or $g_2 = 1$;
(3) computability: $e(g_1, g_2)$ is efficiently computable.

$\mathbb{S} = (q, G, G_T, e)$ is a bilinear map group system composed of the objects as described above.

We introduce notations used in the two schemes.

S_i: the i-th cloud storage server, $1 \leq i \leq t$.
F: the original file, denoted as a sequence of n blocks $f_1, f_2, \ldots, f_n \in Z_q^*$.
F': the encrypted file, obtained by encrypting the original file F with an encryption key K, denoted as a sequence of n blocks $c_1, c_2, \ldots, c_n \in Z_q^*$.
F_i: the file replica stored at the server S_i, denoted as a sequence of n blocks $m_{i1}, m_{i2}, \ldots, m_{in} \in Z_q^*$.
Ω: pseudorandom function (PRF), defined as: $Z_q^* \times \{1, 2, \ldots, n\} \rightarrow Z_q^*$.
ϕ: pseudorandom function, defined as: $Z_q^* \times \{0, 1\}^* \rightarrow Z_q^*$.
π: pseudorandom permutation(PRP), defined as: $Z_q^* \times \{1, 2, \ldots, n\} \rightarrow \{1, 2, \ldots, n\}$.
h: cryptographic hash function, defined as: $Z_q^* \rightarrow G$.
H: cryptographic hash function, defined as: $G \rightarrow Z_q^*$.
h^*: one-way hash function.
v_s: the solution of the puzzle.
$puzzle(cp, x_1, x_2, \ldots, x_n, v_s)$: the puzzle in [19], given $cp, x_1, x_2, \ldots, x_n$, find an solution v_s such that $cp = h^*(x_1, x_2, \ldots, x_n, v_s)$.

3   Two Proxy Multiple-Replica Provable Data Possession Schemes

The proxy multiple-replica provable data possession system involves three different network entities: Client, CSS, and Proxy. The detailed description of the three network entities is given in the following:

Client: Stores multiple replicas at multiple servers;

CSS: Provides data storage services and maintains multiple replicas of the client's data;

Proxy: Be delegated to check possession of client's multiple replicas, is capable of checking possession of multiple replicas.

Let $\mathbb{S} = (q, G, G_T, e)$ be a bilinear map group system, where G and G_T are two groups of a large prime order q. Let g be a generator of G. The original file F is divided into n blocks of equal lengths. The security parameter is denoted by κ.

3.1   Single-Proxy Multiple-Replica Provable Data Possession Scheme

In this section, we give a description of the first proxy multiple-replica provable data possession scheme. In the first scheme, the client delegates the integrity checking task to some proxy. The first scheme consists of 8 phases: $KeyGen$, $ReplicaGen$, $TagGen$, $Challenge1$, $GenPuzzle$, $Challenge2$, $Gen\,Proof$ and $Check\,Proof$.

$KeyGen$: The client chooses a random number $x \in Z_q^*$ and computes $X = g^x$. Its private key is $sk_c = x$ and its public key is $pk_c = X$. The client generates an encryption key K, which is used to encrypt the original file F. The proxy chooses a random number $y \in Z_q^*$, and computes $Y = g^y$. Its private key is $sk_p = y$ and its public key is $pk_p = Y$. The proxy can check the client's remote data possession by using its private key y.

$TagGen$: The client encrypts the original file F into F' by using the encryption key K. Let $F' = (c_1, c_2, \ldots, c_n)$, each block c_i has the same length. The client computes $\varepsilon = H(g^{xy})$ and $W_i = \Omega_\varepsilon(i)$, for $1 \leq i \leq n$. Then the client computes the tag $T_i = (h(W_i)u^{c_i})^x$ for each block c_i.

$ReplicaGen$: The client uses the encrypted file F' to generate t different file replicas $\{F_i\}_{i=1}^t$, and then stores the replica F_i on the corresponding server S_i. More precisely, for the server S_i, let $F_i = \{m_{i1}, m_{i2}, \ldots, m_{in}\}$, where $m_{ij} = c_j + r_{ij}$, $1 \leq j \leq n$, and $r_{ij} = \phi_\varepsilon(i||j)$ is the pseudo-random number generated by the client. After that, the client sends F_i to S_i, and sends $\{T_i\}_{i=1}^n$ to all t servers $\{S_1, S_2, \ldots, S_t\}$.

$Challenge1$: The proxy challenges the server S_i to check the integrity of the replica F_i. The proxy chooses a random number N_p, and sends (ID_p, N_p) to the server S_i.

$GenPuzzle$: The server S_i checks whether ID_p is valid. If ID_p is valid and N_p is fresh, S_i determines the difficulty of the client puzzle based on the current loading. S_i chooses v_s and generates a random number N_s, then computes $cp = h^*(ID_p, N_p, N_s, v_s)$. At last, S_i sends (cp, N_s, ID_p) to proxy.

$Challenge2$: Upon receiving the message (cp, N_s, ID_p), the proxy applies a brute-force method to solve the puzzle. Once the proxy finds an answer v_s^* such that $cp = h^*(ID_p, N_p, N_s, v_s^*)$, the proxy chooses a d-element subset $I = (s_1, s_2, \ldots, s_d)$ of set $[1, 2, \ldots, n]$, where $s_1 < s_2 < \ldots < s_d$. The proxy chooses a random number v_j for each $j \in I$. Then, the proxy sends $\{v_s^*, (j, v_j)\}_{j \in I}$ to the server S_i.

$GenProof$: After receiving the message $\{v_s^*, (j, v_j)\}_{j \in I}$, S_i checks whether the equation $cp = h^*(ID_p, N_p, N_s, v_s^*)$ holds. If it does not hold, S_i stops and outputs FALSE. If it holds, S_i chooses a random number γ, and calculates $R = e(u, X)^\gamma$. Then, S_i computes $\mu' = (\sum_{j \in I} v_j m_{ij}) + \gamma$ and $T = \prod_{j \in I} T_j^{v_j}$. After that, S_i sends $\{\mu', T, R\}$ to the proxy.

$Check\,Proof$: After receiving $\{\mu, T, R\}$ from S_i, the proxy computes $\varepsilon = H(g^{xy})$ and the pseudo-random number $r_{ij} = \phi_\varepsilon(i\|j)$, for each $j \in I$, then computes $\mu = \mu' - \sum_{j \in I} r_{ij}$. The proxy checks the verification equation

$$R \cdot e(T, g) \stackrel{?}{=} e(\prod_{j \in I} (h(\Omega_\varepsilon(j)))^{v_j} \cdot u^\mu, X).$$

If it holds, the proxy outputs TRUE, otherwise, the proxy outputs FALSE. The correctness of the verification is elaborated as follows:

$$
\begin{aligned}
&R \cdot e(T, g) \\
&= e(u, X)^\gamma \cdot e(\prod_{j \in I} (h(\Omega_\varepsilon(j)) \cdot u^{c_j})^{x \cdot v_j}, g) \\
&= e(u^\gamma, X) \cdot e(\prod_{j \in I} (h(\Omega_\varepsilon(j)) \cdot u^{c_j})^{v_j}, X) \\
&= e(u^\gamma, X) \cdot e(\prod_{j \in I} (h(\Omega_\varepsilon(j)))^{v_j} \cdot u^{\sum_{j \in I} v_j c_j}, X) \\
&= e(\prod_{j \in I} (h(\Omega_\varepsilon(j)))^{v_j} \cdot u^{(\sum_{j \in I} v_j c_j) + \gamma}, X) \\
&= e(\prod_{j \in I} (h(\Omega_\varepsilon(j)))^{v_j} \cdot u^\mu, X).
\end{aligned}
$$

In the first scheme, only the client and the proxy can compute $\varepsilon = H(g^{xy})$ and have the capability to check possession of the multiple replicas. The first scheme uses the client puzzle protocol in [19,23] to defend against DoS/DDOS attacks. The puzzle is based on finding partial part of a pre-image of a hash function, and the server requires the proxy to find the solution of a puzzle before the server generates a proof.

3.2   Multi-proxy Multiple-Replica Provable Data Possession Scheme

In this section, we describe the second scheme. In the second scheme, the client delegates the remote data integrity checking task to a group, which consists of multiple proxies. All the group members can verify the integrity of multiple replicas stored in the cloud storage servers. The second scheme consists of 9 phases: *KeyGen, GroupKeyGen, TagGen,* ReplicaGen, *Challenge*1, *GenPuzzle, Challenge*2, *Gen* Pr *oof* and *Check* Pr *oof*.

KeyGen: The client's public-secret key pair is $(pk_c = g^x, sk_c = x)$. The client generates an encryption key K, which is used to encrypt the original file F. Assume l proxies form a group $U = \{P_1, P_2, \ldots, P_l\}$. Each P_i randomly chooses a signing key pair (pk_{p_i}, sk_{p_i}).

GroupKeyGen: Group members $\{P_1, P_2, \ldots, P_l\}$ utilize the group key agreement scheme [20] to share a group secret key Gsk. $\{P_1, P_2, \ldots, P_l\}$ form a circle structure, with $P_0 = P_l$, $P_1 = P_{l+1}$.

Step 1: Each P_i generates a random number $x_i \in Z_q^*$, computes $X_i = g^{x_i}$ and a signature $Sig_{sk_{p_i}}(X_i)$ on X_i. Each P_i sends X_i and $Sig_{sk_{p_i}}(X_i)$ to his neighbours P_{i-1}, P_{i+1}.

Step 2: Each P_i computes its left key $K_i^L = X_{i-1}^{x_i}$, right key $K_i^R = X_{i+1}^{x_i}$, $Y_i = K_i^R / K_i^L$, and a signature $Sig_{sk_{p_i}}(Y_i)$ on Y_i. Each P_i broadcasts Y_i and $Sig_{sk_{p_i}}(Y_i)$.

Step 3: Each P_i generates a group secret key Gsk as:

$$Gsk = K_1^R K_2^R K_3^R \ldots K_n^R = g^{x_1 x_2 + x_2 x_3 + x_3 x_4 + \ldots x_n x_1},$$

where $K_{i+j}^R = K_i^R Y_{i+1} Y_{i+2} \ldots Y_{i+j}$, for $j = \{1, 2, \ldots, n - 1\}$. Each P_i computes the group public key $Gpk = g^{Gsk}$.

TagGen: The client uses the encryption key K to encrypt the original file F into F'. Let $F' = (c_1, c_2, \ldots, c_n)$. The client computes $\varepsilon = H(g^{x \cdot Gsk})$ and $W_i = \Omega_\varepsilon(i)$, $1 \leq i \leq n$. Then the client computes the tag $T_i = (h(W_i)u^{c_i})^x$ for each block c_i.

ReplicaGen: The client uses the encrypted file F' to generate t different file replicas $\{F_i\}_{i=1}^t$, and then stores the replica F_i on the corresponding server S_i. More precisely, for the server S_i, let $F_i = \{m_{i1}, m_{i2}, \ldots, m_{in}\}$, where $m_{ij} = c_j + r_{ij}$, $1 \leq j \leq n$, and $r_{ij} = \phi_\varepsilon(i||j)$ is the pseudo-random number generated by the client. After that, the client sends F_i to S_i, and sends $\{T_i\}_{i=1}^n$ to all t servers $\{S_1, S_2, \ldots, S_t\}$.

*Challenge*1: All the proxies in the group U can challenge the servers to check the integrity of the client's replicas. Assume the proxy P_k $(1 \leq k \leq l)$ to challenge the server S_i. P_k chooses a random number N_{p_k}, and sends (ID_{p_k}, N_{p_k}) to the server S_i.

GenPuzzle: The server S_i checks whether ID_{p_k} is valid. If ID_{p_k} is valid and N_{p_k} is fresh, S_i determines the difficulty of the client puzzle based on the current loading. S_i chooses v_s and generates a random number N_s, then computes $cp = h^*(ID_{p_k}, N_{p_k}, N_s, v_s)$. At last, S_i sends (cp, N_s, ID_{p_k}) to P_k.

*Challenge*2: Upon receiving the message (cp, N_s, ID_{p_k}), P_k applies a brute-force method to solve the puzzle. Once P_k finds an answer v_s^* such that

$cp = h^*(ID_{p_k}, N_{p_k}, N_s, v_s^*)$, P_k chooses a d-element subset $I = (s_1, s_2, \ldots, s_d)$ of set $[1, 2, \ldots, n]$, where $s_1 < s_2 < \ldots < s_d$. P_k chooses a random number v_j for each $j \in I$. After that, P_k sends $\{v_s^*, (j, v_j)\}_{j \in I}$ to the server S_i.

$Gen\, Proof$: After receiving the message $\{v_s^*, (j, v_j)\}_{j \in I}$ from P_k, S_i checks whether the equation $cp = h^*(ID_{p_k}, N_{p_k}, N_s, v_s^*)$ holds. If it does not hold, S_i stops and outputs FALSE. If it holds, S_i chooses a random number γ, and calculates $R = e(u, X)^\gamma$. Then, S_i computes $\mu' = (\sum_{j \in I} v_j m_{ij}) + \gamma$ and $T = \prod_{j \in I} T_j^{v_j}$. After that, S_i sends $\{\mu', T, R\}$ to P_k.

$Check\, Proof$: After receiving $\{\mu', T, R\}$ from S_i, P_k computes $\varepsilon = H(g^{x \cdot Gsk})$ and the pseudo-random number $r_{ij} = \phi_\varepsilon(i||j)$, for each $j \in I$, then computes $\mu = \mu' - \sum_{j \in I} r_{ij}$. P_k checks the verification equation

$$R \cdot e(T, g) \stackrel{?}{=} e(\prod_{j \in I} (h(\Omega_\varepsilon(j)))^{v_j} \cdot u^\mu, X).$$

If it holds, P_k outputs TRUE, otherwise, P_k outputs FALSE. The correctness of the verification is similar as that of the first scheme.

In the second scheme, all the proxies in the group U can verify the integrity of multiple replicas stored in the cloud.

4   Security Analysis and Performance

In this section we analyze the security and performance of the two schemes.

4.1   Security Analysis

In the two schemes, the blocks of encrypted file F' are masked with random numbers which can be computed by the client or the proxy (proxies). The two schemes use the masking technique to resist the collusion attack. In the two schemes, a cloud storage server cannot generate a valid proof unless the server possesses its replica, which is shown in Theorem 1.

Theorem 1. If the server passes integrity verification, it must indeed possess all the data blocks challenged by the proxy.

The security of the two schemes is highly similar to the security of the single replica scheme proposed in [21]. Therefore, the detailed security proof is omitted here.

4.2   Computation and Communication

In this section, we analyze the computation and communication costs of the two schemes.

Computation: The two schemes have the same computation costs in $TagGen$, $Challenge1$, $Challenge2$, $Check\,Pr\,oof$ and $Gen\,Pr\,oof$ phases. The computation cost of hash operations and modular arithmetic operations is negligible [22], we only consider the computation costs of pairing operations and exponentiation operations. In the two proposed schemes, suppose the client generates t different file replicas, and each replica is divided into n blocks of equal lengths. In the $TagGen$ phase, the client needs to perform $2n + 1$ exponentiations. In the $Gen\,Pr\,oof$ phase, the server S_i needs to do one pairing and $d + 1$ exponentiations. In the $Check\,Pr\,oof$ phase, the proxy needs to perform 2 pairings, and $d + 2$ exponentiations. In the $GroupKeyGen$ phase of the second scheme, each proxy needs to do 4 exponentiations and 2 signatures. The computation cost of an exponent operation is denoted by $[E]$, and the computation cost of a bilinear map is denoted by $[B]$. Table 1 presents the computation costs of the two schemes. In Table 1, because S_i determines the difficulty of the client puzzle, the computation cost for solving a client puzzle is neglected.

Communication: The two schemes have the same communication costs in the $Challenge1$, $Challenge2$, $GenPuzzle$ and $Gen\,Pr\,oof$ phases. In the $Challenge2$ phase, the proxy sends $2d + 1$ elements in Z_q^* to the cloud storage server S_i. In the $Gen\,Pr\,oof$ phase, the server S_i sends one element in Z_q^* and one element in G, one element in G_T to the proxy.

Table 1. Computation costs of the two schemes

	The first scheme	The second scheme
TagGen	$(2n + 1)[E]$	$(2n + 1)[E]$
GroupKeyGen		$4l \cdot [E]$
GenProof	$[B] + (d + 1)[E]$	$[B] + (d + 1)[E]$
CheckProof	$2[B] + (d + 2)[E]$	$2[B] + (d + 2)[E]$

5   Conclusion

In this paper, we extend Wang's scheme and propose two proxy multiple-replica provable data possession schemes. The two schemes use client puzzles to prevent DoS/DDoS attacks. In the first scheme, the client delegates the integrity checking task to only one proxy. In the second scheme, the client delegates the integrity checking task to a group, which consists of multiple proxies. In the two schemes, only the client and the proxy (proxies) can guarantee that multiple replicas are correctness. Finally, we analyze the security and performance of the two schemes.

Acknowledgments. This work was supported by the National Natural Science Foundation of China (61202239, 61370126, and 61170189), the Specialized Research Fund for the Doctoral Program of Higher Education (20111102130003), the Program for the Top Young Talents of Higher Learning Institutions of Hebei (BJ201414).

References

1. Ateniese, G., Burns, R., Curtmola, R., et al.: Provable data possession at untrusted stores. In: 14th ACM conference on Computer and communications security, pp. 598-609. ACM (2007)
2. Juels, A., Kaliski Jr., B.S.: PORs: Proofs of retrievability for large files. In: 14th ACM conference on Computer and communications security, pp. 584-597. ACM (2007)
3. Ateniese, G., Di Pietro, R., Mancini, L.V., et al.: Scalable and efficient provable data possession. In: 4th international conference on Security and privacy in communication netowrks, pp. 1-10. ACM (2008)
4. Shacham, H., Waters, B.: Compact proofs of retrievability. In: Pieprzyk, J. (ed.) ASIACRYPT 2008. LNCS, vol. 5350, pp. 90–107. Springer, Heidelberg (2008)
5. Boneh, D., Lynn, B., Shacham, H.: Short signatures from the Weil pairing. J. Cryptol. $17(4)$, 297–319 (2004)
6. Wang, H.: Proxy provable data possession in public clouds. IEEE Trans. Ser. Comput. $6(4)$, 551–559 (2013)
7. Erway, C., Kupcu, A., Papamanthou, C., et al.: Dynamic provable data possession. In: 16th ACM conference on Computer and communications security, pp. 215–222. ACM (2009)
8. Wang, Q., Wang, C., Ren, K., et al.: Enabling public auditability and data dynamics for storage security in cloud computing. IEEE Trans. Parallel Distrib. Syst. $22(5)$, 847–859 (2011)
9. Merkle, R.C.: Protocols for public key cryptosystems. In: 2012 IEEE Symposium on Security and Privacy, pp. 122-122. IEEE Computer Society (1980)
10. Zhu, Y., Hu, H., Ahn, G.J., et al.: Cooperative provable data possession for integrity verification in multicloud storage. IEEE Trans. Parallel Distrib. Syst. $23(12)$, 2231–2244 (2012)
11. Wang, H., Zhang, Y.: On the knowledge soundness of a cooperative provable data possession scheme in multicloud storage. IEEE Trans. Parallel Distrib. Syst. $25(1)$, 264–267 (2014)
12. Yang, K., Jia, X.: An efficient and secure dynamic auditing protocol for data storage in cloud computing. IEEE Trans. Parallel Distrib.Syst. $24(9)$, 1717–1726 (2013)
13. Ni, J., Yu, Y., Mu, Y., et al.: On the security of an efficient dynamic auditing protocol in cloud storage. IEEE Trans. Parallel Distrib.Syst. $25(10)$, 2760–2761 (2013)
14. Liu, C., Chen, J., Yang, L., et al.: Authorized public auditing of dynamic big data storage on cloud with efficient verifiable fine-grained updates. IEEE Trans. Parallel Distrib. Syst. $25(9)$, 2234–2244 (2013)
15. Curtmola, R., Khan, O., Burns, R., et al.: MR-PDP: Multiple-replica provable data possession. In: The 28th International Conference on Distributed Computing Systems ICDCS 2008, pp.411-420. IEEE (2008)
16. Hao, Z., Yu, N.: A multiple-replica remote data possession checking protocol with public verifiability. In: 2010 Second International Symposium on Data, Privacy and E-Commerce (ISDPE), pp. 84-89. IEEE (2010)
17. Barsoum, A.F., Hasan, M.A.: Integrity verification of multiple data copies over untrusted cloud servers. In: 12th IEEE/ACM International Symposium on Cluster, Cloud and Grid Computing (CCGRID 2012), pp. 829-834. IEEE Computer Society (2012)

18. Xiao, D., Yang, Y., Yao, W., et al.: Multiple-file remote data checking for cloud storage. Comput. Secur. **31**(2), 192–205 (2012)
19. Hwang, M.S., Chong, S.K., Chen, T.Y.: DoS-resistant ID-based password authentication scheme using smart cards. J. Syst. Softw. **83**(1), 163–172 (2010)
20. Dutta, R., Barua, R.: Provably secure constant round contributory group key agreement in dynamic setting. IEEE Trans. Inf. Theory **54**(5), 2007–2025 (2008)
21. Wang, C., Chow, S.S.M., Wang, Q., et al.: Privacy-preserving public auditing for secure cloud storage. IEEE Trans. Comput. **62**(2), 362–375 (2013)
22. Barreto, P.S.L.M., Galbraith, S.D., OhEigeartaigh, C., et al.: Efficient pairing computation on supersingular Abelian varieties. Des. Codes Crypt. **42**(3), 239–271 (2007)
23. Aura, T., Nikander, P., Leiwo, J.: DOS-Resistant authentication with client puzzles. In: Christianson, B., Crispo, B., Malcolm, J.A., Roe, M. (eds.) Security Protocols 2000. LNCS, vol. 2133, pp. 170–177. Springer, Heidelberg (2001)

Privacy-Preserving Regression Modeling and Attack Analysis in Sensor Network

Jianjun Wu$^{(\boxtimes)}$ and Fengjuan Zhang$^{(\boxtimes)}$

Computer Science, University of Science and Technology of China, Hefei, China
{wjianjun,fjzhang}@mail.ustc.edu.cn

Abstract. With the advancements of sensing technologies, participatory data fusion and analysis have raised more and more attention as it provides a promising way enabling the public benefit from this process. However, it is increasingly becoming a challenging problem about how to construct a statistical model for this kind of particular phenomenon under the premise of protecting the data privacy. In this paper, we present a method to build a linear regression model describing a phenomenon observed in the distributed network. At the same time, the private data of each node is preserved.

We propose a data aggregation algorithm to fuse the indispensable data for constructing linear regression equation even though the data is private and scattered in the whole network. We also point out that the aggregate node can not only conduct regression modeling, but can conduct some complex statistical analyses based on the aggregation result as well. In addition, we investigate the ability and the degree of the algorithm to preserve the data privacy. We mainly focus on the ability to protect the aggregation result of a community composed of sensor nodes. The experiment shows that the aggregation result can not be disclosed to people unless all of the sensor nodes in the community collude with each other.

1 Introduction

Distributed sensor network can be applied in various areas (i.e. habitat monitoring, environment observation, health applications, etc.). Users of the sensor network can gain an insight into a particular phenomenon via in-network data aggregation. Data aggregation is a promising method to save energy and provide accurate observation in sensor networks [3]. However, some data may be sensitive and private. In this situation, data aggregation may cause threats to the privacy of sensor nodes.

There are a number of nodes in a sensor network, and each of them have some data related with their local state, which is called local data in this paper. Local data is sensitive to be disclosed to others. We aim at enabling users to conduct accurate linear regression modeling and preserving each node's privacy at the same time. The main contributions of this paper are summarized as follows:

- we propose a improved data aggregation algorithm based [14], through which we can get the global data related with the global state of the sensor network.

© Springer International Publishing Switzerland 2015
W. Qiang et al. (Eds.): CloudCom-Asia 2015, LNCS 9106, pp. 354–366, 2015.
DOI: 10.1007/978-3-319-28430-9_27

- we give details of how the user conduct complex modeling and analysis according to the aggregation result, which is just the global data.
- we investigate the ability of the aggregation algorithm to protect the aggregation result when facing a collusive attack, which is based on MLE (Maximum Likelihood Estimate) and can be formulated as a constrained optimization problem.

The rest of the paper is organized as follows. Most relevant researches are presented in Sect. 2. In Sect. 3, we introduce some preliminaries about linear regression and network model. In Sect. 4, we propose our data aggregation algorithm. In Sect. 5, we present how to conduct modeling and some complex regression analyses. Details on attack analysis are presented in Sect. 6. The paper is concluded in Sect. 7.

2   Related Work

In this section, we summarize the most relevant existing researches. [2] presents the first provably secure sensor network data aggregation protocol, which can prevent a portion of the sensor nodes controlled by an adversary from skewing the final aggregate result. [10] epitomizes the privacy preserving issue in WSN. It discusses two main categories of privacy-preserving, data-oriented and context-oriented. It also reviews some existing techniques for privacy-preserving data aggregation. [7] presents the algorithm CPDA and SMART to preserve the privacy of each sensor. CPDA adopts adding random numbers to the raw data, and SMART divides the raw data into slices and then sends the slices to different nodes. [6] proposes a series of compromise-resilient schemes, which is built on the preloaded secret in the sensors. A sensor report the sum of the original data and the secret shared with the sink node.

[13] PriSense focuses on people-centric urban sensing systems. It shares the similar idea with [7], but differs in the application scenario. It presents three cover-selection schemes to cope with the dynamic nature of urban sensing systems. iPDA [8] utilizes node-disjoint aggregation trees to verify the integrity of the aggregation results, and data privacy is achieved through data slicing and assembling techniques. In iPDA, the data integrity-protection and data privacy-preservation mechanisms work synergistically. [12] considers two cases, vertically and horizontally partitioned data. When the global data is vertically partitioned, in order to compute the global regression model without revealing the private data, the author develops an iteration strategy based on Powell's algorithm [11] to minimize the norm of residual error. [5] studies some multivariate statistical analysis methods in secure 2-party computation (S2C) framework, and develops a set of basic protocols for conducting matrix product and matrix inverse, which can be used for secure linear regression and classification. But both of them [5, 12] require nodes with strong computation ability, and bring about heavy communication load, which is not suitable for sensor network scenarios.

[1] proposes a technique to collect some features of private data and construct the regression model of an node while protecting its privacy, and it studies the

reconstruction of privacy data. But it doesn't investigate its application in sensor network where are lots of nodes. [14] proposes an aggregation algorithm to fuse the necessary data for regression modeling and model fitting, which can be used in the sensor network. As it requires the cluster to have a big size, which results in a high-dimension problem it couldn't deal with. Our method can deal with the high-dimension problem by building an aggregation tree and dividing the data into segments. Furthermore, we can conduct some complex regression analyses besides regression modeling and model fitting. We also study the reconstruction issue of the aggregation result.

3   Problem Formulation and Network Model

Linear regression is widely used in many fields. We assume that there is linear correlation relationship between some explanatory variables $x_1, x_2, \cdots, x_{p-1}$ and the response variable y. That is to say:

$$y = \beta_0 + \beta_1 x_1 + \beta_2 x_2 + \ldots + \beta_{p-1} x_{p-1} = \mathbf{x}\beta$$

After we obtain m records, we have,

$$\mathbf{Y} = \mathbf{X}\beta$$

where

$$\mathbf{X} = \begin{bmatrix} 1 & x_{11} & \cdots & x_{1,p-1} \\ 1 & x_{21} & \cdots & x_{2,p-1} \\ \vdots & \vdots & \ddots & \vdots \\ 1 & x_{m,1} & \cdots & x_{m,p-1} \end{bmatrix}, \mathbf{Y} = \begin{bmatrix} y_1 \\ y_2 \\ \vdots \\ y_m \end{bmatrix}$$

The least squared estimator(LSE) $\widehat{\beta}$ of the regression coefficient β and the unexplained variation SSE are respectively

$$\widehat{\beta} = (\mathbf{X}^T\mathbf{X})^{-1}\mathbf{X}^T\mathbf{Y} \tag{1}$$

$$SSE = (\mathbf{X}\widehat{\beta} - \mathbf{Y})^T(\mathbf{X}\widehat{\beta} - \mathbf{Y}) \tag{2}$$

We assume that the sensor network consists of a user and n sensor nodes. In order to reduce communication cost and achieve the goal of privacy-preserving, we build a tree structure rooted at the user to implement the data aggregating algorithm, which is called aggregation tree. We can adopt EADAT [4] to build the aggregation tree. Each non-leaf node aggregates its own data with the data transmitted from all its children nodes and then it transmits the aggregated result to its parent. Note that nodes sharing the same parent consist of a community. Such an aggregation tree is shown in Fig. 1.

Each sensor node can get some local data related with its own state. The user wants to obtain the global data to construct linear regression model. We call the model derived from the global data *global model*. Let $\mathbf{X}^{(i)}, \mathbf{Y}^{(i)}$ denote the local data of node $V^{(i)}, i = 1, 2, \cdots, n$. Then the global data is

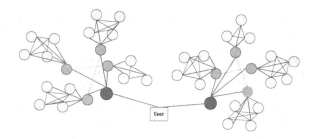

Fig. 1. Aggregation tree

$$\mathbf{X}^{(glb)} = \begin{bmatrix} \mathbf{X}^{(1)} \\ \mathbf{X}^{(2)} \\ \cdots \\ \mathbf{X}^{(n)} \end{bmatrix}, \mathbf{Y}^{(glb)} = \begin{bmatrix} \mathbf{Y}^{(1)} \\ \mathbf{Y}^{(2)} \\ \cdots \\ \mathbf{Y}^{(n)} \end{bmatrix}$$

The regression coefficient of the *global model* is

$$\hat{\beta}^{glb} = ((\mathbf{X}^{(glb)})^T \mathbf{X}^{(glb)})^{-1} (\mathbf{X}^{(glb)})^T \mathbf{Y}^{(glb)}$$

In order to protect the privacy of each node, the user can not obtain $\mathbf{X}^{(glb)}$, $\mathbf{Y}^{(glb)}$, but the user still wants to get the regression coefficient of the *global model* and conduct analysis on the *global model*.

4   Data Aggregation Algorithm

Based on PDA [7], we propose a data aggregation algorithm below. We minimize the traffic in the network and try our best to avoid transmitting redundant data. We only transmit the indispensable data for coefficient estimation of the *global model*.

Step 1: Generating feature data
Each node $V^{(i)}$, collects local data consisting of $n^{(i)}$ records according to its local state, and constructs its own matrix $\mathbf{X}^{(i)}, \mathbf{Y}^{(i)}$. Then each of them computes $\mathbf{XTX}^{(i)} = (\mathbf{X}^{(i)})^T \mathbf{X}^{(i)}, \mathbf{XTY}^{(i)} = (\mathbf{X}^{(i)})^T \mathbf{Y}^{(i)}$. Note that $\mathbf{XTX}^{(i)}$ is symmetric, so we only need to transmit the upper triangular of it, which is different from [14]. Then $V^{(i)}$ generates its feature data $\mathbf{S}^{(i)}$. Here $\mathbf{S}^{(i)}$ is a $\frac{p^2+3p+2}{2}$-dimensional vector, and each component of $\mathbf{S}^{(i)}$ is initialized as follows respectively.

$$\mathbf{XTX}_{11}^{(i)}, \mathbf{XTX}_{12}^{(i)}, \cdots, \mathbf{XTX}_{1,p}^{(i)}, \mathbf{XTX}_{22}^{(i)}, \mathbf{XTX}_{23}^{(i)}, \cdots,$$

$$\mathbf{XTX}_{2,p}^{(t)}, \mathbf{XTX}_{33}^{(i)}, \cdots, \mathbf{XTX}_{3,p}^{(i)}, \cdots, \mathbf{XTX}_{p,p}^{(i)}, \mathbf{XTY}_{1}^{(i)},$$

$$\mathbf{XTY}_{2}^{(i)}, \cdots, \mathbf{XTY}_{p}^{(i)}, \sum_{k=1}^{n^{(i)}} (y_{k}^{(i)})^2$$

The subscript $(.)_{ij}$ represents the element located at the $i - th$ row and $j - th$ column of the matrix.

Step 2: Aggregating data in each community

This step will be initiated by all of the leaf-communities consisting of the leaf nodes of the tree. Now we assume a leaf-community $\mathcal{B}$ consists of leaf nodes $\{V^{(1)}, V^{(2)}, \cdots, V^{(K)}\}$. Each node has got K different random positive numbers $\rho_1, \rho_2, \ldots, \rho_K$ from its parent node $V^{(p)}$. We adopt segmentation method to deal with the high dimension problem, which occurs when $\frac{p^2+3p+2}{2} > K$ and encountered by [14]. Segmentation enables aggregation can be done on a small scale of community'size, which is helpful for the energy saving in WSNs and the random numbers ρ_i can be reused repeatedly. First, M-dimensional vector $\mathbf{f}^{(i)}(j), i, j = 1, 2, \cdots, K$ is computed by node $V^{(i)}$, and every component is initialized as follows.

$$\begin{cases} \mathbf{f}_1^{(i)}(j) = \sum_{l=1}^{K} \mathbf{S}_l^{(i)} \rho_j^{l-1} \\ \mathbf{f}_2^{(i)}(j) = \sum_{l=K+1}^{2K} \mathbf{S}_l^{(i)} \rho_j^{l-K-1} \\ \cdots, \cdots, \cdots, \\ \mathbf{f}_M^{(i)}(j) = \sum_{l=q}^{MK} \mathbf{S}_l^{(i)} \rho_j^{l-q}, \\ j = 1, 2, \cdots, K \end{cases} \tag{3}$$

where $M = \lceil (p^2+3p+2)/2K \rceil, q = (M-1)K+1$. When $l > P = (p^2+3p+2)/2$, we let $\mathbf{S}_l^{(i)} = r_{l-P}^{(i)}$, where $r_1^{(i)}, r_2^{(i)}, \cdots$ are random numbers generated by the node $V^{(i)}$.

Afterwards, node $V^{(i)}$ sends $\mathbf{f}^{(i)}(j)$ to node $V^{(j)}$, and it keeps $\mathbf{f}^{(i)}(i)$ for itself. After node $V^{(j)}$ get all the values $\mathbf{f}^{(i)}(j), i = 1, 2, \cdots, K$, it computes $\mathbf{F}^{(j)}$;

$$\mathbf{F}^{(j)} = \sum_{i=1}^{K} \mathbf{f}^{(i)}(j) \tag{4}$$

Node $V^{(j)}$ sends $\mathbf{F}^{(j)}$ to its parent. Then the parent node solves the following M equations.

$$\begin{bmatrix} 1 & \rho_1 & \cdots & \rho_1^{K-1} \\ 1 & \rho_2 & \cdots & \rho_2^{K-1} \\ \vdots & \vdots & \ddots & \vdots \\ 1 & \rho_K & \cdots & \rho_K^{K-1} \end{bmatrix} \mathbf{Z}_t = \begin{bmatrix} \mathbf{F}_t^{(1)} \\ \mathbf{F}_t^{(2)} \\ \vdots \\ \mathbf{F}_t^{(K)} \end{bmatrix} \quad t = 1, \cdots, M \tag{5}$$

and let $\mathbf{Z} = \begin{bmatrix} \mathbf{Z}_1^T & \mathbf{Z}_2^T & \cdots & \mathbf{Z}_M^T \end{bmatrix}$. Each node divides its feature data into M segments. Then each segment will be perturbed by some random numbers according to Eq. (3) and get a perturbed vector $\mathbf{f}^{(i)}(j)$. After each pair of nodes has exchanged their perturbed vector, each of them adds up all its received perturbed vectors and sends the sum $\mathbf{F}^{(j)}$ to the parent. The parent solves M similar equation systems, and each of them corresponds to one segmentation. In other words, the solution of $t - th$ equation system is just the sum of each node's $t - th$ segmentation. That is

$$\mathbf{Z}_t = \left[\sum_{i=1}^{K} \mathbf{S}_{(t-1)K+1}^{(i)} \quad \sum_{i=1}^{K} \mathbf{S}_{(t-1)K+2}^{(i)} \quad \cdots \quad \sum_{i=1}^{K} \mathbf{S}_{tK}^{(i)}\right]^T$$

The parent $V^{(p)}$ assembles the solutions to $\mathbf{Z}$ and extracts the anterior $(p^2 + 3p + 2)/2$ components to form $\mathbf{S}^{(children)}$, let $\mathbf{S}^{(p)} = \mathbf{S}^{(p)} + \mathbf{S}^{(children)}$. Now the leaf-community $\mathcal{B}$ has completed the aggregation procedure and the parent has got the aggregation result of this community. All other leaf-communities also conduct the same procedure simultaneously. When all the leaf-communities have completed this step, the same procedure will be initiated again by the communities consisting of these parents, until the root-community, whose parent node is the root, has also completed this procedure.

Step 3: Reassembling data at the user side
When the root-community has completed the aggregation, let $\mathbf{S}$ denotes the aggregation result obtained by the user. The user declares some variables, a matrix $\mathbf{A}_{p \times p}$, a vector $\mathbf{b}_{p \times 1}$, one float variable T. They are initialized as follows:

$$\begin{cases} \mathbf{A}_{kl} = \mathbf{S}_{(2p-k+2)(k-1)/2+l-k+1} \\ \mathbf{A}_{lk} = \mathbf{A}_{kl}, \\ (k = 1, 2, \ldots, p, l = k, \cdots, p), \\ \mathbf{b}_k = \mathbf{S}_{p(p+1)/2+k}, (k = 1, 2, \ldots, p) \\ T = \mathbf{S}_{(p^2+3p+2)/2} \end{cases} \tag{6}$$

5 Privacy-Preserving Regression Analysis

The user can conduct some analysis only according to $\mathbf{A}, \mathbf{b}, T$. [14] points out that the user can conduct coefficient estimation, error computation and variable significance test, etc. Besides these, we will show that some complex analyses such as stepwise regression and multicollinearity detection can be conducted as well. According to Eq. (6), we can get following conclusions.

$$\mathbf{A} = \sum_{i=1}^{n} (\mathbf{X}^{(i)})^T \mathbf{X}^{(i)} = (\mathbf{X}^{glb})^T \mathbf{X}^{glb} \tag{7}$$

$$\mathbf{b} = \sum_{i=1}^{n} (\mathbf{X}^{(i)})^T \mathbf{Y}^{(i)} = (\mathbf{X}^{glb})^T \mathbf{Y}^{glb} \tag{8}$$

$$T = \sum_{i=1}^{n} \sum_{k=1}^{n^{(i)}} \left(y_k^{(i)}\right)^2 = (\mathbf{Y}^{glb})^T \mathbf{Y}^{glb} \tag{9}$$

$$\widehat{\beta}^{glb} = \mathbf{A}^{-1}\mathbf{b} \tag{10}$$

Pay attention that *glb* indicates that the statistic is related to the *global model*.

We expect that our regression model can describe the phenomenon accurately and inexpensively. We want to build the regression model with the optimal subset of complete explanatory variables rather than all of them. What's more important is that we want to reduce the error resulting from overfitting.

Assuming $X_V = \{x_{j_1}, x_{j_2}, \ldots, x_{j_k}\}$ is a subset, we define the complete explanatory variable set as: $X_{all} = \{x_1, x_2, \ldots, x_{p-1}\}$. We apply a screening policy to find the optimal subset and avoid huge computation cost. Stepwise regression is a widely-used iterative method among these screening policies. At each step of the stepwise regression algorithm, we add a new variable into the old variable subset (denoted by old) or delete a variable from the old variable subset. Then we'll get a new variable subset(denoted by new). We conduct a test to decide which variable should be added or deleted. Suppose the candidate variable which should be added or deleted is x_k, we define:

$$F_k = \frac{SSE_{old} - SSE_{new}}{SSE_{new}/(n - |old| - 2)} \tag{11}$$

SSE_{old} is the unexplained variation of the model built with the old variable set. $|old|$ represents the number of variables in the old variable set. At each step of the stepwise regression, we can decide whether add or delete a variable according to this statistic value. In the following, we will illustrate that user can compute this statistic only according to $\mathbf{A}, \mathbf{b}, T$. We also point out that we can get all the models built with arbitrary variable subset without more information.

Definition 1. Let $V = \{i_1, i_2, \ldots, i_p\}, U = \{j_1, j_2, \ldots, j_q\}$. The operator $\nabla_{V \times U}(\mathbf{A})$ represents all the elements of matrix $\mathbf{A}$ located at $i_1 - th$ row, $i_2 - th$ row, $\ldots$, $i_p - th$ row, and $j_1 - th$ col, $j_2 - th$, $\ldots, j_q - th$ column. They form a new matrix in their original order. Operator $\nabla_V(\mathbf{b})$ represents a new vector formed by the $i_1 - th, i_2 - th, \ldots, i_p - th$ components of vector $\mathbf{b}$ in their original order.

Theorem 1. For $\forall k, 0 \leq k \leq p - 1$, arbitrary $V = \{j_1, j_2, \ldots, j_k\} \subseteq \{1, 2, , p - 1\}$. If we construct the regression model with the subset $X_V = \{x_{j_1}, x_{j_2}, \ldots, x_{j_k}\} \subseteq \{x_1, x_2, \ldots, x_{p-1}\}$. Let $V_0 = \{1, j_1 + 1, j_2 + 1, \ldots, j_k + 1\}$. Then the LSE of the coefficient of the model is:

$$\widehat{\beta}_{X_V}^{glb} = (\nabla_{V_0 \times V_0}(\mathbf{A}))^{-1} \nabla_{V_0}(\mathbf{b})$$

Proof: In the following procedure, the auxiliary mark all represents all the rows or columns.

$$(\nabla_{all \times V_0}(\mathbf{X}^{glb}))^T \nabla_{all \times V_0}(\mathbf{X}^{glb})$$

$$= \left[(\nabla_{all \times V_0}(\mathbf{X}^{(1)}))^T \cdots (\nabla_{all \times V_0}(\mathbf{X}^{(n)}))^T \right] \begin{bmatrix} \nabla_{all \times V_0}(\mathbf{X}^{(1)}) \\ \vdots \\ \nabla_{all \times V_0}(\mathbf{X}^{(n)}) \end{bmatrix}$$

$$= \sum_{i=1}^{n} (\nabla_{all \times V_0}(\mathbf{X}^{(i)}))^T \nabla_{all \times V_0}(X^{(i)})$$

$$= \sum_{i=1}^{n} \begin{bmatrix} \sum_{k=1}^{n^{(i)}} \mathbf{X}_{k,1}^{(i)} \mathbf{X}_{k,1}^{(i)} & \cdots & \sum_{k=1}^{n^{(i)}} \mathbf{X}_{k,1}^{(i)} \mathbf{X}_{k,j_k+1}^{(i)} \\ \vdots & \ddots & \vdots \\ \sum_{k=1}^{n^{(i)}} \mathbf{X}_{k,j_k+1}^{(i)} \mathbf{X}_{k,1}^{(i)} & \cdots & \sum_{k=1}^{n^{(i)}} \mathbf{X}_{k,j_k+1}^{(i)} \mathbf{X}_{k,j_k+1}^{(i)} \end{bmatrix}$$

$$= \sum_{i=1}^{n} \nabla_{V_0 \times V_0} ((\mathbf{X}^{(i)})^T \mathbf{X}^{(i)}) \tag{12}$$

$$= \nabla_{V_0 \times V_0} \left(\sum_{i=1}^{n} ((\mathbf{X}^{(i)})^T \mathbf{X}^{(i)}) \right)$$

$$= \nabla_{V_0 \times V_0}(\mathbf{A})$$

And we have,

$$(\nabla_{all \times V_0}(\mathbf{X}^{glb}))^T \mathbf{Y}^{glb}$$

$$= \nabla_{V_0} \left(\sum_{i=1}^{n} ((\mathbf{X}^{(i)})^T \mathbf{Y}^{(i)}) \right) \tag{13}$$

$$= \nabla_{V_0}(\mathbf{b})$$

If we construct the model with the subset X_V, the LSE of the coefficient is computed as follows:

$$\widehat{\beta}_{X_V}^{glb} = \left((\nabla_{all \times V_0}(\mathbf{X}^{glb}))^T \nabla_{all \times V_0}(\mathbf{X}^{glb}) \right)^{-1} (\nabla_{all \times V_0}(\mathbf{X}^{glb}))^T \nabla_{all}(\mathbf{Y}^{glb})$$

$$= (\nabla_{V_0 \times V_0}(\mathbf{A}))^{-1} \nabla_{V_0}(\mathbf{b}) \tag{14}$$

We can get Theorem 1 similarly. Due to space limitation, we omit the details.

Theorem 1. *If we construct the model with the subset* $X_V = \{x_{j_1}, x_{j_2}, \ldots, x_{j_k}\}$ $\subseteq \{x_1, x_2, \ldots, x_{p-1}\}$. *Let* $V_0 = \{1, j_1 + 1, j_2 + 1, \ldots, j_k + 1\}$. *the unexplained variation is*

$$SSE_{X_V}^{glb} = T - (\nabla_{V_0}(\mathbf{b}))^T (\nabla_{V_0 \times V_0}(\mathbf{A}))^{-1} \nabla_{V_0}(\mathbf{b})$$

Note $SSE_{\emptyset}^{glb} = T - \mathbf{b}_1^2/\mathbf{A}_{11} = \sum_{i=1}^{n} \sum_{k=1}^{n^{(i)}} \left(y_k^{(i)} - \overline{y^{glb}} \right)^2$. With the help of Theorem 1, we can conduct stepwise regression. At each step we can compute F_k and the detailed procedure is listed in **Algorithm 1**.

6 Attack Analysis

Each node in the sensor network may be curious about the privacy of other nodes or have been breached so that the attacker can threaten other's privacy. For example, in the people-centric urban sensing system, each sensor (such as smart phone) can record the data transmitted to it and has the ability to derive some information about the privacy of other nodes. In our aggregation algorithm, node $V^{(i)}$ sends its perturbed vector $\mathbf{f}^{(i)}(j)$ to its brother node $V^{(j)}$, which implicates some formation of its feature data $S^{(i)}$. Then its brother nodes may have the

---

Algorithm 1. The algorithm for privacy-preserving stepwise regression

---
Require:
    $\mathbf{A}, \mathbf{b}, T, \alpha_E, \alpha_D$;
Ensure:
1: $old = \emptyset; complete = \{x_1, x_2, \cdots, x_{p-1}\}, n = \mathbf{A}_{11}$
2: **for** each $x_k \in complete - old$ **do**
3:    $new = old \bigcup \{x_k\}$;
4:    compute F_k according to Eqn.(11) and theorem 1;
5: **end for**
6: let $k_0 = arg\ max_k\{F_k\}$;
7: $p = P(F(1, n - |old| - 2) \geq F_{k_0})$;
8: **if** $p \geq \alpha_E$ **then**
9:    **return** old ;
10: **else**
11:    $old = old \bigcup \{x_{k_0}\}$;
12: **end if**
13: **for** each $x_k \in old$ **do**
14:    $new = old - \{x_k\}$;
15:    compute F_k according to Eqn.(11) and theorem 1;
16: **end for**
17: let $k_0 = arg\ min_k\{F_k\}$;
18: $p = P(F(1, n - |old| - 1) \geq F_{k_0})$;
19: **if** $p \geq \alpha_D$ **then**
20:    $old = old - x_{k_0}$;
21: **end if**
22: goto 2;

---

ability to derive $S^{(i)}$ in some way. Since $S^{(i)}$ is the feature data of node $V^{(i)}$'s private data, it is possible for brother nodes to derive the privacy of node $V^{(i)}$. On the other hand, the parent of a community can get the aggregation result of the community, and then it may try to deduce the privacy of its children nodes. So we have to investigate the ability of the aggregation algorithm to preserve the privacy of nodes.

After the aggregation algorithm performed, the parent node gets the aggregation result of the community. If the parent node adopts maximize likelihood estimation to reconstruct the private data of their children nodes, it has to solve an optimal problem subject to non-linear equalities, which is very difficult to solve and will bring about huge computation load. For example, if the user wants to reconstruct the original data of all the sensor nodes, it has to solve an optimal problem which is non-convex and has $\sum_{i=1}^{n} n^{(i)}p$ optimal variables. The parent also has to face a difficult permutation problem. Details about this point have been discussed in [1]. We should remember that, except the leaf nodes, all the other nodes will aggregate their own private data with the data of all their children nodes, which will bring about more difficulties to reconstruct. So we can conclude that the privacy of each node is very difficult to be disclosed to its parent when we adopt the aggregation algorithm.

In the following, we demonstrate that the aggregation result of a community won't be disclosed to its children nodes. Suppose there is a community composed of node $\{V^{(1)}, V^{(2)}, \cdots, V^{(c)}\}$, and node $\{V^{(1)}, V^{(1)}, \cdots, V^{(N)}\}$ collide with each other to deduce the aggregation result of the node $\{V^{(N+1)}, V^{(N+2)}, \cdots, V^{(c)}\}$. If they succeed, then the aggregation result of this community is disclosed to them. If they adopt the maximize likelihood estimation to deduce the aggregation result, the problem can be formulated as follows:

$$min \sum_{k=1}^{|\mathbf{S}|} \left(\sum_{i=N+1}^{c} \mathbf{S}_k^{(i)} - MLS_k \right)^2$$

where MLS_k is the maximum likelihood value of the random variable $\sum_{i=N+1}^{c} \mathbf{S}_k^{(i)}$. According to our aggregation algorithm, the collusive nodes can get the following equalities:

$$
\begin{cases}
f_1^{(N+1)}(t) = \sum_{l=1}^{c} \mathbf{S}_l^{(N+1)} \rho_t^{l-1} \\
f_2^{(N+1)}(t) = \sum_{l=c+1}^{2c} \mathbf{S}_l^{(N+1)} \rho_t^{l-c-1} \\
\cdots, \cdots, \cdots, \\
f_M^{(N+1)}(t) = \sum_{l=(M-1)c+1}^{Mc} \mathbf{S}_l^{(N+1)} \rho_t^{l-(M-1)c-1}
\end{cases}
$$
$$\cdots, \cdots, \cdots$$
$$t = 1, 2, \cdots, N;$$

The equalities above are about each node's feature data. If we want to get the aggregation result of them, we need to merge these equalities. Let $\sum_{i=N+1}^{c} \mathbf{S}_k^{(i)} = \mathbf{S}_k'$, and we can get:

$$
min \sum_{k=1}^{|\mathbf{S}'|} \left(\mathbf{S}_k' - MLS_k' \right)^2
$$
$$
\begin{cases}
\sum_{l=1}^{c} \mathbf{S}_l' \rho_t^{l-1} = \sum_{i=N+1}^{c} f_1^{(i)}(t) \\
\sum_{l=c+1}^{2c} \mathbf{S}_l' \rho_t^{l-c-1} = \sum_{i=N+1}^{c} f_2^{(i)}(t) \\
\cdots, \cdots, \cdots, \\
\sum_{l=(M-1)c+1}^{Mc} \mathbf{S}_l' \rho_t^{l-(M-1)c-1} = \sum_{i=N+1}^{c} f_M^{(i)}(t)
\end{cases}
\qquad (15)
$$
$$t = 1, 2, \cdots, N;$$

It is an optimization problem only subjects to linear equality constraint, the attacker can adopt Lagrange multiplier method to solve it. The Hesse matrix of this problem is $2E$, which means that it is positive definite. In addition it is a convex optimization problem, so the solution is just the global minimum value. Then we get a linear equation system. The matrix formulation of this equation system is displayed in the following.

$$
\begin{bmatrix} 2\mathbf{E} & \mathbf{V}^T \\ \mathbf{V} & 0 \end{bmatrix} \begin{bmatrix} \mathbf{S}' \\ \lambda \end{bmatrix} = \begin{bmatrix} 2\mathbf{MLS}' \\ \mathbf{F}' \end{bmatrix}
\qquad (16)
$$

where

$$\mathbf{V} = \begin{bmatrix} \mathbf{E}_{11}\mathbf{V}_1 & \mathbf{E}_{12}\mathbf{V}_1 & \cdots & \mathbf{E}_{1M}\mathbf{V}_1 \\ \mathbf{E}_{11}\mathbf{V}_2 & \mathbf{E}_{12}\mathbf{V}_2 & \cdots & \mathbf{E}_{1M}\mathbf{V}_2 \\ \cdots & \cdots & \ddots & \cdots \\ \mathbf{E}_{11}\mathbf{V}_N & \mathbf{E}_{12}\mathbf{V}_N & \cdots & \mathbf{E}_{1M}\mathbf{V}_N \end{bmatrix}, \mathbf{MLS}' = \begin{bmatrix} MLS'_1 \\ MLS'_2 \\ \cdots \\ MLS'_{|S|} \end{bmatrix},$$

$$\mathbf{F}' = \begin{bmatrix} \sum_{i=N+1}^{c} f_1^{(i)}(1) \\ \vdots \\ \sum_{i=N+1}^{c} f_M^{(i)}(1) \\ \sum_{i=N+1}^{c} f_1^{(i)}(2) \\ \vdots \\ \sum_{i=N+1}^{c} f_M^{(i)}(N) \end{bmatrix}, \mathbf{V}_i = \begin{bmatrix} 1 & \rho_i & \rho_i^2 & \cdots & \rho_i^{c-1} \\ 0 & 0 & 0 & \cdots & 0 \\ 0 & 0 & 0 & \cdots & 0 \\ \cdots\cdots\cdots & & \ddots & \cdots \\ 0 & 0 & 0 & \cdots & 0 \end{bmatrix},$$

$\mathbf{E}_{ij}$ is an elementary transformation, which swaps the $i-th$ row and $j-th$ row of the unit matrix. After getting the solution of above problem, the collusive nodes add the solutions and the feature data of themselves to get the *estimated* aggregation result of the whole community.

$$\mathbf{S}^{(eglb)} = \sum_{i=1}^{N} \mathbf{S}^{(i)} + \mathbf{S}'$$

In order to measure the difference between the *estimated* aggregation result and the real one, we suppose that these malicious nodes will construct the *estimated* model with $\mathbf{S}^{(eglb)}$ according to Eqs. (6) and (10). Then we compare the *estimated* model with the *real* model derived from the real aggregation result based on SSE.

We conduct the experiment on the data set housing [9], which concerns the task of predicting housing values in the area of Boston. It has 13 explanatory variables, so the length of the feature data is 120. We build the aggregation tree with height 2 and 8 leaf nodes, where the user is the root, so $M = 15$. We subsample from housing [9] and get a new data set with 500 records. We divide the new data set into training and test set. We use the product of mean of all the collusive nodes' feature data and $c - N$ as the **MLS**. In fact in many cases, the equation system (16) has a very large condition number resulting in an inaccurate solution, so we adopt QR decomposition to solve it.

Figure 2 shows the fitness level of *estimated* model and the *real* one on the training set when there is only one macilious node. The fitness error of the *real* model is 8922.1, while that of the *estimated* model is 1713316.4.

With the number of malicious nodes increasing, the *estimated* model gets close to the *real* model. After we carry out the attack 7 times, we get Fig. 3. We can find that the more the collusive nodes, the closer the *estimated* model get to the *real* one. Because the feature data is perturbed by random number, we can find some fluctuation between 7 curves. When there are 7 collusive nodes,

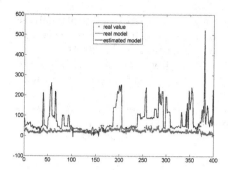

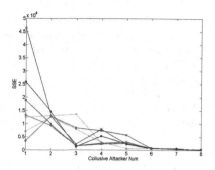

Fig. 2. Fitness level under single malicious node

Fig. 3. Convergence of the fitness SSE under multiple malicious node

we find the mean of the 7 *estimated* models' SSE is 43668, which is 5 times as the real one.

We can find that the *estimated* model deduced by a single malicious node or even 7 collusive nodes is quite different from the real one. So we can make a conclusion that the aggregation algorithm has a strong ability to protect the aggregation result.

7   Conclusion

To preserve the privacy, we propose a data aggregation algorithm to aggregate the necessary information for constructing linear regression model in the sensor network. We point out that the user can conduct some complex analyses without more information transmitted. Having done theoretical analyses and experiment, we find that our method can protect the privacy of each node and the aggregation result of each community very well. With the help of our work, the privacy-preserving modeling and analyses for linear regression can really come into use.

References

1. Ahmadi, H., Pham, N., Ganti, R., Abdelzaher, T., Nath, S., Han, J.: Privacy-aware regression modeling of participatory sensing data. In: Proceedings of the 8th ACM Conference on Embedded Networked Sensor Systems. pp. 99–112. ACM (2010)
2. Chan, H., Perrig, A., Song, D.: Secure hierarchical in-network aggregation in sensor networks. In: Proceedings of the 13th ACM conference on Computer and communications security. pp. 278–287. ACM (2006)
3. Chen, X., Makki, K., Yen, K., Pissinou, N.: Sensor network security: a survey. IEEE Commun. Surv. Tutorials **11**(2), 52–73 (2009)
4. Ding, M., Cheng, X., Xue, G.: Aggregation tree construction in sensor networks. In: 2003 IEEE 58th Vehicular Technology Conference VTC 2003-Fall, vol. 4, pp. 2168–2172. IEEE (2003)

5. Du, W., Han, Y., Chen, S.: Privacy-preserving multivariate statistical analysis: Linear regression and classification. In: Proceedings of the 4th SIAM International Conference on Data Mining. vol. 233. Lake Buena Vista, Florida (2004)
6. Feng, T., Wang, C., Zhang, W., Ruan, L.: Confidentiality protection for distributed sensor data aggregation. In: The 27th IEEE Conference on Computer Communications INFOCOM 2008 IEEE (2008)
7. He, W., Liu, X., Nguyen, H., Nahrstedt, K., Abdelzaher, T.: Pda: Privacy-preserving data aggregation in wireless sensor networks. In: 26th IEEE International Conference on Computer Communications INFOCOM 2007, pp. 2045–2053. IEEE (2007)
8. He, W., Nguyen, H., Liu, X., Nahrstedt, K., Abdelzaher, T.: IPDA: an integrity-protecting private data aggregation scheme for wireless sensor networks. In: IEEE Military Communications Conference MILCOM 2008, pp. 1–7. IEEE (2008)
9. L, T.: http://www.liaad.up.pt/ltorgo/regression/datasets.html
10. Li, N., Zhang, N., Das, S.K., Thuraisingham, B.: Privacy preservation in wireless sensor networks: a state-of-the-art survey. Ad Hoc Netw. **7**(8), 1501–1514 (2009)
11. Powell, M.: An efficient method for finding the minimum of a function of several variables without calculating derivatives. Comput. J. **7**(2), 155–162 (1964)
12. Sanil, A., Karr, A., Lin, X., Reiter, J.: Privacy preserving regression modelling via distributed computation. In: Proceedings of the Tenth ACM SIGKDD International Conference on Knowledge Discovery And Data Mining, pp. 677–682. ACM (2004)
13. Shi, J., Zhang, R., Liu, Y., Zhang, Y.: Prisense: privacy-preserving data aggregation in people-centric urban sensing systems. In: 2010 IEEE Proceedings of INFOCOM, pp. 1–9. IEEE (2010)
14. Xing, K., Wan, Z., Hu, P., Zhu, H., Wang, Y., Chen, X., Wang, Y., Huang, L.: Mutual privacy-preserving regression modeling in participatory sensing. In: 2013 IEEE Proceedings of INFOCOM, pp. 3039–3047. IEEE (2013)

Research of Terminal Transparent Encryption Storage Mechanism for Multi-cloud Disks

Chaowen Chang[✉], Shuai Wang, Yutong Wang, and PeiSheng Han

Zhengzhou Institute of Information Science and Technology,
Zhengzhou, Henan, China
changchaowen5@163.com, {921269551,bigcat123}@qq.com,
wyt87380345@126.com

Abstract. In the view of data storage security for the existing cloud disk, this paper presents a terminal transparent encryption storage mechanism which is irrelevant to existing cloud storage security mechanism. This mechanism can solve the problem of the user's trusting in the cloud disk storage fundamentally. By the method of formatting into N size fixed files, this mechanism of cloud disk can be seen as a RAW storage device, which stores data into the physical sector. In Multi-cloud disks mode, a dynamic selection strategy based on partitioning factor is proposed to promote the efficiency of file storage by uneven parallel cloudy disk access. To prove the feasibility of data recovery, an interrupt retransmission mechanism and an encoding redundancy backup mechanism are put forward, respectively. By utilization of transferring the control of data security from the cloud storage to the client side, the effective terminal transparent encryption storage mechanism for multi-cloud disks is guaranteed, which not only solves the problem of the user's trusting in the cloud disk storage fundamentally and concerns of the cloud storage operator reliability but also enhances the security and reliability of the file storage and improves the file access rate as well as storage rate of document transparent encryption.

Keywords: Cloud disk · Transparent encryption · Control · Security · Redundancy backup

1 Introduction

With the constantly advancing cloud computing technology [1–3], cloud storage [4] is becoming a new storing way applied deeply in every aspect of daily life. Currently, there are numerous cloud storage platforms over the world [5].The relatively prominent representatives overseas are Simple Storage Service (S3), Nirvanix, and so forth, which can provide a lot of low-cost file storage service; domestic prominent representatives are Baidu Cloud disk, Kingsoft Disk, Sina Plate, Nuts Clouds, Cool Disk, and 360 Cloud. Cloud disk is the product of the continuous information technology development, and more users choose it to store privacy data (address book, photos, etc.); it can provide users with real-time data storing, reading, and other services via the Internet, which is the next-generation network storage service platform with mobility, flexibility, and on-demand functions; it can effectively control the cost of inputs and meet the

© Springer International Publishing Switzerland 2015
W. Qiang et al. (Eds.): CloudCom-Asia 2015, LNCS 9106, pp. 367–385, 2015.
DOI: 10.1007/978-3-319-28430-9_28

diverse needs of business users continuously, and it is more flexible and convenient compared with traditional physical media; users can easily obtain the required data from it via the Internet without carrying data storing physical medium [3].

According to survey, the most important feature of cloud disk data storage is the following: all users' data is stored in the storage service provider of cloud disk; security and reliability of stored data are ensured by the storage service providers of cloud disk [6, 7]; thus, security risks of stored data are coming up; users lose the absolute control over their own data and only can rely on the secure storage mechanisms provided by the storage service providers of cloud disk to ensure the security of stored data [8], but users' harbor doubts about the security capabilities of them which will impede the shift of data and business to the cloud storage service platforms and security issues of data storage are the biggest obstacle in promotion and application of cloud storage disk [9]. Users' trust problem to cloud disk storage service providers cannot be solved ultimately relying only on moral constraints or management rules [10], it must be addressed on technology; that is, users must take control of the stored data.

For the above security problems in cloud storage disk, the literature [11] proposes encryption tools to solve the integrity and consistency issues of users' data and studies data recoverability mechanism; but it cannot fundamentally solve the security issues of users' data with poor ease. Literature [12] proposes data security storage system model and strategy for cloud storage in which way the control right of data is put entirely to user terminal; data is completely disrupted in user terminal before storage by using symmetric encryption technology and nonstandard coding techniques of firmware or hardware to ensure that the stored data cannot be cracked by unauthorized users, so as to ensure security and effectiveness of the stored data, but it does not analyse the efficiency of reading data and the reliability of users' data; however, data encryption and decryption methods can greatly affect the efficiency of reading data. The encrypted file system proposed in literature [13–15] can solve the problem of confidentiality of user's stored data in a certain extent, but it is only considered about confidentiality, and it has great limitation in solving the problem of security of data storage. Through a combination of RAID redundancy ideologies, literature [16] presents a distributed storage system, using the distributed RAID technology to store data after redundant management of stored data, which can guarantee the availability of stored data, but it is mainly designed for the storage mechanism of storage platform, and user terminals do not participate in the process to protect the reliability of stored data; for users, it is still incredible and data confidentiality cannot be protected.

Considering the ideas of cloud disk storage security in literatures above, this paper designs a terminal transparent encryption storage system for Multi-cloud disk (TTEFMC) independent of disk storage, and the system combines organically virtual disk technology, transparent encryption and decryption technology, and redundant backup technology to achieve terminal secure, rapid, transparent encryption and decryption and effective management of stored data; the control right of data belongs to user terminal, through the transfer of control of data to reduce user's excessive reliance on cloud disk storage service providers, dispelling the user's security concerns to cloud disk storage service providers. In the process of system design, this paper research focuses on storage mapping management system, data block storage mechanism, data encryption and decryption mechanism, retransmission mechanism, and coding inter-ruption redundant backup mechanism.

2 System Design

TTEFMC abstracts each cloud disk applied by users into RAW storage device and virtualizes each RAW storage device into a local disk and then formats it; the disk storage space is unified allocated by the virtual file system, and virtual disk partition tables, file directory, and file allocation table (FAT) will be stored in the USB Key; only with USB Key, users can obtain the required data; it not only improves the usability of the system but also further enhances data storage security. Virtual file system receives system calls from user layer and data access from core layer, and it redirects relevant data access to remote cloud services provider in accordance with certain mapping relationship; the terminal is the leading role in TTEFMC, through a series of terminal security operations, and users will have absolute control of the stored data. Users complete the file blocks transparent encryption and decryption storage through the operation of the virtual disk file; users' data is stored in remote cloud disk rather than the local computer, but it feels like operating the user's local computer disk and does not change the user's operating habits [17]. Figure 1 is a diagram for the overall design of TTEFMC, in solving the existing problems of cloud disk data storage security, this system improves the security of data and efficiency of file reading through storing users' file blocks into each cloud disk applied by users; it also utilizes the interrupt retrans-mission mechanism and encoding redundancy backup mechanism to improve the availability of data; that is, when a cloud disk breaks down or loses data, the system can still restore the required files by other cloud disk data. Multi-cloud disks block storage for users' files not only improves the efficiency of file access but also shares effectively

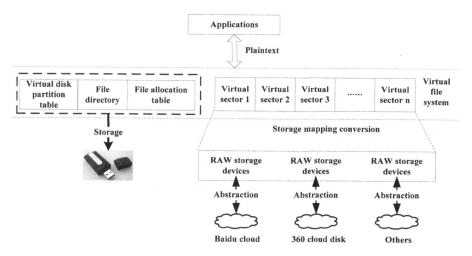

Fig. 1. Design of the system

the security risks of the single cloud disk data storage, and it also enhances the reliability of data storage and significantly improves the performance of the system.

2.1   Storage Mapping Management Mechanism

In TTEFMC, virtual storage mapping principle of the virtual file system is particularly important, the size of the virtual disks is not directly related to the size of cloud disk, and users can create virtual disks according to appropriate virtual disk. Storage mapping management mechanism is responsible for managing the mappings between the data blocks corresponding to the files, file attributes, access control information, the size of stored files, and so on; it mainly includes the mapping of files to the virtual sectors and the virtual sectors to each cloud disk; virtual storage of files is achieved by creating the virtual disk partition tables, file directory, and file allocation table (FAT) through storage mapping management mechanism. The design of storage mapping management mechanism of this system is based on the file management mechanism of the FAT file system. Figure 2(a) shows the file management mechanism of the FAT file system. Compared with the FAT file system, Fig. 2(b) shows the storage mapping management mechanism of the virtual file system; it adds the conversion table, from logical block to the file block (LFB), between FCB and FAT table; it achieves the storage mapping of the virtual sectors to each cloud disk by LFB table and virtual storage of users' files by utilizing 3-level mapping management mechanism.

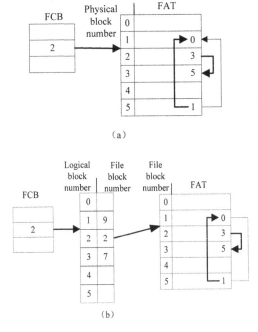

Fig. 2. (a) Management mechanism of FAT file system; (b) Storage mapping management mechanisms of the virtual file system

In the process of virtual storage, the structure of the virtual sector addresses the sum of the corresponding volume number and the offset address of the data block. As the virtual sector can be mapped to any cloud disk by the network without any impact by geographical location, volume mapping table and the block mapping table are required to implement virtual mapping. The specific virtual mapping process is shown in Fig. 3. The virtual file system divides each file into a plurality of data blocks, which stores the data block corresponding to each file on each cloud disk. Each file has a directory entry in the file directory that the corresponding directory entry and index node of this file can be found by searching the directory and then the actual physical location of the storage file is found. Throughout the system, the files are stored in accordance with a uniform manner. Since virtual file system only communicates with the virtual disk, the association of the file with the virtual sector and the virtual sector with the cloud disk is completely transparent for the file system. Therefore, it is conducive to the management and utilization of the stored data and provides very good protection for the stored data.

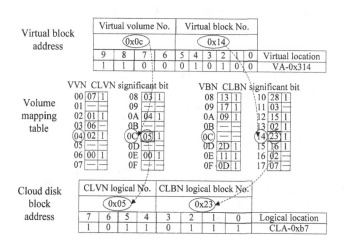

Fig. 3. Virtual storage mapping process

2.2 Data Block Storage Mechanism

The main idea of the data block storage is dynamically to divide the storage file into a plurality of data blocks with the same size according to the file size in order to store every data block to each cloud disk applied by users and improve the upload and download rate of data by parallel transmission [18]. Also storing the data block to each cloud disk can improve the security of the stored files.

A Dynamic Selection Strategy Based on Partitioning Factor. In the ideal state, while storing a user file, the first data block can be encrypted and transferred to the first cloud disk, the second data block to the second cloud disk, ..., and the Nth data block to the Nth cloud disk. When a user accesses the data stored in the cloud disk, the system will use parallel reading mode and the client terminal will start multiple threads; namely, users can simultaneously read the data from the cloud disk 1–N. Thus, the data access rate from the cloud disk has been raised $N-1$ times.

In practice, the file size must be considered while storing the data block since the systems establish communication links with multiple cloud disks when multiple cloud disks simultaneously upload or download a smaller file, which will bring extra connection overhead. As a result, the file uploaded or downloaded speed is not necessarily faster than that of a single cloud disk. Meanwhile, the load state of the multiple cloud disks in the system must be considered, as some disks may store relatively more data; namely, they are in an overload state while some cloud disks may store less data, which causes the storage resource utilization rate of the whole system to be out of balance.

Storage space can be effectively used with the utilization of fixed-size partition strategy, but it is likely to cause one data block to differ greatly from others, and the unevenness of data blocks will affect the overall performance of the system. Above the analysis of the partition strategy explains that users' block size is not the same for different file types when storing data in blocks. Therefore, all the divided blocks with the same size certainly are unable to meet the requirements of various file types, which will greatly restrict the file transfer efficiency and seriously affect the quality enhancement of the storage services by cloud disk. A reasonable data partition strategy can effectively improve the utilization of cloud disk storage space, as well as the speed of data read-write, which will greatly impact the performance of the system. For the analysis above, this paper proposes a dynamic selection strategy based on partitioning factor, and the basic idea is as follows: in accordance with the number of cloud disks and file sizes, the storage file dynamically is equally partitioned, while, for small files below a threshold, there is no partitioning. While the huge size file must be partitioned in order to achieve better transmission efficiency, and for the files to be partitioned, the partition granularity should be as large as possible, then the data blocks are uniformly stored to each cloud disk. Thus, it can take full advantage of the storage space of each cloud disk while reducing the amount of file metadata. If the partition is unreasonable, the download efficiency may be very low, sometimes even lower than that of downloading data from a single cloud disk.

When multiple cloud disks are storing the file *File*, the original *File* needs to be divided into a plurality of data blocks; then the specific methods of data partitioning are as follows.

First, make the following assumptions.

(i)   The file size is S.
(ii) Divide the file into n blocks and $n \in N^+$.
(iii) The cloud disk number is N and $N > 1$.
(iv) The data blocks, respectively, are $file_1, file_2, file_3, \ldots, file_n$.
(v)   The data block size is Lj; namely, the partition threshold Z_i of the partition factor $f = \{Z_0, Z_1, Z_2, Z_3, \ldots \ldots, Z_p\}$, where $j \in \{1, 2, 3, \ldots \ldots, n\}$, $i \in \{0, 1, 2, 3, \ldots \ldots, p\}$, $Z_i \in N^+$, and $Z_0 < Z_1 < Z_2 < Z_3 \ldots \ldots < Z_p$.

The starting and ending position of the data blocks $file_1, file_2, file_3, \ldots, file_n$ in the original file are, respectively $(0, L_1 - 1), (L_1, (L_1 + L_2) - 1), (L_1 + L_2, (L_1 + L_2 + L_3) - 1), \ldots, (\sum_{j=1}^{n-1} Lj, S)$. Thus, $file_1 \bigcup file_2 \bigcup file_3 \bigcup file_4 \bigcup \ldots \bigcup file_n = File$ and the intersection of any two data blocks from $file_1, file_2, file_3, \ldots, file_n$ is empty.

The specific description of dynamic selection strategy based on partitioning is shown as follows. Define a set of partitioning factor $f = \{Z_0, Z_1, Z_2, Z_3, \ldots\ldots, Z_p\}$ to get the more flexible block size.

(i) When the size of *File* $S \leq Z_0$, the file will be stored directly after copying and backup without partition.

(ii) When the size of *File* $S > Z_0$, the system will send the polling message R to each cloud disk, and the message R contains the utilization rate λ of each cloud disk and their connection status. At the same time, the system will receive the feedback Q from each cloud disk and obtain the number of cloud disks and the utilization rate λ of each cloud disk through the message Q. Meanwhile, S will orderly calculate with the threshold value Z_i of each block in f to obtain the file block number q. If S can be divisible by Z_i, then $q = S/Z_i$. If not, $q = [S/Z_i] + 1$. The value of $[S/Z_i]$ is less than or equal to the largest integer value of S/Z_i.

Case 1. Orderly calculate S with the threshold value Z_i of each block in f, and if there exists $q \leq N$, partition the File with the threshold value Z_i which meets the condition that the number of data blocks q is the closest to that of cloud disks N. When the data size of last data block is not enough to partition, then it is uniformly filled with "0", according to the cloud disk utilization rate λ to determine which cloud disks are used to store data, and the smaller λ is more likely to be the selected cloud disk for data storage. At last, the q blocks are sequentially stored to the selected disk cloud.

Case 2. All the values sequentially calculated S with the threshold value Z_i of each block in f are greater than the number of cloud disks namely $q > N$, using the greatest threshold Zp to partition the stored *File*, and each cloud disk can store j data blocks, $j = [S/(N \times Zp)]$.

If the remaining data $S - j \times (N \times Z_p) < Z_0$ store them after copying and backup without partition. Otherwise, $S - j \times (N \times Z_p) > Z_0$, $S - j \times (N \times Z_p)$ will calculate sequentially with threshold value Z_i of each block in f to obtain the number of the remaining data blocks d, and if $S - j \times (N \times Z_p)$ can be divisible by Z_i, then $d = [S - j \times (N \times Z_p)]/Z_i$. If not, then $d = [(S - j \times (N \times Z_p))/Z_i] + 1$. Therefore, the remaining data blocks will be allocated and stored to each cloud disk in accordance with the method above in Case 1.

For large files, utilizing the data partitioning strategy and the parallel data transmission will greatly reduce the transfer time and take full advantage of the storage space and service of each cloud disk, so as to provide satisfactory services to users. The dynamic selection strategy based on partitioning factor takes into full account the utilization of cloud disk storage space, while also considering the amount of metadata and the network latency. Dynamic data partitioning can take full advantage of cloud disk storage space and effectively reduce the number of TCP connections by reducing the amount of metadata. Through dynamic data partition strategy, the storage file is partitioned into multiple data blocks and dispersedly stored to different disks, which is beneficial to parallel access to user's data and greatly improves the read-write efficiency.

Data Access Efficiency Analysis. Storing user's data blocks to each cloud disk by utilizing dynamic selection strategy based on partitioning factor can effectively improve the efficiency of files access. Next, the data access efficiency will be analyzed in detail. Parameters involved in analysis procedure of data access efficiency are shown in Table 1.

Table 1. Parameter description.

Parameters	Description
T_Q	Time for client sending a request transmission
T_C	Time for client connecting to the cloud disks
S	Storage file size
Z_p	Maximum of data blocks
L_i	Size of data blocks
B	Network bandwidth of file transfer process
N	Number of cloud disks

The total time required for storing users' files to a single disk is

$$T = T_Q + T_C + \frac{S}{B} \tag{1}$$

In case 1, the total time required for block transmitting files is

$$T_1 = T_Q + T_C + \frac{L_i}{B} \tag{2}$$

Further, $T_1 - T = (Li/B) - (S/B) < 0$.
In case 2, the total time required for block transmitting files is

$$T_2 = T_Q + 2T_C + \frac{Z_P}{B} \times j + \frac{L_i}{B}, j = \left\lceil \frac{S}{N \times Z_p} \right\rceil \tag{3}$$

Further,
$T_2 - T = TC + (Zp/B) \times j + (Li/B) - (S/B) = TC + (Zp \times j + Li - S)/B.$ Consider

$$\frac{Z_P \times j + L_i - S}{B} < 0 \tag{4}$$

In the partitioning transfer process of large files, the time of establishing the communication links between clients and cloud disks will be far less than that of file transmission; thus, $TC < < (S - Zp \times j - Li)/B$, namely, $T_2 - T < 0$.

In case 1 and 2, the total time required for block transmitting files is less than that for storing users' files to a single disk, so utilizing the reasonable data partitioning strategy can effectively improve the efficiency of file access.

2.3   Data Encryption and Decryption Mechanism

The operation of data encryption and decryption mainly prevents the information from leaking out by unauthorized users after encryption and decryption of the block data. Data encryption and decryption mechanism is as follows. After intercepting the IRP request of writing data to cloud disk, the filter driver will compare the user privilege for the current session with the access control privilege of the awaiting file, to determine the session user's subsequent action. If they are consistent, call a corresponding encryption algorithm to encrypt plaintext in IRP and finally store the encrypted data in cloud disk via network. When IRP request is reading data from cloud disk, after authority comparison the filter driver will call a corresponding decryption algorithm to decrypt the received ciphertext and then write the plaintext into IRP data segment and return to the application, which implements the encryption and decryption to the stored data [19].

Data encryption and decryption operate in the kernel mode, which is embedded with encryption algorithm, through intercepting, filtering IRP_MJ_WRITE and IRP_MJ_READ, and other related routines to complete data encryption and decryption [20]. Among them, the data encryption is completed in the IRP_MJ_WRITE routine of the filter driver, and the decryption is operated in IRP_MJ_READ routine. Attached to the volume device object (VDO), the filter driver can filter all IRP and fast I/O calls for this volume, such as creating, opening, reading, and writing requests. The operation mechanism of data encryption and decryption is shown in Fig. 4, and the main course consists of two parts: the user space part and the kernel space part. The user space part is primarily responsible for transmitting control commands to the kernel space and appointing the encryption rules based on user's needs, such as the protection of file directory and file type, while it controls start and stop of the data encryption and decryption module. When application-layer control module sends the requests of data operation, the system call will notify the I/O manager to transfer the operation requests into IRP which can be identified by the kernel driver and sent to the data encryption and decryption module. After reading the encryption rules, it will judge the operation to determine whether the current request should be processed by the filter driver, and then it will process encryption and decryption according to the request type. Data encryption and decryption module is mainly responsible for storing transparent encryption and decryption of data. Key management module is mainly responsible for the access of key information, and the system uses a mobile device (USB Key, etc.) to store keys to ensure security and effectivity of the key information.

Files can be encrypted and decrypted automatically through the data encryption and decryption module, and they are stored in each cloud disk in the form of ciphertext and delivered to users in the form of plaintext while they are being used, but the plaintext of files is presented to users temporarily. In cloud disk, the actual storage format of the files remains ciphertext. Protected files through user's processing are encrypted and decrypted automatically without user controlling, and the encryption and decryption are completely transparent for user without feeling the conversion between ciphertext and plaintext.

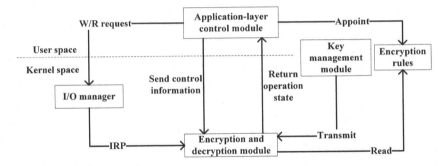

Fig. 4. Operation mechanism of data encryption and decryption

2.4   System Reliability Design

Interruption Retransmission Mechanism. Data block is stored in each cloud disk, and the system will simultaneously open multiple threads to improve the efficiency of reading files. Each thread establishes communication link with cloud disk, when the client is reading a file, multithreads will receive the data block of files at the same time. But during the data transmission, the data transfer may be interrupted due to power outages, leaving off the network and other unforeseen circumstances. Then, the data interruption retransmission mechanism will be demanded. The stored data in system is "atomic", and the successfully transmitted data is stored in the receiver and the remaining data is stored in the sender. There is an interruption in the last receiving of the nth block by receiver, and then the receiver will receive $n - 1$ data blocks in default and the nth data block will be discarded. Every time receiving a data block, the receiver will send a confirmation message to the sender, and it is assumed that the data blocks which the sender has transmitted are file $=$ file1 $+$ file2 $+$ file$n$ $+$ $\cdots$ $+$ file$n$. Then, sender will retransmit data blocks from file$n$, and the receiver will receive and store data blocks from file$n$ again.

Code Redundancy Backup Mechanism. Data code redundancy backup refers to the fact that the raw data is divided into N parts, through the code redundancy, and the equal N data blocks, respectively, generate N data blocks with redundancy. Through redundancy backup, when a cloud disk applied by user breaks down, it will not affect the normal user's access requests with the loss of data, and the system can recover the lost data according to other cloud disks' data blocks. Compared with the security mechanism of a single cloud disk, the reliability improves an order of magnitude.

Redundant management and storage access of copy backup mechanism [21] are relatively simple, but, in order to ensure the availability of stored data, they will consume a large number of storage resources. However, through blocking and encoding the raw data, any certain amount of code data block can be accessed, so that the data decoding and reconstructing can be completed and the original data can be restored. The size of redundant data blocks is the same with the original ones, and the required memory resources to store the redundant data can also be greatly reduced. Figure 5 shows the flowchart of code redundancy backup and recovery.

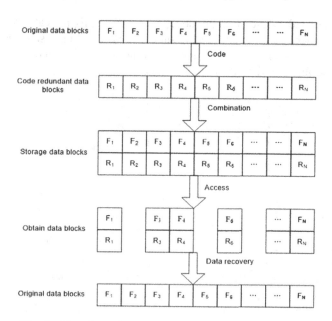

Fig. 5. Flowchart of code redundancy backup and recovery

(i) Redundancy Treatment. In $N + i$ code redundancy backup mode, the N copies of data blocks and the redundant data of each data block combine to store. The redundant data of every data block in $N + 0$ code redundancy backup mode is an XOR of all data blocks; thus, the redundant data block is $R = F1 \oplus F2 \oplus F3 \oplus F4 \oplus \cdots \oplus FN$. The redundant data R is also the check field of storing data, so it is required to generate check fields in the $N + 0$ code redundancy backup mode every time storing data and store the field into the USB Key to test the integrity of stored data. $N + i$ code redundancy backup mode is shown in Table 2, in which $i \in (0, N - 1)$. As i increases, data backup and recovery capabilities will continue to develop, but the problems of the amount of data calculation will increase. However, this can achieve data recovery after two or more of the damaged cloud disks.

Table 2. N + i code redundancy backup mode.

Mode	Data block						
	F_1	F_2	F_3	...	F_j	...	F_N
$N + 0$	R_{01}	R_{02}	R_{03}	...	R_{0j}	...	R_{0N}
$N + 1$	R_{1N}	R_{11}	R_{12}	...	$R_{1(N-1+j)}$	...	$R_{1(N-1)}$
$N + 2$	$R_{2(N-1)}$	R_{2N}	R_{21}	...	$R_{2(N-2+j)}$	...	$R_{2(N-2)}$
$N + 3$	$R_{3(N-2)}$	$R_{3(N-1)}$	R_{3N}	...	$R_{3(N-3+j)}$	...	$R_{3(N-3)}$
$\vdots$	$\vdots$	$\vdots$	$\vdots$	$\vdots$	$\vdots$	$\vdots$	$\vdots$
$N + i$	$R_{i(N-i+1)}$	$R_{i(N-i+2)}$	$R_{i(N-i+3)}$	...	$R_{i(N-i+j)}$	...	$R_{i(N-i)}$

In $N + i$ code redundancy backup mode, $F_1, F_2, F_3, \ldots, F_N$ indicates N equal size of data blocks, R_{ij} denotes the jth redundant subblock right shifting i steps, referred to as M_j, where $i \in (0, N-1)$, $j \in (1, N)$, i represents the mode of redundancy backup, and j represents the jth redundant data block, where

$$
\begin{aligned}
M_j = R_{ij} = {} & F_{(j+i)\bmod N} + F_{(j+i+1)\bmod N} \\
& + F_{(j+i+2)\bmod N} + F_{(j+i+3)\bmod N} + F_{(j+i+4)\bmod N} \\
& + F_{(j+i+4)\bmod N} + \cdots + F_{[j+(N-2)]\bmod N} \\
& + F_{[j+(N-1)]\bmod N}
\end{aligned}
\tag{5}
$$

But in $N + 0$ code redundancy backup mode, $R_{01} = R_{02} = R_{03} = R_{04} = \cdots = R_{0j} = \cdots = R_{0N} = R$. From Table 2, the essence of code redundancy backup is the memory block that combined the data block F_{j+i} with the redundant data block R_{ij} obtained by XOR. If any n ($n \leq N/2$) data blocks are lost, the original data can be restored by use of the remaining data blocks and redundant blocks.

List Treatment. To achieve recovery of the original data through the list data, divide the raw data into N parts, generate information through redundancy treatment, generate the stored data blocks with combination of the N copies data and the redundant data, and list-treat the stored data blocks to recover the stored files. Figure 6 is the flowchart of list treatment.

Stored data after data code redundancy backup in the client terminal, which can effectively prevent data leakage and damage on the third-party cloud disk storage platform. Divide original data into N data blocks, select the appropriate code redundancy backup mode for the N data blocks, and obtain N redundant data blocks. The loss of data blocks in transmission results in obtaining $m \leq N$ data blocks, but, as long as $m \geq N/2$, the recovery mechanism can recover the original files using any m data blocks.

Recoverability of Storage Data. Mathematical induction proves the feasibility of original data recovery, and there are two cases of continuous lost and discontinuous lost.

(i) Data Block Continuously Lost. Assume that n data blocks are lost continuously, when n = 1, that is missing data block Fj, under $N + 0$, $N + 1$ mode, using the redundancy Mj, R under data block Fj + 1, then it is easy to get $Fj = Mj \oplus R$; Assuming when $n = k(k > 1)$, that is, missing data blocks $F_j, F_{j+1}, \ldots, F_{j+k-1}$, under $N+0, N+k$ mode, using the redundancy $M_j, M_{j+1}, \ldots, M_{j+k-1}$ data blocks under data blocks $F_{j+k}, F_{j+k+1}, \ldots, F_{j+2k}$ and data blocks $F_{j+k}, F_{j+k+1}, \ldots, F_{j+2k-1}$, thus:

$$
\begin{aligned}
F_{j+k-1} &= F_{j+k} \oplus F_{j+k+1} \oplus \ldots \oplus F_{j+2k-1} \oplus M_{j+k-1} \oplus R \\
F_{j+k-2} &= F_{j+k} \oplus F_{j+k+1} \oplus \ldots \oplus F_{j+2k-2} \oplus M_{j+k-1} \oplus M_{j+k-2}
\end{aligned}
$$

$$\cdots\cdots$$

$$
F_j = F_{j+k} \oplus M_{j+1} \oplus M_j
$$

$$\tag{6}$$

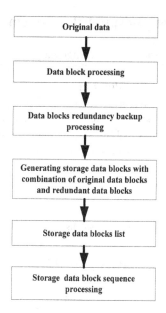

Fig. 6. Flowchart of list treatment

So,

$$F_{j+k} = F_{j+k-1} \oplus F_{j+k+1} \oplus \ldots \oplus F_{j+2k-1} \oplus M_{j+k-1} \oplus R,$$
$$= F_{j+k+1} \oplus F_{j+k+2} \tag{7}$$

When $n = k+1$, under $N+0, N+k+1$ mode, using the redundancy $M_j, M_{j+1}, \ldots, M_{j+k-1}$ under data blocks $F_{j+k}, F_{j+k+1}, \ldots, F_{j+2k}$ and data blocks $F_{j+k}, F_{j+k+1}, \ldots, F_{j+2k-1}$, lost data blocks $F_j, F_{j+1}, \ldots, F_{j+k-1}, F_{j+k}$ can recover the original data. So, the assumption is reliable.

When missing $n(n \leq N/2)$ data blocks, that is the loss of data blocks $F_j, F_{j+1}, \ldots, F_{j+n-1}$, the lost data can be recover only if there has the redundancy $M_j, M_{j+1}, \ldots, M_{j+n-1}$ under data blocks $F_{j+n}, F_{j+n+1}, \ldots, F_{j+2n}$ and data blocks $F_{j+n}, F_{j+n+1}, \ldots, F_{j+2n-1}$, under $N+0, N+n$ mode as follows:

$$F_{j+n-1} = F_{j+n} \oplus F_{j+n+1} \oplus \ldots \oplus F_{j+2n-1} \oplus M_{j+n-1} \oplus R$$
$$F_{j+n-2} = F_{j+n} \oplus F_{j+n+1} \oplus \ldots \oplus F_{j+2n-2} \oplus M_{j+n-1} \oplus M_{j+n-2}$$
$$\ldots\ldots \tag{8}$$
$$F_j = F_{j+n} \oplus M_{j+1} \oplus M_j$$

(ii) Data Block Discontinuous Lost. That is equal to a plurality of continuous missing data blocks, which can also use the above method and conclusions to recover the original data.

3   Implementation Mechanism of the System

TTEFMC is based on public cloud as the storage medium, which unitizes the interfaces with the public cloud communication engine. In a computer with the structure of the file system driver supported by Windows operating platform, it virtualizes the stored files into files in the disks in order to be user-friendly for the users. When reading or writing the virtual disk files, it achieves the secure storage of files through a filter driver for file encryption or decryption. TTEFMC includes three parts: public cloud communication engine, encryption filter driver, and virtual disks. Figure 7 is a diagram of TTEFMC.

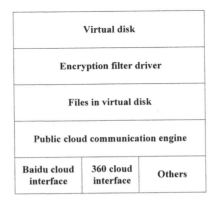

Fig. 7. Implementation diagram of the system

Public cloud communication engine provides unified cloud disk storage interfaces to achieve the physical storage of public cloud.

Encryption filter driver achieves a transparent encryption and decryption when users read or write files, and the encryption key will change every time.

Using virtual disk technology to virtualize each cloud disk into a local disk makes the system user-friendly for users.

The main workflow of the system is as follows. After identity authentication, it will establish a communication connection to users' cloud disks and provide data storage services. The client terminal calls the virtual disk drivers in the virtual file system, and then users' cloud virtual disks are virtualized into disks, and users may need to generate files, modify files, or delete files in the virtual disk. According to the size of users' file blocks, the system uses redundancy backup mechanism to back up the data blocks. Then, filter driver encrypts each block of data blocks. After encrypting the stored files, public cloud communications engine will store the files into cloud disks and save the virtual storage mapping table of the files to the USB Key. When the user needs to access the stored data, virtual file system reads the virtual storage mapping table stored in the USB Key at first. Then, it will integrate the dispersed data which is transmitted in parallel based on the table and provide the users with the complete encrypted data.

Finally, the encryption filter driver decrypts the encrypted data and the decrypted data will be presented to the users. This process is completely transparent to the users and the users do not need to change their operation habits. The client parallel transmits users' file blocks to each cloud disk. This is helpful to improve the efficiency of data access. Transmitting the stored data to each cloud disk after backup is helpful to enhance the availability and reliability of the stored data. Even if one of the cloud disks breaks down, users can also access data from other cloud disks normally. Blocking and the encrypting backup operations improve the security of the entire system.

In this system, loading virtual disks after identity authentication is the first protection of secure storage by the terminal. Transparent encryption and decryption based on the file filter driver are the second protection. When the filter driver detects the read request from IRP, multiple threads will start to receive data blocks and decrypt the data after reorganization and then show the plaintext data to the users. When the filter driver detects the write request from IRP, the system will encrypt users' data with the encryption module and start multiple threads to transmit the encrypted data blocks to each cloud disk. Encrypting an decrypting the stored data in the operating system kernel layer can protect all the users' data and this method is safe, efficient, and stable [23, 24]. Transmitting users' data to each cloud disk after backup is the third protection. When a cloud disk breaks down, users can still recover lost data from other cloud disks to ensure proper access to users' data.

4 System Verification

The system is installed in a computer of file system driving structure supported by Windows operating platform. Due to business reasons, most API interfaces of domestic cloud Disks has not open to the public yet. Even if a few cloud service providers open the API interface, the application process is complicated and has a long cycle; therefore, in the course of the experiment, NAS network storage environment is used to simulate cloud storage for experimental verification. Each NAS network storage server simulates each cloud disk, and each network storage server virtualizes into local disk through virtualization technology, in the meantime the stored files are encrypted transmitted to every network storage server after blocked through the safety measures, such as the data block storage mechanism, data encryption mechanism, code redundancy backup mechanism and so on, and through the terminals' treatment of storage file to ensure safe and reliable data storage. The detailed information of system verification environment is as follows.

(i) Terminal

> Machine Type: Lenovo B460
> Operating System Version: Windows 7 32 Bit
> CPU Type: Intel(R) Core(TM) i3 CPU M 350 @ 2.27 GHz 2.27 GHz
> Memory size: 2.00 GB

(ii) Network File Storage System

> NAS Network Storage Server Type: HP Storage Works X1600 (AP804A)
> NAS Network Storage Server Type Number: 3

(iii) Network Transmission Rate: 100 Mb/s

The file transmission efficiency is focused tested through the system verification process. For easy analysis and comparison, the single network storage server storing files and three sets of network storage server parallel storing files are experimented respectively. Table 3 gives comparison of storing file efficiency between the single network storage server and three sets in the 100 Mb/s of network transmission rates, where eight different sizes of documents were selected to analyze, and each file is tested 10 times for an average value. The results are clearly shown that the performance of single network storage server and 3 sets of network storage sever.

Table 3. Comparison of files access time

File size	Single network storage server (S)		3 sets of network storage server (S)	
	Store	Download	Store	Download
532 KB	0.06	0.05	0.07	0.06
1.27 MB	0.12	0.10	0.14	0.11
3.98 MB	0.38	0.33	0.19	0.16
8.12 MB	0.64	0.54	0.28	0.26
10.98 MB	0.84	0.78	0.37	0.31
20.92 MB	1.74	1.61	0.68	0.58
50.70 MB	3.17	3.62	1.42	1.29
120.0 MB	7.63	7.63	3.21	3.03

Test results show that the efficiency of 3 network storage servers parallel access file is significantly higher than the single network storage server, although after the data block, file transparent encryption and decryption, coding redundancy backup and other security operation by client, the file access efficiency is decreased, the loss of access efficiency caused by the system is acceptable compared to the safety provided. When the small (below 2 M) file is transmitted, the access wastage is less than 1 s, and it is similar to that of the single network storage server; but with the increasing of file storage size, the access time raises to straight up; however users access large file without long time use the system, and it does not produce the problem of document long processing delay, so its performance is significantly higher than that of a single access network storage server. Through the experimental verification, the system basically meets the expected design requirements, and the simple operation, user's well experience, and strong safety and reliability, greatly enhance the access speed of the file. Therefore, this system not only guarantees the secure of storage data, but also do not the lost much of access efficiency; transformation of the terminal environment greatly improves the management capabilities on remote storing data for user.

5   Conclusion and Future Work

Compared with the existing cloud disk storage mechanism, TTEMFMC transfers the control right from cloud disks to user terminal, which increases user's capability data storage management and control. This mechanism can be implemented to ensure the terminal security of the stored data, timeless transparent encryption or decryption, and effective management and also can enhance the confidentiality, integrity, and availability of users' data [25].

Analysis shows that TTEMFMC can enhance the security and reliability of the stored files, improve the rate of access, and effectively overcome the effect of file transparent encryption and decryption on storage rate. In the future, we will focus on transparent encryption and decryption of files and parallel transmission in order to further improve the rate of storage.

By the method of formatting into N size fixed files, the proposed system storage mechanism is already very satisfactory, while there are still many possible ways to further improve the proposed mechanism, such as the sound fuzzy method; see [26] and references therein. As is well known, networked systems, complex networks, sensor networks, and multiagent systems are common systems in daily life, and research on which has become increasingly active in recent years, due primarily to their wide applications in many fields; see, for example, [27]. Naturally, how to apply our storage mechanism to optimize resources in different systems is still a thoughtful issue. On the other hand, disturbance (such as white noise and periodic narrowband noise) and incomplete information (such as data-packet dropouts and missing measurements.) inevitably exist in many kinds of systems and networks; see, for example, [28, 29], which will undoubtedly influence the feasibility and security of the suggested mechanism. Therefore, accuracy and robustness of the designed mechanism in complicated background area promising and valuable research direction in the future.

References

1. Jian-Hua, Z., Nan, Z.: Cloud computing-based data storage and disaster recovery. In: Proceedings of the International Conference on Future Computer Science and Education (ICFCSE 2011), Xi'an, China, August 2011, pp. 629–632 (2011)
2. Vaquero, L.M., Rodero-Merino, L., Caceres, J.: A break in the clouds: towards a cloud definition. ACM SIGCOMM Comput. Commun. Rev. 39(1), 50–55 (2009)
3. Armbrust, M., Fox, A., Griffith, R., et al.: A view of cloud computing. Commun. ACM 53 (4), 50–58 (2010)
4. Grossman, R.L., Gu, Y., Sabala, M., Zhang, W.: Compute and storage clouds using wide area high performance networks. Future Gener. Comput. Syst. 25(2), 179–183 (2009)
5. Luo, J.Z., Jin, J.H., Song, A.B.: Cloud computing: architecture and key technologies. J. Commun. 32(7), 3–21 (2011)
6. Wu, J.Y., Fu, J.Q., Ping, L.D.: Study on the P2P cloud storage system. Acta Electronica Sinica 39(5), 1100–1107 (2011)

7. Chow, S.S., Chu, C.-K., Huang, X., Zhou, J., Deng, R.H.: Dynamic secure cloud storage with provenance. In: Naccache, D. (ed.) Cryphtography and Security: From Theory to Applications. LNCS, vol. 6805, pp. 442–464. Springer, Heidelberg (2012)

8. Brinkmann, A., Eddert, S., Meyer Auf Der Heide, F.: Dynamic and redundant data placement. In: 27th International Conference on Distributed Computing Systems (ICDCS 2007), Toronto, pp. 87–95 (2007)

9. Feng, D.G., Zhang, M., Zhang, Y.: Study on cloud computing security. J. Softw. 22(1), 77–83 (2011)

10. Yang, J., Wang, H.H., Wang, J.: Survey on some security issues of cloud computing. J. Chin. Comput. Syst. 33(3), 472–479 (2012)

11. Cachin, C., Alex, I.K., Shraer, E.: Trusting the cloud. ACM SIGACT News 40(2), 455–461 (2009)

12. Lin, Q.Y., Gui, X.L., Shi, D.Q., Wang, X.P.: Study of the secure access strategy of cloud storages. J. Comput. Res. Dev. 48, 240–243 (2011)

13. Blaze, M.: Cryptographic file system for Unix. In: Proceedings of the 1st ACM Conference on Computer and Communications Security, Fairfax, VA, USA, November 1993, pp. 9–16 (1993)

14. Fu, K., Kaashoek, M.F., Mazieres, D.: Fast and secure distributed read-only file system. In: Proceedings of the 4th Symposium on Operating Systems Design and Implementation (OSDI), San Diego, CA, pp. 181–196 (2000)

15. Wright, C.P., Martino, M.C., Zadok, E.: NCryptfs: a secure and convenient cryptographic file system. In: Proceedings of the Annual USENIX Technical Conference, San Antonio, pp. 197–210 (2003)

16. Storer, W.M., Greenan, M.K., Miller, E.: POTSHARDS: a secure, recoverable, long-term archival storage system. ACM Trans. Storage 5(2), 1–35 (2009)

17. Zhao, Y.L., Dai, Z.X., Wang, Z.G.: Research on storage system architecture of the intelligent network disk (IND). Chin. J. Comput. 31(5), 858–867 (2008)

18. Wang, Z., Luo, W.M., Yan, B.P.: Optimal mechanism of parallel downloading. J. Softw. 20(8), 2255–2268 (2009)

19. Tao, M.: Design and Implementation of File Encryption System Based on Windows IFS Filter Driver. Chengdu Electronics Technology University, Chengdu (2012)

20. Zheng, L., Ma, Z.F., Gu, M.: Techniques of file system filter driver-based and security-enhanced encryption system. J. Chin. Comput. Syst. 28(7), 1181–1184 (2007)

21. Shvachko, K., Kuang, H.R., Radia, S.: The hadoop distributed file system. In: Proceedings of the IEEE 26th Symposium on Storage Systems and Technologies (MSST 2010), Piscataway, NJ, USA, May 2010, p. 1 (2010)

22. Subashini, S., Kavithaa, V.: A survey on security issues in service delivery models of cloud computing. J. Netw. Comput. Appl. 34(1), 1–11 (2011)

23. Kaufman, L.M.: Data security in the world of cloud computing. IEEE Secur. Priv. 2009(7), 61–64 (2009)

24. Yumerefendi, A.R., Chase, J.S.: Strong accountability for network storage. ACM Trans. Storage 3(3), 1–33 (2007). (article 11)

25. Maheshwari, U., Vingralek, R., Shapiro, W.: How to build a trusted database system on untrusted storage. In: Proceedings of the Symposium on Operating System Design and Implementation, San Diego, Calif, USA, pp. 10–20 (2000)

26. Zhang, S.J., Wang, Z.D., Ding, D.R., Shu, H.S.: Fuzzy filtering with randomly occurring parameter uncertainties, interval delays, and channel fadings. IEEE Trans. Cybern. 44(3), 406–417 (2014)

27. He, X., Wang, Z.D., Liu, Y.R., Zhou, D.H.: Least-squares fault detection and diagnosis for networked sensing systems using a direct state estimation approach. IEEE Trans. Industr. Inf. **9**(3), 1670–1679 (2013)

28. Wang, Z.D., Ding, D.R., Dong, H.L., Shu, H.: H_∞ consensus control for multi-agent systems with missing measurements: the finite-horizon case. Syst. Control Lett. **62**(10), 827–836 (2013)

29. Hu, J., Wang, Z., Shen, B., Gao, H.: Quantised recursive filtering for a class of nonlinear systems with multiplicative noises and missing measurements. Int. J. Control **86**(4), 650–663 (2013)

Novel Intrusion Detection System for Cloud Computing: A Case Study

Ming-Yi Liao[1]([✉]), Zhi-Kai Mo[1], Mon-Yen Luo[2], Chu-Sing Yang[1], and Jiann-Liang Chen[3]

[1] Department of Electrical Engineering, Institute of Computer and Communication Engineering, National Cheng Kung University, Tainan, Taiwan, ROC
myliao@ee.ncku.edu.tw
[2] Department of Computer Science and Information Engineering, National Kaohsiung University of Applied Sciences, Kaohsiung, Taiwan, ROC
[3] Department of Electrical Engineering, National Taiwan University of Science and Technology, Taipei, Taiwan, ROC

Abstract. Because of the growth in cloud computing and manturity of virtualization technology, many enterprises are virtualizing their servers to increase server utilization and lower costs. However, the complex network topology arising from virtualization makes clouds vulnerable, and security breaches have occurred on cloud computing platforms in recent years. Therefore, a comprehensive mechanism for detecting and preventing malicious traffic is necessary. We propose a network intrusion detection system that is based on a virtualization platform. This system, developed from a multipattern based network traffic classifier, collects packets from the virtual network environment and analyzes their content by using deep packet inspection for identifying malicious network traffic and intrusion attempts. We improve the intrusion detection features of the network traffic classifier and deploy it on a Xen virtualization platform. Our system can be combined with the Linux Netfilter framework to monitor inter-virtual-machine communications in the virtualization platform. It efficiently inspects packets and instantly protects the cloud computing environment from malicious traffic.

Keywords: Cloud computing · Deep packet inspection · Intrusion detection system

1 Introduction

With the rise of cloud computing and maturity of virtualization technology, many enterprises are virtualizing their servers to enhance server utilization and reduce costs. However, the security events that have occurred on cloud computing platforms in recent years not only cause economic loss but also affect user privacy. Potential security risks in virtualized environments include disclosure of server configuration, disclosure of virtual machine (VM) information by a VM manager, and using VMs as

© Springer International Publishing Switzerland 2015
W. Qiang et al. (Eds.): CloudCom-Asia 2015, LNCS 9106, pp. 386–398, 2015.
DOI: 10.1007/978-3-319-28430-9_29

bouncers for attacking other targets. To secure cloud computing networks, a comprehensive mechanism that detects and prevents malicious traffic is essential. A critical task in cloud security is providing virtual host, and hypervisor protection, anti-malware and intrusion prevention solution. Inter host communication among VMs cannot be monitored using security systems outside the host. Some solutions entail establishing an agent in every guest VM, causing redundant overhead because all agents run simultaneously. To secure cloud computing networks, we propose a network intrusion detection system (IDS) that is based on virtualization platforms. Our system can be combined with the Netfilter framework [1] for efficiently inspecting packets and instantly protecting the cloud computing environment from malicious traffic. The remainder of this paper is organized as follows. In Sect. 2, we introduce the study background and related research. Section 3 describes the implementation of our system architecture in an IDS. Section 4 describes the deployment of the IDS in a Xen virtualization platform. Section 5 describes our experiments and discusses the evaluation results, and Sect. 6 presents the conclusions.

2   Related Works

Libpcap [2] is a widely used system-independent interface for user-level packet capture that provides a portable framework for low-level network monitoring. Some IDSs and packet classifiers, such as Snort [3] and Bro [4], use libpcap. Another approach is to capture packets through a communication application program interface between the kernel and user-space in Linux [5].

Port-based classification is traditionally used in packet classification [6]. Statistics-based classification is an approach that indirectly analyzes packets [7, 8]. This method performs traffic classification without accessing sensitive data. Usually, statistics-based methods provide features of network flow for machine learning and heuristic algorithms [9]. Each feature can be considered a dimension of a vector space [10]. Data clustering algorithms are applied for data training and modeling. Deep Packet Inspection (DPI) technique matchs a predefined signature for traffic classification and intrusion detection [11]. Regular expression engines process a regular expression statement by constructing a deterministic finite automaton (DFA) [12, 13] with a time and memory cost of $O(2^m)$ for a regular expression of size m running DFA on a string of size n in time $O(n)$. In protocol decoding, packet payloads are parsed according to the header of the application layer [14], and connections are tracked using finite state automata. The classifier [15] is a DPI system based on the Linux Netfilter framework.

Snort is an open source network IDS that detects packets through signature-based methods. Bro is a passive, open source network traffic analyzer. It supports many application-layer protocols, including DNS, FTP, HTTP, IRC, SMTP, SSH, and SSL.

Some open source platforms, and commercial solutions use virtualized servers for cloud computing. The Kernel-based Virtual Machine (KVM) [16] is an open source VM built in the Linux kernel. The KVM supports Intel VT and AMD-V virtualization, and combines the hypervisor and operating system kernel for reduced redundancy and enhanced efficiency. Xen [17] is an open source VM developed by the University of Cambridge, and runs at most 128 complete operating systems in a host.

Restricting user access to the critical resources in a virtual environment [18] is one facilitative approach to prevent the network in cloud computing from malicious threats. Another approach involves binding users to different security groups according to their anomaly levels [19]. To prevent sniffing and spoofing in virtual networks, a network model that isolates a group of VMs proposed [20]. VMs are split into shared subnetworks and are assigned unique IDs, and these IDs prevent communication between subnetworks. In addition, firewall is established between the subnetworks to prevent spoofing. A VM monitor based IPS [21] is proposed for detecting intrusion attempts in the Xen virtualization platform.

3   IDS Implementation

3.1   Multipattern Based Packet Classifier Workflow

The classifier is connected to the Netfilter and processes packets through the PROMISC hook, using connection tracking for efficient packet inspection. When packets pass through the classifier module, a sequence of inspections is triggered. First, whether each packet is an IP packet is determined. Subsequently, the classifier searches connection tracking data to ensure that the connection associated with this packet was identified previously. If the connection is unidentified, the classifier parses the packet payload and begin pattern matching. The classifier queries the rules in two steps because variable pattern matching is more time intensive than matching processes. If no match is identified after fixed and arithmetic pattern matching, variable pattern matching is applied. Finally, the classifier matches the rule again. If the matching is successful, information about the application type is written to the connection tracking data. However, when the classifier matches the first packet in a connection, connection tracking of all subsequent packets in the connection is ignored. Therefore, some exploitative behaviors and malicious connections appearing in the general protocol cannot be identified. Hence, we propose an improvement to this classifier, that entails modifying the packet matching procedure.

3.2   Multipattern Based Classifier with State Checking Function

As discussed in Sect. 3.1, connection tracking ignores certain exploitative and malicious connections in the general protocol. The Heartbleed exploit [24], a serious vulnerability in the widely used OpenSSL cryptographic software library, illustrates the vulnerability of connection tracking. It was discovered by a team of security engineers at Codenomicon and Neel Mehta of Google Security. This vulnerability arising from an implementation problem in some versions of the OpenSSL library, enables anyone on the Internet to read system memory. Attackers could steal user passwords and secret keys used to encrypt the traffic. Heartbleed exploits the fact that the OpenSSL software does not check the data length in the heartbeat protocol (RFC 6520), which is used to negotiate and monitor service availability. The heartbeat protocol includes message payload and payload length, and a heartbeat request is sent to obtain a corresponding response message. The receiver responds with a message that carries a copy of the requested payload. When an attacker sends a request where the payload length is greater than the actual length, the vulnerable OpenSSL server responds with data of the length specified by the request, causing a memory leak.

The heartbeat protocol is an extension of TLS/SSL, and the exploit occurs after a TLS connection is established. The conventional classifier detects the TLS connection by matching the hello packet but ignores the subsequent packets; therefore, it cannot detect the Heartbleed exploit. Another example is CVE-2014-3466 [25], which occurs in GnuTLS, a secure communications library that implements the SSL and TLS protocols. GnuTLS does not check the length of the session ID in the ServerHello message; this vulnerability enables attackers to overflow stack buffer and execute arbitrary code, which cannot be detected by the conventional classifier. We modified the packet matching procedure and added functions for checking the packet state and format. A sequence of functions is designed for detecting exploits and malware. Of the four signature matching patterns, arithmetic matching is used to calculate the data length and determine whether the length equals the value of the specific bytes; thus, it is similar to protocol decoding. We extend this functionality to intrusion detection, for detecting the Heartbleed exploit and botnets. If all packets are processed, the processing costs increase. Therefore, each function is assigned to a corresponding protocol. An easy solution is to construct a list of functions and search specific functions according to the application protocol of the packets; however, searching entails more costs.

We use connection tracking for assigning state-checking functions. When a packet is identified, information about the service type and matched rule is returned from the rule set, and connection tracking writes this information to the corresponding hash table. Therefore, we set a pointer in the rule entry to the corresponding functions so that packets with identified service types can be processed through predefined state-checking functions. Thus, packets are checked according to a specific protocol, reducing packet processing costs. Figure 1 illustrates the procedure for assigning state-checking functions by using connection tracking; HTTP and TLS are two example protocols.

The work flow of the classifier is in Fig. 2. When a packet enters the classifier, the classifier searches the corresponding entry in the connection tracking table. If this connection was not tracked, the classifier performs signature matching to identify the packet. If the classifier recognizes the packet service type, it writes the information to

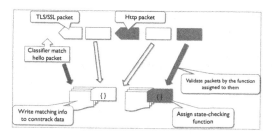

Fig. 1. Assigning state-checking functions to specific packets

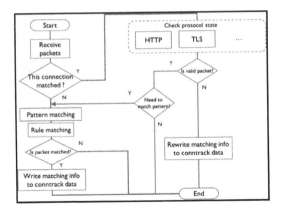

Fig. 2. Flow diagram of the classifier with state-checking function

the connection tracking entry. The subsequent packets belonging to the connection are processed using state-checking functions for protocol validation. Malicious traffic appearing in the general protocol must be detected using string matching; therefore, the state-checking function should determine packet types that match patterns. For example, HTTP botnets request pages and send commands using HTTP request; the state-checking function of HTTP identifies the request packets and forwards them for pattern matching. If a packet in a flow is considered malicious, the classifier rewrites the connection tracking data of the flow and assigns another state-checking function to it. The other packets in the flow are processed using state-checking function of HTTP. This approach comprehensively detects intrusions.

Here, we analyze the running time of our modified classifier. We first calculate the running time of packet matching without connection tracking. We assume that there are n packets in a flow. The running time of arithmetic pattern matching is t_a, and t_p is the sum of running times for fixed and variable pattern matching and rule matching. The running time for the state-checking function is t_s. The time for matching all packets is obtained as follows.

$$T_{all} = n(t_a + t_p).$$

With connection tracking, only m packets must be matched by the classifier. The time consumed is represented as follows.

$$T_{ct} = m(t_a + t_p).$$

We add the mechanism of the state-checking function to connection tracking and increase the running time (T'_{ct}), which is represented as follows.

$$T'_{ct} = m(t_p + t_a) + t_s(n - m)$$
$$= nt_s + (t_{p=} + t_a - t_s).$$

Because the state-checking function and arithmetic pattern matching employ the same methods for checking packets, we assume that they consume the same amount of time. Practically, the state-checking function for a specific application is a subset of the arithmetic patterns; therefore, the running time of the state-checking function is less than that of arithmetic pattern matching. If we assume $t_s = t_a$, we can rewrite T'_{ct}:

$$T'_{ct} = nt_a + mt_p.$$

The running time of the classifier with the state-checking function is less than that for matching all packets. We calculate their time consumption ratio:

$$\frac{T'_{ct}}{T_{all}} = \frac{nt_a + mt_p}{nt_a + nt_p}.$$

In related research, t_p is approximately 66.9 ms and t_a is approximately 5.3 ms in the worst case. For example, if there are ten packets and the first packet is matched using string pattern matching, the ratio is 0.166, which is 66 % higher than that of the conventional classifier; nevertheless, the time consumption of the modified classifier is much lower than that for matching all packets. In the subsequent section, we prove that the proposed design is reasonable in combination with the state-checking function and connection tracking.

4 Deployment

We deploy the classifier on a Xen platform and test the proposed design in several network scenarios to demostrate its suitability in cloud computing environments. The classifier is a kernel module connected to the Netfilter and processes packets from and to the VM detecting intrusions; however, comprehensively securing virtual networks in virtualization platforms is not easy. Networking in Xen can be constructed using the Linux bridge or Open vSwitch in Domain 0. We set up the classifier in combination with the virtual switch or the bridge, as explained subsequently.

4.1   Open vSwitch

Open vSwitch is a multilayer virtual switch designed to enable effective network automation through programmatic extensions. If the Xen network is connected by a virtual switch, the function of the bridge module in Domain 0 is replaced. All packets transmited to and from VMs are forwarded by the virtual switch; therefore, the Netfilter does not receive these packets, and the classifier must run in the offline mode. We enable the mirror port to collect traffic from all VMs and create a virtual interface for receiving packets mirrored from the virtual switch. Figure 3 shows the architecture of the virtual network with Open vSwitch. If more than one physical machine is used, the virtual switches on each host mirror its ports to a central traffic detection system installed in the classifier, as shown in Fig. 4.

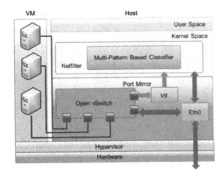

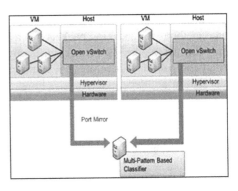

Fig. 3. Multipattern based classifier with Open vSwitch

Fig. 4. Mirroring ports outside the multipattern based classifier

4.2 Linux Bridge

Linux provides virtual bridges for connecting several Ethernet segments independent of the protocol. The Linux bridge by default connects the network interface of Domain 0 and Domain Us in the Xen network. The Bridge-Netfilter is a Netfilter component that provides several modules for packet manipulation at the data link layer. In addition, it can be connected to user-defined modules. The five hooks in both layers are the same, but the BROUTING hook in the Bridge-Netfilter redirects packets between interfaces with different subnets. To filter the packets passing through the bridge, connecting the classifier to the Bridge-Netfilter is an easy approach. However, the classifier requires the connection tracking module (for enhanced packet matching efficiency), which the Bridge-Netfilter does not support. We use sysctl to modify the configuration to enable packets in the bridge to pass through the PREROUTING hook of the Netfilter and return to the Bridge-Netfilter so that they can beprocessed by the classifier, as depicted in Fig. 5.

However, the NAT table of the Netfilter is traversed only by the first packet in each connection. Previous approaches overcome this problem by defining a PROMISC hook

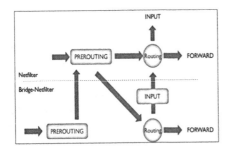

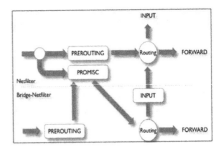

Fig. 5. PREROUTING hook of the Netfilter is enabled in the bridge

Fig. 6. The entrance of PREROUTING is substituted with that of PROMISC

in the Netfilter. We substitute the entrance of the PREROUTING hook with that of PROMISC. Figure 6 shows the path modified for the classifier in the Bridge-Netfilter.

5 Evaluation

We conducted evaluation experiments to prove that our system is a suitable IDS for cloud computing environments. To test packet loss rate and bit rate, we set up the classifier on a Xen virtualization platform and operated VMs to generate packets. Moreover, we tested the accuracy of our classifier by replaying traffic samples of malware and exploits. The experimental environment is described in the subsequent sections.

5.1 Environment

Table 1 details our environment. We constructed a Xen system and installed software on VMs. The Linux kernel was rebuilt from the source to enable installing our modified classifier. The iptables are required to enable the classifier; hence, we updated the iptables for compatibility with a newer kernel. The Linux bridge and Open vSwitch were not enabled simultaneously because of compatibility concerns.

Table 1. Experimental environment

Component	Software
Dom0	CentOS 6.4, Linux kernel 3.12.4, iptables 1.4.21, Snort 2.9.6.1
DomU	CentOS 6.4
Xen	Xen 4.3.1

5.2 Packet Loss Rate and Throughput Testing

The experimental scenario is illustrated in Fig. 7. We used two VMs; one floods user data protocol (UDP) traffic to the other, which receives and counts the packets. We gathered statistics from the network interfaces of the VMs to determine the number of packets sent and received. For comparison, we individually tested the modified classifier and Snort. Snort was set in the inline mode for detecting traffic passing through the bridge. Because the classifier contains 397 rules, we enabled only 383 exploit-kit rules in Snort. We used Hping [26] to generate UDP traffic, and varied the source and destination ports of each packet from 1 to 65,535 to prevent connection tracking from ignoring most packets. We tested packet sizes ranging from 64 bytes to 1,512 bytes and observed the variation. The experimental results are presented in Figs. 8 and 9. In Fig. 8, the Snort packet loss rate fluctuates between 17 % and 36 % and is higher than that of the classifier, of which the loss rate is < 1 %. Figure 9 reveals that our classifier affords a higher bit rate than Snort does.

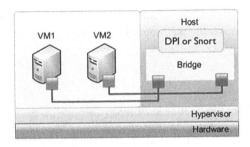

Fig. 7. Connecting VMs by using the Linux bridge

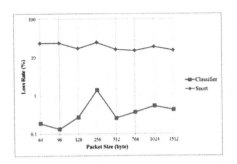
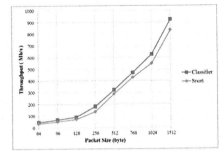

Fig. 8. Packet loss rate of traffic between VMs   **Fig. 9.** Throughput of traffic between VMs

In another experiment, we flood UDP traffic from a PC outside the virtualization platform to a VM. The test environment is depicted in Fig. 10.

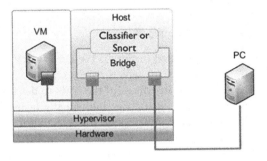

Fig. 10. Connecting a VM and a PC by using the Linux bridge

As in the earlier experiment, we conducted separate experiments with the classifier and Snort. As Fig. 11 shows, the packet loss rate of Snort decreases from 17 % to 7 %, and the packet loss rate of the classifier does not exceed 0.01 %. In Fig. 12, the bit rate of Snort is higher than that of our classifier, possibly because packets pass through the classifier at a lower speed to ensure transmission. Snort runs in the user space and

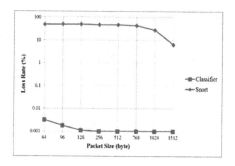
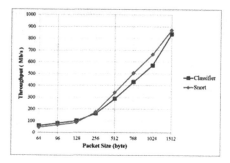

Fig. 11. Packet loss rate of traffic from a PC to a VM

Fig. 12. Throughput of traffic from a PC to a VM

receives packets from the kernel by AF_PACKET, which causes severe packet loss. These experiments prove that our classifier remains stable during packet transmissions.

5.3   Accuracy Testing

We collected five types of malicious traffic a worm, two botnets, and two exploits to test the accuracy of the classifier. Table 2 describes the flow count, packet count, and the corresponding rules in the classifier for these samples. These samples are stored in the PCAP files and are split into smaller files according to TCP streams. We shuffled these files and replayed them to generate malicious traffic.

Table 2. Traffic samples of malware and exploits

Name	Type	Flow count	Packet count	Byte count	Rules in Classifier
BlackSun	Botnet	47	566	182311	37
Zeus	Botnet	22	3644	3214907	24
Vobfus	Worm	9	1044	1170678	4
CVE-2014-0160 (Heartbleed)	Exploit	10	400	330860	1
CVE-2014-3466	Exploit	10	100	11530	1

The accuracy test results of the conventional and modified classifiers as listed in Tables 3 and 4, respectively. Comparing these results reveals that a higher number of BlackSun and TLS exploits are detected by the modified classifier than by the conventional classifier. The conventional classifier cannot detect exploits inside TLS, such as the Heartbleed exploit and CVE-2014-3466, because they are ignored by connection tracking. A higher number of BlackSun packets are detected by the modified classifier because these packets are commands following the normal HTTP requests in a connection. The other matched packets include HTTP and FTP packets. Some packets are

Table 3. Flows matched by the classifier

Name	Flow count	Packet count	Byte count
BlackSun	37	425	128487
Zeus	21	3643	3164178
Vobfus	9	1044	1157502
CVE-2014-0160 (Heartbleed)	0	0	0
CVE-2014-3466	0	0	0
TLS	20	410	186390
http	8	110	45085
ftp	2	27	2330

Table 4. Flows matched by the modified classifier

Name	Flow count	Packet count	Byte count
BlackSun	38	451	141601
Zeus	21	3643	3164178
Vobfus	9	1044	1157502
CVE-2014-0160 (Heartbleed)	10	310	176260
CVE-2014-3466	10	100	10130
TLS	0	0	0
http	7	84	31971
ftp	2	27	2330

not shown in the results because they are malicious TCP packets that are not received by the Netfilter. The HTTP packets are requests for elements in the pages of BlackSun, and the FTP packets are file transmissions without significant signatures. Thus, we demonstrate that our modification enhances the functionality of intrusion detection.

6 Conclusion

We propose a novel IDS that is an improvement of multipattern based network packet classifier. We modified the classifier procedure and added protocol validation for detecting exploits that appear in the general protocol and malicious commands that follows normal requests in a connection. Therefore, we used connection tracking to assign state-checking functions to flow packets. Experimental results show that the modified classifier succeeds in detecting a higher number of exploits and malicious traffic than the conventional version does. To detect inter-VM communication in the virtualization platform, we set up our classifier on the virtual network of a Xen system. The bridge-Netfilter was modified to enable the PROMISC hook of the Netfilter for packet processing because supports connection tracking. The classifier connected to the PROMISC hook receives packets and returns them to bridge after pattern matching. We experimentally demonstrated that our classifier performs more accurately than another IDS does. Because only the hypervisor of the Xen platform is modified, the modified classifier can be set up on other virtualization platforms, such as KVM. In future work, we aim to manage several IDSs when more than one is deployed, with experimental testing in a practical scenario being our objective.

Acknowledgement. This research was supported by a grant from the Ministry of Science and Technology, Taiwan, Republic of China, under Grants MOST-103-2221-E-006-145-MY3 and MOST-103-2811-E-006-049, for which we are grateful.

References

1. Netfilter. http://www.netfilter.org
2. Libpcap. http://www.tcpdump.org
3. Snort. http://www.snort.org
4. The Bro Network Security Monitor. http://www.bro.org
5. Wehrle, K., Pählke, F., Ritter, H., Müller, D., Bechler, M.: The Linux Networking Architecture: Design and Implementation of Network Protocols in the Linux Kernel (2004)
6. Qi, Y., Xu, L., Yang, B., Xue, Y., Li, J.: Packet classification algorithms: from theory to practice. In: INFOCOM 2009, pp. 648–656. IEEE (2009)
7. Finsterbusch, M., Richter, C., Rocha, E., Muller, J.: A survey of payload-based traffic classification approaches. In: IEEE Communications Surverys & Tutorials (2012)
8. Sicker, D.C., Ohm, P., Grunwald, D.: Legal issues surrounding monitoring during network research. In: Proceedings of the 7th ACM SIGCOMM on Internet Measurement (2007)
9. Rotsos, C., Van Gael, J., Moore, A.W., Ghahramani, Z.: Probabilistic graphical models for semi-supervised traffic classification. In: Proceedings of the 6th International Wireless Communications and Mobile Computing Conference, pp. 752–757 (2010)
10. Piskac, P., Novotny, J.: Using of time characteristics in data flow for traffic classification. In: Proceedings of the Autonomous Infrastructure, Management, and Security: Managing the Dynamics of Networks and Services, pp. 173–176 (2011)
11. Lin, P.C., Li, Z.X., Lin, Y.D., Lai, Y.C., Lin, F.: Profiling and accelerating string matching algorithms in three network content security applications. IEEE Commun. Surv. Tutorials **8**(2), 24–37 (2006)
12. Liu, C., Wu, J.: Fast deep packet inspection with a dual finite automata. IEEE Trans. Comput. **62**(2), 310–321 (2013)
13. Wang, X., Jiang, J., Tang, Y., Liu, B., Wang, X.: StriD2FA: scalable regular expression matching for deep packet inspection. In: 2011 IEEE ICC Conference, pp. 1–5 (2011)
14. Risso, F., Baldi, M., Morandi, O., Baldini, A., Monclus, P.: Lightweight, payload-based traffic classification: an experimental evaluation. In: IEEE ICC Conference (2008)
15. Liao, M.Y., Luo, M.Y., Yang, C.S., Chen, C.H., Wu, P.C., Chen, Y.C.: Design and evaluation of deep packet inspection system: a case study. IET Netw. **1**, 2–9 (2012)
16. KVM. http://www.linux-kvm.org/page/Main_Page
17. Xen. http://www.xenproject.org/
18. Khoudali, S., Benzidane, K., Sekkaki, A.: Inter-VM packet inspection in cloud computing. In: Communications, Computers and Applications (MIC-CCA) (2012)

19. Lee, J.H., Park, M.W., Eon, J.H., Chung, T.M.: Multilevel intrusion detection system and log management in cloud computing. In: ICACT (2011)
20. Wu, H., Yi, D., Winer, C., Li, Y.: Network security for virtual machine in cloud computing. In: ICCIT 2010 (2010)
21. Jin, H., Xi, G.F., Zou, D.Q., Wu, S., Zhao, F., Li, M., Zheng, W.: A VMM-based intrusion prevention system in cloud computing environment. J. Supercomputing **66**, 1133–1151 (2013)

Author Index